Lecture Notes in Physics

176

Gauge Theory and Gravitation

Proceedings of the International Symposium
on Gauge Theory and Gravitation (g & G)
Held at Tezukayama University
Nara, Japan, August 20 – 24, 1982

Edited by K. Kikkawa, N. Nakanishi, and H. Nariai

Springer-Verlag
Berlin Heidelberg New York 1983

Editors

Keiji Kikkawa
Department of Physics, Osaka University
Toyonaka 560, Japan

Noboru Nakanishi
Research Institute for Mathematical Sciences
Kyoto University
Kyoto 606, Japan

Hidekazu Nariai
Research Institute for Theoretical Physics
Hiroshima University
Takehara 725, Japan

ISBN 3-540-11994-9 Springer-Verlag Berlin Heidelberg New York
ISBN 0-387-11994-9 Springer-Verlag New York Heidelberg Berlin

Printed in Germany

Printing and binding: Beltz Offsetdruck, Hemsbach/Bergstr.
2153/3140-543210

PREFACE

The history of gauge theory began with Weyl's pioneering attempt to unify the general theory of relativity and electromagnetic theory. After the establishment of quantum physics, however, the two theories proceeded along quite different routes. Whereas general relativity remained a classical theory and applied solely to phenomena of very large scales, electromagnetic theory brought a great triumph in quantum electrodynamics for microscopic phenomena.

Though quantum electrodynamics was a prototype of quantum field theory, the gauge principle itself was regarded as a special artifice realizing renormalizability in the 1950s. The extension of the gauge principle to a non-Abelian symmetry was proposed by C.N. Yang and R.L. Mills in 1954 and also by R. Utiyama in 1956, but the non-Abelian gauge theory seemed to have no physical reality at that time. This situation changed drastically in the 1970s, when the concept of spontaneous symmetry breakdown was incorporated into the theory. It is now firmly believed that both electroweak and strong interactions are described by non-Abelian theories.

In contrast to the rapid progress of particle physics, the development of the theory of gravitation was rather modest, and remained isolated from the rest of physics. In recent years, however, it has become an increasingly accepted view that gravity should be included in quantum physics, and that the theory of gravitation is indispensable in explaining cosmic phenomena around black-hole spacetime and in the universe itself. Thus, it is widely believed that particle physics and the theory of gravitation must be unified from the standpoint of the gauge principle in a generalized sense.

In Japan, research on theories of gravitation has long been supported by the General Relativity and Gravitation (GRG) Research Group, together with a sub-group (relating relativistic astrophysics and cosmology) in the Nuclear Astrophysical Group. Because of the interest of particle physicists in gravity, the GRG group recently expanded to include such scientists, and a new GRG Research Group was organized in 1981.

Professor Ryoyu Utiyama, President of Tezukayama University, is one of the most distinguished theoretical physicists in Japan. As early as 1956, he made a pioneering contribution to gauge theory and gravitation by showing that the general theory of relativity and the non-Abelian gauge theory could be understood on the same footing. Since then, Professor Utiyama has done a lot of important work on the gauge-theoretical formulation of quantum field theory. He served as an organizer of the GRG group and was professor at Osaka University until 1980.

The International Symposium on Gauge Theory and Gravitation was held at Tezukayama University, Nara, Japan, on 20-24 August 1982 to pay tribute to Professor Utiyama's brilliant research and to foster the development of gauge theory and gravitation. The symposium was supported by the Physical Society of Japan and the International Committee of the GRG, and sponsored by Tezukayama University, the Japan Society for the Promotion of Science, the Yamada Science Foundation, the Nishina Memorial Foundation, the JEC Fund and the Kinki Nippon Rail Line Company. The success of the symposium was made possible by the cordial cooperation of all participants and organizers. Many thanks are due to Mrs. Y. Tsuji and to the graduate students of Osaka University for their secretarial assistance.

December 20, 1982 Editors Keiji Kikkawa

Noboru Nakanishi

Hidekazu Nariai

CONTENTS

GEOMETRICAL ASPECT OF GAUGE THEORY AND GRAVITATION

Chairman: R. Utiyama

Chairman: F.W. Hehl

GAUGE THEORIES I

Chairman: K. Kikkawa

GAUGE THEORIES III

QUANTUM GRAVITY

SUPERSYMMETRY AND SUPERGRAVITY

The Symposium was financially supported by

Tezukayama University
Japan Society for the Promotion of Science
Yamada Science Foundation
JEC FUND*
Kinki Nippon Rail Line

and cooperated by

Physical Society of Japan
International Committee on General Relativity and Gravitation
Research Institute for Fundamental Physics, Kyoto University
Research Institute for Theoretical Physics, Hiroshima University
High Energy Theory Group, Osaka University

———— . ————

Chairman

R. Utiyama — Tezukayama

Advisary Committee		Organization Committee	
Y. Fujii	Tokyo	T. Eguchi	Tokyo
K. Fujikawa	INS	Z.F. Ezawa	Tohoku
H. Hirakawa	Tokyo	R. Fukuda	Kyoto
K. Kikkawa	Osaka	A. Hosoya	Osaka
Z. Maki	Kyoto	M. Kemmoku	Nara
N. Nakanishi	Kyoto	T. Kugo	Kyoto
T. Nakano	Osaka City	E. Takasugi	Osaka
S. Nariai	Hiroshima	A. Tomimatsu	Hiroshima
K. Nishijima	Tokyo	A. Ukawa	INS
H. Sato	Kyoto	T. Yoneya	Tokyo
H. Sugahara	KEK	M. Yoshimura	KEK

*
This International Symposium is executed with the assistance of a grant from the Commerative Association for Japan World Exposition.

The Participants of g&G Symposium, Tezukayama University
in Nara, Japan

A MICRO-deSITTER SPACETIME WITH CONSTANT TORSION: A NEW VACUUM SOLUTION OF THE POINCARÉ GAUGE FIELD THEORY

Peter Baekler and Friedrich W. Hehl
Institute for Theoretical Physics
University of Cologne
D-5000 Cologne 41, West Germany

ABSTRACT

We study the Poincaré gauge field theory with Lagrangians of the type (curvature scalar + torsion^2)/l^2 + curvature^2/$\varkappa$. Here l is the Planck length and $\varkappa$ the coupling constant of the hypothetical Lorentz gauge bosons. We find a new vacuum solution with a deSitter metric and with constant microscopic torsion $\sim\sqrt{\varkappa}/\ell$ and curvature $\sim \varkappa/\ell^2$. Its curvature displays double duality properties.

CONTENTS

1. INTRODUCTION AND SUMMARY

Soon after Yang and Mills /1/ developed the SU(2)-gauge concept in the context of the conserved isotopic spin current, Utiyama /2/ extended this idea to any semi-simple Lie-symmetry group and demonstrated furthermore that general relativity (GR) can be understood as a gauge theory related in some sense to local Lorentz invariance. A most lucid presentation of Utiyama's gauge theoretical ideas can be found in his more recent paper /3/. This approach to GR can be traced back to an earlier attempt of Weyl /4/.

Subsequently Sciama /5/ and Kibble /6/ have shown that, in a gauge theoretical set-up, the ad-hoc assumption of a symmetric connection of spacetime, used by Utiyama, should be dropped. Thus one arrives at a gravitational theory formulated in the framework of a Riemann-Cartan spacetime U_4. It is most natural to interpret the resulting theory as a gauge theory with local Poincaré invariance, cf. Cartan /7/. We will call it a Poincaré gauge field theory+ (PG). The present article is based on our four Erice lectures on this subject /32/.

Einstein "deduced" GR and its Riemannian spacetime V_4 by studying the motion of forcefree point-particles in non-inertial frames of reference and by applying the equivalence principle. In a gauge theoretical approach one studies the Lagrangian of a fermionic field, typically of a Dirac field, in non-inertial frames of reference and applies again a principle of equivalence++: A Riemann-Cartan spacetime U_4 is the outcome.

It is the heuristic power of the Einsteinian type of procedure as applied to a fermionic field which lends support to the PG. For fermions a U_4 is a much more natural "habitat" than a V_4. The latter is naturally adapted to point-particles. -

In Sect. 2 we will formulate the two field equations of the PG. In Sect. 3 the most general polynomial Lagrangian in torsion and curvature will be displayed yielding quasi-linear 2nd-order field equations. For a detailed motivation and derivation in the context of both of these sections see /32/.

Subsequently we are are going to look for exact vacuum solutions of the field equations with non-vanishing torsion. Such solutions with spherical symmetry and reflection invariance were found earlier for specific Lagrangians of the PG by Neville /36/, Baekler /37/, Baekler, Hehl and Mielke /38/, by Benn, Dereli and Tucker /39,40/ and by Baek-

---

+) Cf. also Hayashi and Shirafuji /8/ and Hayashi and Nakano /9/. There is a vast literature on this field. References can be found by starting, say, from the articles of Ne'eman /10/, Trautman /11/ and Hehl, Nitsch and Von der Heyde /12/ in the GRG-Einstein Centennial Volume (Held /13/). More recent work includes the articles of Hennig and Nitsch /14/, Mielke /15/, Ne'eman /16/, Nieh and Yau /17/, see also /18,19/, Schweizer /20/, Szczyrba /21,22/, Thirring /23/, Tseytlin /24/, Wallner /25,26/, Yasskin /27/, and, on the equations of motion, Audretsch /28/ and Rumpf /29,30/; cf. also Drechsler /31/.

++) Cf. Von der Heyde /34/, Mack /35/, and Utiyama /3/. In the gravitational part of Utiyama's paper on p. 2218 $\Gamma_\varsigma = 0$ (locally) is postulated in order "to derive the Riemannian Γ from a gauge theoretical approach". If one put $A_\varsigma = 0$ (locally) instead, one would arrive at a Riemann-Cartan spacetime. Utiyama writes in a letter to one of the authors furthermore: "I know that the requirement $A_\varsigma = 0$, locally (D) is more general than the requirement $\Gamma_\varsigma = 0$, locally (E). Einstein's equivalence principle should be replaced with the requirement (D) which is applicable even to microscopic cases. The condition (D) fits the orthodox view of the general theory of gauge fields.... this requirement is a generalized principle of equivalence." We completely agree.

ler /42/. J.D. McCrea /43/ found corresponding cylindrically symmetric solutions and Adamowicz /44/ plane wave solutions, see also Chen and Chern /45/. Mielke /46/ developed a general method for the generation of exact solutions.

In Sect. 4 we turn to spherical symmetry with reflection invariance. Earlier work on this subject, besides the articles cited in the last paragraph, has been done by Ramaswamy and Yasskin /47/, Baekler and Yasskin /48/, Nieh and Rauch /49,50/, Rauch, Shaw and Nieh /51/ and Rauch /52/, see also Tsamparlis /53,54/. In the papers /49,50,51/ the assumption of reflection invariance has been relaxed.

In Sect. 5 we specialize the tetrad such as to find a spatially homogeneous time-dependent solution. First, in Sect. 6, we execute this program for the purely quadratic model Lagrangian (6.1). For this purpose we use the two LISP-based algebraic manipulation schemes REDUCE /55/ and ORTOCARTAN /56,57/, cf. d'Inverno /58/. Our REDUCE-programs rely heavily on similar programs written by J.D. McCrea, cf. /43/. We present the new constant curvature and constant torsion solution in eqs. (6.13) and (6.14) and show that it is weakly double self dual.

In Sect. 7 it is shown that our new micro-deSitter solution with constant torsion is, by a suitable adaption of the constants, see. eq. (6.14)', also a solution of the field equations of the general polynomial Lagrangian (3.1), provided the fairly weak constraints (7.1) and (7.2) on the corresponding coupling constants are fulfilled.

2. THE TWO FIELD EQUATIONS OF THE POINCARÉ GAUGE FIELD THEORY

The independent gravitational potentials of the Riemann-Cartan spacetime of the Poincaré gauge field theory (PG) are the tetrad coefficients $e_i{}^\alpha$ and the connection coefficients $\Gamma_i{}^{\alpha\beta} = -\Gamma_i{}^{\beta\alpha}$. Here i,j ... = 0 ... 3 are holonomic (world) indices and α, β ... = 0 ... 3 anholonomic (Lorentz) indices. The tetrads are chosen orthonormal, i.e. the local metric $\eta_{\alpha\beta}$ coincides with the Minkowski metric diag. (-+++). The tetrad coefficients $e_i{}^\alpha$ can be interpreted as translational and the connection coefficients $\Gamma_i{}^{\alpha\beta}$ as Lorentz (or rotational) gauge potentials. The corresponding field strength tensors are torsion

$$F_{ij}{}^\alpha := 2 D_{[i} e_{j]}{}^\alpha = 2\left(\partial_{[i} e_{j]}{}^\alpha + \Gamma_{[i|\beta}{}^\alpha e_{|j]}{}^\beta\right) \tag{2.1}$$

and curvature

(2.2) $$F_{ij\alpha}{}^{\beta} = 2\left(\partial_{[i}\Gamma_{j]\alpha}{}^{\beta} + \Gamma_{[i|\varepsilon}{}^{\beta}\,\Gamma_{|j]\alpha}{}^{\varepsilon}\right),$$

respectively. The operator D_i represents the covariant exterior derivative. Let be given a matter field represented by a Poincaré spinor-tensor $\Psi(x^i)$ and referred to a local tetrad; its indices are suppressed. The action function W of a matter field Ψ that is interacting with the Poincaré gauge fields, can be put into the form

(2.3) $$W = \int d^4x\, e\left[L(\eta_{\alpha\beta}, \gamma^{\alpha} \ldots \Psi, D_\alpha\Psi) + V(\varkappa_1, \varkappa_2, \ldots, \eta_{\alpha\beta}, F_{\alpha\beta}{}^{\gamma}, F_{\alpha\beta\gamma}{}^{\delta})\right].$$

Here L is the special relativistic matter Lagrangian of, say, a Dirac field minimally coupled to the geometry, γ^{α} = Dirac matrices, $e: = \det e_i{}^{\alpha}$, and $\mathcal{L}: = eL$, whereas $\mathcal{V}: = eV$ is the gauge field Lagrangian depending on some coupling constants $\varkappa_1, \varkappa_2 \ldots$, on the local metric, and on the anholonomic components of torsion and curvature, respectively.

Let be given the field momenta by

(2.4) $$\mathcal{H}_{\alpha}{}^{ij} = 2\frac{\partial \mathcal{V}}{\partial F_{ji}{}^{\alpha}}, \qquad \mathcal{H}_{\alpha\beta}{}^{ij} = 2\frac{\partial \mathcal{V}}{\partial F_{ji}{}^{\alpha\beta}},$$

the momentum current of the gauge fields by

(2.5) $$\mathcal{E}_{\alpha}{}^{i} := e^{i}{}_{\alpha}\mathcal{V} - F_{\alpha j}{}^{\gamma}\,\mathcal{H}_{\gamma}{}^{ji} - F_{\alpha j}{}^{\gamma\delta}\,\mathcal{H}_{\gamma\delta}{}^{ji},$$

and the spin current of the gauge fields by

(2.6) $$\mathcal{E}_{\alpha\beta}{}^{i} := \mathcal{H}_{[\beta\alpha]}{}^{i},$$

then the field equations of the PG read

(2.7) $$D_j\,\mathcal{H}_{\alpha}{}^{ij} - \mathcal{E}_{\alpha}{}^{i} = e\,\Sigma_{\alpha}{}^{i},$$

(2.8) $$D_j\,\mathcal{H}_{\alpha\beta}{}^{ij} - \mathcal{E}_{\alpha\beta}{}^{i} = e\,\tau_{\alpha\beta}{}^{i}.$$

The sources on the right hand sides are the canonical momentum current and the canonical spin current of the matter field. For simplicity we will concentrate in this article on <u>vacuum</u> solutions of the field equations, i.e. we will put the material sources in (2.7,2.8) equal to zero later on.

The field equations are supplemented by the two Bianchi identities for torsion and curvature, respectively:

(2.9) $$D_{[i}F_{jk]}{}^{\beta} \equiv F_{[ijk]}{}^{\beta},$$

(2.10) $$D_{[i}F_{jk]\alpha}{}^{\beta} \equiv 0.$$

3. THE MOST GENERAL LAGRANGIAN YIELDING QUASI-LINEAR 2ND ORDER FIELD EQUATIONS

The guiding principle for the construction of the gauge field Lagrangian V is that we allow at most second derivatives of the gauge potentials $e_i^{\ \alpha}$ and $\Gamma_i^{\ \alpha\beta}$, and these second derivatives should appear in the field equations only linearly (hypothesis of quasi-linearity). If we further assume V to be polynomial in the field strengths, we find

$$
\begin{aligned}
(3.1)\qquad V = {} & \frac{1}{\mu l^4} + \frac{1}{l^2}\Big[\frac{1}{2\chi}F + \frac{1}{4}F_{\alpha\beta}{}^{\gamma}\big(d_1 F^{\alpha\beta}{}_{\gamma} + d_2 F_{\gamma}{}^{\beta\alpha} + d_3 \delta_{\gamma}^{\beta} F^{\alpha\mu}{}_{\mu}\big)\Big] \\
& - \frac{1}{4\varkappa}F_{\alpha\beta\gamma\delta}\Big[F^{\alpha\beta\gamma\delta} + f_1 F^{\alpha\gamma\beta\delta} + f_2 F^{\gamma\delta\alpha\beta} + f_3 \eta^{\alpha\delta}F^{\beta\gamma} \\
& + f_4 \eta^{\alpha\delta}F^{\gamma\beta} + f_5 \eta^{\alpha\delta}\eta^{\beta\gamma}F\Big] .
\end{aligned}
$$

The U_4-Ricci tensor is defined by $F_{\alpha\beta} := F_{\gamma\alpha\beta}{}^{\gamma}$, its contraction, the curvature scalar, by $F := F_{\gamma\mu}{}^{\mu\gamma}$. The Planck length is denoted by l, $1/\mu l^4$ represents the cosmological constant, and χ, $\varkappa$, f_A and d_A are dimensionless coupling constants. In (3.1) we hypothesized that, in addition to the usual gravitational potential $e_i^{\ \alpha}$ ("gravitons", weak Einstein gravity with coupling constant l^2), a propagating Lorentz gauge potential ("tordions", strong Yang-Mills gravity with coupling constant $\varkappa$) does exist in nature.+

By using (2.4) we derive the corresponding gauge field momenta, which are linear in torsion and curvature, respectively:

$$
(3.2)\qquad \mathcal{H}_{\alpha}{}^{ij} = \frac{e}{l^2}\big(-d_1 F^{ij}{}_{\alpha} + d_2 F_{\alpha}{}^{[ij]} + d_3 e^{[i}{}_{\alpha}F^{j]\gamma}{}_{\gamma}\big) ,
$$

$$
\begin{aligned}
(3.3)\qquad \mathcal{H}_{\alpha\beta}{}^{ij} = {} & \frac{e}{\chi l^2} e^{i}{}_{[\alpha} e^{j}{}_{\beta]} + \frac{e}{\varkappa}\big(F^{ij}{}_{\alpha\beta} + f_1 F^{i}{}_{[\alpha}{}^{j}{}_{\beta]} + f_2 F_{\alpha\beta}{}^{ij} \\
& + f_3 e^{[i}{}_{[\beta}F^{j]}{}_{\alpha]} + f_4 e^{[i}{}_{[\beta}F_{\alpha]}{}^{j]} + f_5 F e^{i}{}_{[\beta}e^{j}{}_{\alpha]}\big) .
\end{aligned}
$$

By substituting (3.1),(3.2) and (3.3) into (2.5)-(2.8), we find the two field equations of the PG in their explicit form:

+) This was proposed by Hehl, Ne'eman, Nitsch and Von der Heyde /59/. For a corresponding ansatz in supergravity see Nishino /60/.

(FIRST)
$$\frac{1}{\chi}\left(F_\alpha{}^i - \frac{1}{2}F e^i{}_\alpha + \Lambda e^i{}_\alpha\right) + \frac{1}{e}D_j\left(e d_1 F^{ji}{}_\alpha + e d_2 F_\alpha{}^{[ij]} + e d_3 e^{[i}{}_\alpha F^{j]\mu}{}_\mu\right)$$
$$+ F_{\alpha j}{}^\gamma\left(d_1 F^{ij}{}_\gamma + d_2 F_\gamma{}^{[ji]} + d_3 e^{[j}{}_\gamma F^{i]\mu}{}_\mu\right)$$
$$- \frac{1}{4}e^i{}_\alpha F_{\mu\nu}{}^\lambda\left(d_1 F^{\mu\nu}{}_\lambda + d_2 F_\lambda{}^{\nu\mu} + d_3 \delta^\nu{}_\lambda F^{\mu\gamma}{}_\gamma\right)$$
$$+ \frac{\ell^2}{\varkappa} F_{\alpha j}{}^{\gamma\delta}\left(F^{ji}{}_{\gamma\delta} + f_1 F^{[j}{}_\gamma{}^{i]}{}_\delta + f_2 F_{\gamma\delta}{}^{ji} + f_3 e^{[j}{}_{[\delta} F^{i]}{}_{\gamma]}\right.$$
$$\left. + f_4 e^{[j}{}_{[\delta} F_{\gamma]}{}^{i]} + f_5 F e^j{}_{[\delta} e^i{}_{\gamma]}\right)$$
$$- \frac{\ell^2}{4\varkappa} e^i{}_\alpha F_{\mu\nu\gamma\delta}\left(F^{\nu\mu\gamma\delta} + f_1 F^{\nu\gamma\mu\delta} + f_2 F^{\gamma\delta\nu\mu} + \right.$$
$$\left. f_3 F^{\mu\gamma}\eta^{\nu\delta} + f_4 F^{\gamma\mu}\eta^{\nu\delta} + f_5 F \eta^{\nu\delta}\eta^{\gamma\mu}\right)$$
$$= \ell^2 \Sigma_\alpha{}^i \ ,$$

(SECOND)
$$\frac{\varkappa}{2\chi\ell^2}T_{\alpha\beta}{}^i + \frac{1}{e}D_j\left(e F^{ij}{}_{\alpha\beta} + e f_1 F^{[i}{}_\alpha{}^{j]}{}_\beta + e f_2 F_{\alpha\beta}{}^{ij}\right.$$
$$\left. + e f_3 e^{[i}{}_{[\beta} F^{j]}{}_{\alpha]} + e f_4 e^{[i}{}_{[\beta} F_{\alpha]}{}^{j]} + e f_5 F e^i{}_{[\beta} e^j{}_{\alpha]}\right)$$
$$+ \frac{\varkappa}{\ell^2}\left(d_1 F^i{}_{[\beta\alpha]} - \frac{d_3}{2} e^i{}_{[\alpha} F_{\beta]}{}^\mu{}_\mu + \frac{3}{4} d_2 F_{[\alpha\beta}{}^{i]} + \frac{d_2}{2} F_{\alpha\beta}{}^i\right) = \varkappa \tau_{\alpha\beta}{}^i \ .$$

For brevity we put $\Lambda := -\frac{\chi}{\mu\ell^2}$. With this sign convention, and for $\mu \neq 0$, we formally receive the same Λ as in GR. Observe that one of the constants f_2, f_4 or f_5 can be eliminated by use of the Euler-Gauss-Bonnet theorem. For convenience we choose $f_5 = 0$.

4. SPHERICAL SYMMETRY WITH REFLECTION INVARIANCE

For spherical symmetry and in spherical coordinates, the metric turns out to be

(4.1)
$$ds^2 = -e^{2\mu(T,R)}dT^2 + e^{2\lambda(T,R)}dR^2 + r^2(T,R)(d\vartheta^2 + \sin^2\vartheta\, d\varphi^2)$$

with the unknown functions μ, λ and r. Under the same conditions the torsion tensor has eight independent components, which can be further reduced by imposing spatial reflection invariance: then the only non-zero components of the torsion tensor $F_{\alpha\beta\gamma}$ are

(4.2) $$\begin{aligned} F_{010} &= -f(T,R) \ , \\ F_{011} &= -h(T,R) \ , \\ F_{022} &= F_{033} = -K(T,R) \ , \\ F_{122} &= F_{133} = g(T,R) \ . \end{aligned}$$

$(\ )^{\cdot}$ will denote differentiation with respect to T and $(\ )'$ with respect to R.

In the following we will calculate the anholonomic components of the various geometrical objects involved. For this purpose we specify a tetrad associated with the metric (4.1). We choose the coframe as follows:

(4.3) $$\omega^0 = e^{\mu} dT \ , \quad \omega^1 = e^{\lambda} dR \ , \quad \omega^2 = r\, d\vartheta \ , \quad \omega^3 = r \sin\vartheta\, d\varphi \ .$$

The corresponding tetrad coefficients $e_i{}^{\alpha}$ can be read off from $\omega^{\alpha} = e_i{}^{\alpha} dx^i$.

The anholonomic connection is determined by

(4.4) $$\Gamma_{\alpha\beta\gamma} = -\Omega_{\alpha\beta\gamma} + \Omega_{\beta\gamma\alpha} - \Omega_{\gamma\alpha\beta} + \tfrac{1}{2}(F_{\alpha\beta\gamma} - F_{\beta\gamma\alpha} + F_{\gamma\alpha\beta})$$

with the object of anholonomity

(4.5) $$\Omega_{\alpha\beta\gamma} = \eta_{\gamma\delta}\, e^i{}_{\alpha}\, e^j{}_{\beta}\, \partial_{[i} e_{j]}{}^{\delta} \ .$$

For spherical symmetry (4.4) becomes

(4.6) $$\begin{aligned} \Gamma_{010} &= -(\mu' e^{-\lambda} + f) := V(T,R) \ , \\ \Gamma_{101} &= \dot{\lambda} e^{-\mu} + h := X(T,R) \ , \\ \Gamma_{202} &= \Gamma_{303} = \frac{\dot{r}}{r} e^{-\mu} + K := Y(T,R) \ , \\ \Gamma_{212} &= \Gamma_{313} = \frac{r'}{r} e^{-\lambda} - g := -W(T,R) \ , \\ \Gamma_{323} &= \frac{1}{r} \cot\vartheta \ . \end{aligned}$$

The curvature tensor

(4.7) $$F_{\alpha\beta\gamma}{}^{\delta} = 2\partial_{[\alpha} \Gamma_{\beta]\gamma}{}^{\delta} + 2\Gamma_{[\alpha|\varepsilon}{}^{\delta}\, \Gamma_{|\beta]\gamma}{}^{\varepsilon} + 2\Omega_{\alpha\beta}{}^{\varepsilon}\, \Gamma_{\varepsilon\gamma}{}^{\delta}$$

has the following non-vanishing components

$$
\begin{aligned}
(4.8)\quad F_{0101} &= -e^{-\mu-\lambda}\left[(X e^{\lambda})^{\cdot} - (V e^{\mu})' \right] &&:= A(T,R) ,\\
F_{0220} &= F_{0330} = -\frac{e^{-\mu}}{r}(Yr)^{\cdot} - VW &&:= C(T,R) ,\\
F_{1220} &= F_{1330} = -\frac{e^{-\lambda}}{r}(Yr)' - XW &&:= -D(T,R) ,\\
F_{0221} &= F_{0331} = \frac{e^{-\mu}}{r}(Wr)^{\cdot} + VY &&:= -G(T,R) ,\\
F_{1221} &= F_{1331} = \frac{e^{-\lambda}}{r}(Wr)' + XY &&:= H(T,R) ,\\
F_{2323} &= \frac{1}{r^2} + Y^2 - W^2 &&:= L(T,R) .
\end{aligned}
$$

5. A SPECIFIC TIME-DEPENDENT ANSATZ FOR METRIC AND TORSION

In order to find homogeneous time-dependent and not necessarily isotropic solutions of the field equations, we specialize the ansatz (4.3) to

$$(5.1)\quad \omega^0 = dT ,\quad \omega^1 = \Phi(T,R)\,dR ,\quad \omega^2 = \Phi(T,R)\,d\vartheta ,\quad \omega^3 = \Phi(T,R)\,R\sin\vartheta\, d\varphi$$

yielding the metric

$$(5.2)\quad ds^2 = -dT^2 + \Phi^2(T,R)\left(dR^2 + R^2 d\Omega^2\right) .$$

For the torsion we assume

$$
(5.3)\quad \begin{aligned}
-h(T,R) &= -k(T,R) =: Z(T,R) ,\\
g(T,R) &= f(T,R) \equiv 0 .
\end{aligned}
$$

Then the anholonomic connection reads

$$
(5.4)\quad \begin{aligned}
\Gamma_{101} &= \Gamma_{202} = \Gamma_{303} = \frac{\dot{\Phi}}{\Phi} - Z(T,R) ,\\
\Gamma_{212} &= \Gamma_{313} = \frac{1}{R} ,\quad \Gamma_{233} = \frac{\cot\vartheta}{R} ,
\end{aligned}
$$

and for the anholonomic components of the U_4-curvature $F_{\alpha\beta\gamma\delta}$ we find

$$
(5.5)\quad \begin{aligned}
F_{0101} &= F_{0202} = F_{0303} = \ddot{\Phi}/\Phi - \dot{\Phi}Z/\Phi - \dot{Z} ,\\
F_{0221} &= F_{0331} = -\dot{\Phi}'/\Phi^2 + \dot{\Phi}\Phi'/\Phi^3 = -(1/\Phi)(\dot{\Phi}/\Phi)' ,\\
F_{1202} &= F_{1303} = \dot{\Phi}'/\Phi^2 - \dot{\Phi}\Phi'/\Phi^3 - Z'/\Phi = (1/\Phi)(\dot{\Phi}/\Phi)' \\
&\qquad - Z'/\Phi ,\\
F_{1212} &= F_{1313} = (1/\Phi^2)(\Phi'/\Phi)' + \Phi'/\Phi^3 - \dot{\Phi}^2/\Phi^2 + 2\dot{\Phi}Z/\Phi - Z^2 ,\\
F_{2323} &= \Phi'^2/\Phi^4 + 2\Phi'/\Phi^3 R - \dot{\Phi}^2/\Phi^2 + 2\dot{\Phi}Z/\Phi - Z^2 .
\end{aligned}
$$

A further simplication can be achieved by the separation ansatz (cf. Fennelly and Pavelle /61/)

$$\Phi(T,R) = \frac{f(T)}{R^2} \quad . \tag{5.6}$$

6. THE NEW SOLUTION FOR THE QUADRATIC MODEL LAGRANGIAN

The field equations (FIRST) and (SECOND) of the 10-parameter Lagrangian are much more complicated than the Einstein equations. Therefore, for computational ease, it is advisable to pick a simple model Lagrangian which should have all the characteristic features of the full 10-parameter Lagrangian (3.1). A particular plausible choice is our <u>purely quadratic</u> Lagrangian /59/

$$V_Q = \frac{1}{4\ell^2} F_{\alpha\beta}{}^{\gamma}\left(-F^{\alpha\beta}{}_{\gamma} + 2\delta^{\beta}_{\gamma} F^{\alpha\mu}{}_{\mu}\right) - \frac{1}{4\varkappa} F_{\alpha\beta\gamma\delta} F^{\alpha\beta\gamma\delta} \quad , \tag{6.1}$$

$$\left(\Lambda = 1/\chi = f_A = 0\,, \; d_1 = -1\,, \; d_2 = 0\,, \; d_3 = 2\right) \; .$$

On substituting the data of Sect. 5 into (FIRST) and (SECOND) and using the parameter set of (6.1), we find for the vacuum case

$$\left\{\begin{aligned}
\mathrm{FIRST}(0,0) &= -3\dot{z}^2 + 6\dot{z}\ddot{f}/f - 6\dot{z}z\dot{f}/f - 2z'^2R^4/f^2 - 3\ddot{f}^2/f^2 + 6z\ddot{f}\dot{f}/f^2 + 3\dot{f}^4/f^4 \\
&\quad - 12z\dot{f}^3/f^3 + 15z^2\dot{f}^2/f^2 - 12z^3\dot{f}/f - 6z(\dot{f}/f)\varkappa/\ell^2 + 3z^4 + 3z^2\varkappa/\ell^2 = 0, \\
\mathrm{FIRST}(0,1) &= z'\left(\dot{z} - \ddot{f}/f + z\dot{f}/f\right) = 0, \\
\mathrm{FIRST}(1,0) &= z'\left(-2\dot{z} + 2\ddot{f}/f - 2z\dot{f}/f + \varkappa/\ell^2\right) = 0, \\
\mathrm{FIRST}(2,2) = \mathrm{FIRST}(3,3) &= \dot{z}^2 - 2\dot{z}\ddot{f}/f + 2\dot{z}z\dot{f}/f - 2\dot{z}\varkappa/\ell^2 + \ddot{f}^2/f^2 - z^4 \\
&\quad - 2z\ddot{f}\dot{f}/f^2 - \dot{f}^4/f^4 + 4z\dot{f}^3/f^3 - 5z^2\dot{f}^2/f^2 + 4z^3\dot{f}/f - 4z(\dot{f}/f)\varkappa/\ell^2 + z^2\varkappa/\ell^2 = 0, \\
\mathrm{FIRST}(1,1) &= \dot{z}^2 - 2\dot{z}\ddot{f}/f + 2\dot{z}z\dot{f}/f - 2\dot{z}\varkappa/\ell^2 + 2z'^2R^4/f^2 + \ddot{f}^2/f^2 - \\
&\quad 2\ddot{f}\dot{f}z/f^2 - \dot{f}^4/f^4 + 4z\dot{f}^3/f^3 - 5z^2\dot{f}^2/f^2 + 4z^3\dot{f}/f - z^4 - 4z(\dot{f}/f)\varkappa/\ell^2 + z^2\varkappa/\ell^2 = 0;
\end{aligned}\right. \tag{6.2}$$

$$\left\{\begin{aligned}
\mathrm{SECOND}(0,0,1) &= (1/f)(z'f)^{\cdot} = 0, \\
\mathrm{SECOND}(1,0,1) &= -\ddot{z} - 3\dot{z}\dot{f}/f - 2z'R^3/f^2 + \dddot{f}/f + \ddot{f}\dot{f}/f^2 - z\ddot{f}/f \\
&\quad - 2\dot{f}^3/f^3 + 5z\dot{f}^2/f^2 - 6z^2\dot{f}/f + 2z^3 + z\varkappa/\ell^2 = 0, \\
\mathrm{SECOND}(2,0,2) = \mathrm{SECOND}(3,0,3) &= -\ddot{z} - 3\dot{z}\dot{f}/f + z''R^4/f^2 + z'R^3/f^2 + 2z^3 \\
&\quad + \dddot{f}/f + \ddot{f}\dot{f}/f^2 - z\ddot{f}/f - 2\dot{f}^3/f^3 + 5z\dot{f}^2/f^2 - 6z^2\dot{f}/f + z\varkappa/\ell^2 = 0, \\
\mathrm{SECOND}(2,2,1) = \mathrm{SECOND}(3,3,1) &= z'\left(\dot{f} - zf\right) = 0 .
\end{aligned}\right. \tag{6.3}$$

Inspection of the antisymmetric part of (FIRST) reveals that

(6.4) $Z' = 0$ or $Z(T,R) = Z(T)$.

The trace of (FIRST) yields

(6.5) $\dot{Z} + 3Z\dot{f}/f - Z^2 = 0$

with the solution

(6.6) $$Z(T) = \frac{1}{f^3(T)\left[C_1 - \int \frac{d\tau}{f^3(\tau)}\right]} ,$$

where C_1 is an integration constant.

Eqs. (6.5) and (6.6) fulfill (SECOND) identically, i.e. the function f(T) should be determined by the tracefree part of (FIRST). Substitution of (6.5) into (FIRST) yields an integro-differential equation for the unknown function f(T)

(6.7) $$2Z(T)\left(2\dot{f}/f - Z(T)\right)\left[\ddot{f}/f + \dot{f}^2/f^2 + \varkappa/(2\ell^2)\right] + \ddot{f}^2/f^2 - \dot{f}^4/f^4 = 0$$

with Z(T) given by (6.6).

Eq. (6.7) admits at least one simple solution for f(T): Choose

(6.8) $Z(T) = \alpha\dot{f}/f$,

which implies $C_1 = 0$ in (6.6), and

(6.9) $f(T) = C e^{\beta T}$,

where C, α, and β are constants. Then (SECOND) is fulfilled for arbitrary C and α with β given by

(6.10) $$\beta^2 = -\frac{\varkappa}{2(\alpha-1)(\alpha-2)\ell^2} , \quad (\alpha \neq 1, \alpha \neq 2) .$$

(FIRST), however, restricts the value of α to

(6.11) $\alpha = 3$.

Accordingly, the metric is found to be

(6.12) $$ds^2 = -dT^2 + \frac{C^2 e^{2\beta T}}{R^4}\left(dR^2 + R^2 d\Omega^2\right)$$

By a coordinate transformation, it can be put into an explicitly conformally flat form. Collecting our results, we finally have the exact vacuum solution

(6.13a) $$ds^2 = \frac{1}{\beta^2 t^2} ds^2_{Minkowski} ,$$

(6.13b) $$F_{011} = F_{022} = F_{033} = \alpha\beta = \text{constant} ,$$

$$(6.13c)\qquad \left\| F_{\alpha\beta}{}^{\gamma\delta} = (\alpha-1)\beta^2 \cdot \begin{pmatrix} 1 & & & & & \\ & 1 & & & 0 & \\ & & 1 & & & \\ & & & 1-\alpha & & \\ & 0 & & & 1-\alpha & \\ & & & & & 1-\alpha \end{pmatrix} = \text{constant} \right\|$$

with

$$(6.14)\qquad \alpha = 3\,,\quad \beta = \sqrt{-\varkappa}/2\ell\ .$$

The antisymmetric index pairs (01,02,03,23,31,12) of the curvature matrix are numbered by (1,2...6) (cf. Misner, Thorne and Wheeler /62/ p. 361).

Consequently, for the purely quadratic Lagrangian solutions exist with constant (anholonomic) torsion and constant (anholonomic) U_4-curvature, but with an $F_{\alpha\beta}{}^{\gamma\delta}$ which does <u>not</u> satisfy $G_{\alpha\beta}(U_4) = \Lambda\eta_{\alpha\beta}$.

The U_4-curvature (6.13c) with (6.14) can be split into a Riemannian V_4-piece, caused by the deSitter metric (6.13a), and a purely torsion dependent piece according to

$$(6.15)\qquad F_{\alpha\beta}{}^{\gamma\delta} = \frac{\varkappa}{4\ell^2}\,\mathbb{1} + \frac{3\varkappa}{4\ell^2}\begin{pmatrix} -1 & & & & & \\ & -1 & & & & \\ & & -1 & & & \\ & & & 1 & & \\ & & & & 1 & \\ & & & & & 1 \end{pmatrix} .$$

This decomposition is irreducible under the local Lorentz group (cf. Debever /63,64/ and Lenzen /65/), the deSitter piece corresponds to the U_4-curvature scalar and the second piece to the tracefree symmetric U_4-Ricci tensor. The U_4-Weyl curvature tensor $C_{\alpha\beta}{}^{\gamma\delta}$ vanishes.

As can be read off from (6.15), the U_4-curvature of our new solution is weakly double self dual

$$(6.16)\qquad \overset{+}{F}_{\alpha\beta}{}^{\gamma\delta} := \frac{1}{2}\left(F_{\alpha\beta}{}^{\gamma\delta} + {}^*F^*_{\alpha\beta}{}^{\gamma\delta}\right) = \gamma\,\frac{\varkappa}{2\ell^2}\,\mathbb{1}$$

for $\gamma = 1/2$, where

$$(6.17)\qquad {}^*F^*_{\alpha\beta}{}^{\gamma\delta} := -\frac{1}{4}\,\eta_{\alpha\beta\rho\sigma}\,F^{\rho\sigma}{}_{\mu\nu}\,\eta^{\mu\nu\gamma\delta}$$

is the double dual of the curvature tensor ($\eta_{\alpha\beta\rho\sigma}$ = totally antisymmetric unit tensor). A double duality ansatz such as (6.16) can be used in the first place in order to find exact solutions (Baekler, Hehl and Mielke /38/, Benn, Dereli and Tucker /39/).+)

+) F. Müller-Hoissen has informed us that he rederived our new solution by solving his Friedmann type equations in /66,67/ for the vacuum case, cf. also Minkevich /68/.

7. THE NEW SOLUTION FOR THE GENERAL CASE

In this section we will show that the solution (6.13, 6.14) can be generalized for the 10-parameter Lagrangian (3.1). For this purpose we substitute (5.1), (5.3), (5.6), and (6.8) into (SECOND) and (FIRST). Then (SECOND), together with the double duality ansatz (6.16), yields the two constraints

(7.1) $$d_1 + \frac{d_2}{2} + \frac{d_3}{2} = 0$$

and

(7.2) $$\frac{1}{\chi} + 2\gamma\varrho - \frac{d_3}{2} = 0 \, ,$$

cf. Baekler /42/ ($\varrho := 1 + \frac{1}{2}f_1 + f_2 + f_3 + f_4$). (FIRST) leads to /42/

(7.3) $$\frac{d_3}{2} G_{\alpha\beta}(V_4) + \left(\frac{\Lambda}{\chi} - 3\gamma^2\varrho\frac{\varkappa}{\ell^2}\right)\eta_{\alpha\beta} = 0$$

or, assuming $G_{\alpha\beta}(V_4) = -3\beta^2\eta_{\alpha\beta}$ to be fulfilled, to

(7.4) $$-\frac{3}{2}d_3\beta^2 + \frac{\Lambda}{\chi} - 3\gamma^2\varrho\frac{\varkappa}{\ell^2} = 0 \quad .$$

The constraint (7.1) holds for the "viable set" of torsion-parameters, which, for $F_{\alpha\beta}{}^{\gamma\delta} = 0$, leads to the teleparallelism theory of gravity.+) Eq. (7.4) classifies the solutions into two different groups: One with $d_3 = 0$ and a second one with $d_3 \neq 0$.

Let us consider at first $d_3 = 0$. Then, as can be seen from (7.3), non-trivial solutions are only possible if we include a non-zero cosmological constant into the Lagrangian (see Benn, Dereli and Tucker /39/). The perhaps more interesting case is, however, that with $d_3 \neq 0$. Then (7.4) yields

(7.5) $$\beta^2 = -\frac{2}{3d_3}\left(3\gamma^2\varrho\frac{\varkappa}{\ell^2} - \frac{\Lambda}{\chi}\right) \quad .$$

These solutions can carry, in the case of $\Lambda = 0$, a "cosmological constant" without having a cosmological constant (see Baekler /37/, Baekler, Hehl and Mielke /38/).

For the function f we find, respectively++),

+) Recently interest in the teleparallelism theory increased greatly. The articles of Kopczyński /69/ and of Müller-Hoissen and Nitsch /70/ unify the formalism and give a critical evaluation of the viability of the theory. Recent work includes the articles of Cho /71/, Hayashi and Shirafuji /72/, Meyer /73/, Nitsch /74/, Nitsch and Hehl /75/, Schweizer and Straumann /76/, Schweizer, Straumann and Wipf /77/ and Smalley /78/.

++) For vanishing torsion, the solution $f = C\exp(\beta t)$ was found by

$$f(T) = \begin{cases} C \exp\left[\sqrt{-\frac{2}{3 d_3}\left(3\gamma^2 \varrho \frac{\varkappa}{\ell^2} - \frac{\Lambda}{\varkappa}\right)}\, T\right] & , d_3 \neq 0, \\ C \exp(\beta T), \ \beta = \text{arbitrary const.} & , d_3 = 0. \end{cases} \tag{7.6}$$

Accordingly,

$$\left\| \begin{array}{lll} \alpha = 3, & \beta^2 = -\frac{2}{3 d_3}\left(3\gamma^2 \varrho \varkappa/\ell^2 - \frac{\Lambda}{\varkappa}\right), & d_3 \neq 0, \\ \alpha = 3, & \beta^2 = \text{arbitrary constant}, & d_3 = 0, \end{array} \right\| \tag{6.14'}$$

together with (6.13), represent solutions for the 10-paramter Lagrangian, provided the two constraints (7.1) and (7.2) are fulfilled.

ACKNOWLEDGEMENTS

One of the authors (F.W.H.) would like to thank Prof. R. Utiyama for the invitation to give a lecture at the Nara Symposium on Gauge Theory and Gravitation upon which this paper is based. Furthermore, we appreciate helpful remarks of B. Mashhoon, H.-J. Lenzen, and J. Nitsch. A. Krasiński brought his ORTOCARTAN to Cologne and installed it, and J. Dermott McCrea (Dublin) showed us how to write effective REDUCE programs for the PG. We are very grateful to them as well as to R. Esser of our Computer Center whose help was indispensible in order to bring our programs to work.

LITERATURE

/1/ Yang, C.N., and R. Mills: Phys. Rev. 96, 191 (1954).
/2/ Utiyama, R.: Phys. Rev. 101, 1597 (1956).
/3/ Utiyama, R.: Progr. Theor. Phys. 64, 2207 (1980).
/4/ Weyl, H.: Z. Phys. 56, 330 (1929).
/5/ Sciama, D.W.: In "Recent Developments in General Relativity". Festschrift for Infeld, p. 415, Pergamon, Oxford (1962).
/6/ Kibble, T.W.B.: J. Math. Phys. 2, 212 (1961).
/7/ Cartan, E.: Ann. Ec. Norm. Sup. 40, 325 (1923); 41, 1 (1924).
/8/ Hayashi, K., and T. Shirafuji: Progr. Theor. Phys. 64, 866, 883, 1435, 2222 (1980); 65, 525, 2079 (E) (1981); 66, 318, 741 (E), 2258 (1981).
/9/ Hayashi, K., and T. Nakano: Progr. Theor. Phys. 38, 491 (1967).
/10/ Ne'eman, Y.: In /13/, Vol. 1, p. 309 (1980).
/11/ Trautman, A.: In /13/, Vol. 1, p.287 (1980).
/12/ Hehl, F.W., J. Nitsch and P. von der Heyde: In /13/, Vol. 1, p. 329 (1980).
/13/ Held, A. (ed.): "General Relativity and Gravitation, One Hundred Years After the Birth of Albert Einstein." Two Volumes. Plenum Press, N.Y. (1980).
/14/ Hennig, J., and J. Nitsch: GRG Journal 13, 947 (1981).
/15/ Mielke, E.W.: Habilitation Thesis, Univ. of Kiel (1982).
/16/ Ne'eman, Y.: In "Some Strangeness in the Proportion, A Centennial Symposium to Celebrate the Achievements of Albert Einstein." H. Woolf (Ed.), p. 429, Addison-Wesley, Reading (1980).

Fennelly and Pavelle /61/ in Yang's gauge model of gravity. Moreover, it also solves the first field quation of SKY-gravity (see /38/ and also Goenner and Havas /79/, Mielke /80/, Ni /81/ and Pavelle /82/).

/17/ Nieh, H.T., and M.L. Yau: Ann. Phys. (N.Y.) 138, 237 (1982).
/18/ Nieh, H.T.: J. Math. Phys. 21, 1439 (1980).
/19/ Nieh, H.T., and M.L. Yau: J. Math. Phys. 23, 373 (1982).
/20/ Schweizer, M.A.: Thesis, Univ. of Zürich (1980).
/21/ Szczyrba, W.: In /33/, p. 105 (1980).
/22/ Szczyrba, W.: Phys. Rev. D25, 2548 (1982).
/23/ Thirring, W.: Lecture Notes in Physics (Springer) 116, 272 (1980).
/24/ Tseytlin, A.A.: Phys. Rev. D., to be published (1982).
/25/ Wallner, R.P.: GRG Journal 12, 719 (1980).
/26/ Wallner, R.P.: Thesis, Univ. of Vienna (1982) and references given.
/27/ Yasskin, P.D.: Thesis, Univ. of Maryland (1979).
/28/ Audretsch, J.: Phys. Rev. D24, 1470 (1981); D25, 605 (E) (1982).
/29/ Rumpf, H.: In /33/, p. 93 (1980).
/30/ Rumpf, H.: GRG Journal 14, 773 (1982).
/31/ Drechsler, W.: Phys. Lett. 107B, 415 (1981).
/32/ Hehl, F.W.: "Four Lectures in Poincaré Gauge Field Thoery." In /33/, p. 5 (1980).
/33/ Bergmann, P.G., and V. de Sabbata (Eds.): "Cosmology and Gravitation: Spin, Torsion, Rotation and Supergravity". Proceedings of the Erice-School, May 1979. Plenum, New York (1980).
/34/ Von der Heyde, P.: Lett. Nuovo Cimento 14, 250 (1975).
/35/ Mack, G.: Fortschr. Phys. 29, 135 (1981).
/36/ Neville, D.E.: Phys. Rev. D21, 2770 (1980).
/37/ Baekler, P.: Phys. Lett. 99B, 329 (1981).
/38/ Baekler, P., F.W. Hehl and E.W. Mielke: Preprint IC/80/140, Trieste (1980), In "Proc. 2nd M. Grossmann Meeting on General Relativity", R. Ruffini (Ed.), North-Holland, Amsterdam (1982).
/39/ Benn, I.M., T. Dereli and R.W. Tucker: GRG Journal 13, 581 (1981).
/40/ Benn, I.M., T. Dereli and R.W. Tucker: J. Phys A15, 849 (1982).
/41/ Baekler, P.: Diploma Thesis, Univ. of Cologne (1980).
/42/ Baekler, P.: Preprint, Univ. of Cologne (1982).
/43/ McCrea, J.D.: Preprint, Univ. College, Dublin (1982).
/44/ Adamowicz, W.: GRG Journal 12, 677 (1980).
/45/ Chen, H.-H., and D.-C. Chern: Preprint, National Central Univ., Taiwan (1982).
/46/ Mielke, E.W.: Preprint IC/81/210, Intern. Centre Theor. Physics, Trieste (1981).
/47/ Ramaswamy, S., and P.B. Yasskin: Phys. Rev. D19, 2264 (1979).
/48/ Baekler, P., and P.B. Yasskin: GRG Journal, to be published (1983).
/49/ Nieh, H.T., and R. Rauch: Phys. Lett. 81A, 113 (1981).
/50/ Rauch, R., and H.T. Nieh: Phys. Rev. D24, 2029 (1981).
/51/ Rauch, R., J.C. Shaw and H.-T. Nieh: GRG Journal 14, 331 (1982).
/52/ Rauch, R.T.: Phys. Rev. D25, 577 (1982); D26, 931 (1982).
/53/ Tsamparlis, M.: Phys. Lett. 75A, 27 (1979).
/54/ Tsamparlis, M.: Phys. Rev. D24, 1451 (1981).
/55/ Hearn, A.C.: REDUCE 2 User's Manual, 2nd ed., Univ. of Utah (1973).
/56/ Krasiński, A., and M. Perkowski: The System ORTOCARTAN - User's Manual, 2nd ed., Warsaw (1980).
/57/ Krasiński, A., and M. Perkowski: GRG Journal 13, 67 (1981).
/58/ d'Inverno, R.A.: In /13/, Vol. 1, p. 491 (1980).
/59/ Hehl, F.W., Y. Ne'eman, J. Nitsch and P. Von der Heyde: Phys. Lett. 78B, 102 (1978).
/60/ Nishino, H.: Progr. Theor. Phys. 66, 287 (1981).
/61/ Fennelly, A.J., and R. Pavelle: J. Phys. A12, 227 (1979).
/62/ Misner, Ch. W., K.S. Thorne and J.A. Wheeler: "Gravitation", Freeman, San Francisco (1973).
/63/ Debever, R.: Bull. Acad. Roy. Belgique, Cl.Sc. 42, 313, 608, 1033 (1956).
/64/ Debever, R.: Cahier de Physique 18, 303 (1964).
/65/ Lenzen, H.J.: Diploma Thesis, Univ. of Cologne (1982).
/66/ Müller-Hoissen, F.: Phys. Lett. A, to be published (1982).
/67/ Müller-Hoissen, F.: Preprint, Univ. of Göttingen (1982).

/68/ Minkevich, A.V.: Phys. Lett. 80A, 232 (1980).
/69/ Kopczyński, W.: J. Phys. A15, 493 (1982).
/70/ Müller-Hoissen, F., and J. Nitsch: Preprint, Univ. of Göttingen (1982).
/71/ Cho, Y.M.: Phys. Rev. D14, 2521 (1976).
/72/ Hayashi, K., and T. Shirafuji: Phys. Rev. D24, 3312 (1981) and references given.
/73/ Meyer, H.: GRG Journal 14, 531 (1982).
/74/ Nitsch, J.: In /33/, p. 63 (1980).
/75/ Nitsch, J., and F.W. Hehl: Phys. Lett. 90B, 98 (1980).
/76/ Schweizer, M., and N. Straumann: Phys. Lett. 71A, 493 (1979).
/77/ Schweizer, M., N. Straumann and A. Wipf: GRG Journal 12, 95 (1980).
/78/ Smalley, L.L.: Phys. Rev. D21, 328 (1980).
/79/ Goenner, H., and P. Havas: J. Math. Phys. 21, 1159 (1980).
/80/ Mielke, E.W.: GRG Journal 13, 175 (1981).
/81/ Ni, W.-T.: Phys. Rev. Lett. 35, 319 (1975).
/82/ Pavelle, R.: Phys. Rev. Lett. 34, 1114, 1484 (E) (1975).

NEW GENERAL RELATIVITY

- TRANSLATION GAUGE THEORY -

Takeshi SHIRAFUJI

Physics Department, Saitama University
Urawa, Saitama 338, Japan

1. Introduction

In 1956 Utiyama proposed to introduce the gravitational field as the gauge field of the Lorentz group.[1] He introduced 24 fields by generalizing 6 constant parameters ω_{ij} $(= - \omega_{ji})$ for homogeneous Lorentz transformations to arbitrary functions $\omega_{ij}(x)$. Later Kibble considered all the 10 parameters of the inhomogeneous Lorentz group (Poincaré group), and laid the basis for Poincaré gauge theory with 40 independent field variables.[2]

Hayashi and Nakano proposed to extend translations only, leaving the six parameters ω_{ij} constant.[3] In this translation gauge theory, 16 field variables $c_k{}^\mu$ are introduced as the gauge field, by requiring that the action integral be invariant under the group of extended translations and global Lorentz transformations. This invariance group is the simplest one that includes the group of general coordinate transformations.

The underlying space-time of the translation gauge theory is the Weitzenböck space-time with absolute parallelism. The notion of absolute parallelism was first introduced into physics by Einstein, trying to unify gravitation and electromagnetism.[4] His attempt failed because there was no Schwarzschild solution in his field equation.[5] A purely gravitational theory based on the Weitzenböck space-time was revived by Møller,[6] and its Lagrangian formulation was given by Pellegrini and Plebanski.[7]

The theory of gravitation based on the Weitzenböck space-time was extensively studied by Hayashi and Shirafuji,[8] and it was given the name, new general relativity, since Einstein in 1928, after inventing general relativity, considered absolute parallelism for the first time, and since its main consequences were analogous to those of general relativity so far as macroscopic phenomena were concerned.

2. Fundamental particles and translation gauge group

We start from the action integral in special relativity for the fundamental particles of spin 1/2,

$$S_M = \int d^4x\, L_M(q, \partial_k q) , \tag{2.1}$$

which is invariant under Lorentz transformations,

$$\delta x^k = c^k + \omega^k{}_j x^j , \qquad (\omega_{kj} = - \omega_{jk}) , \tag{2.2a}$$

$$\delta q = (i/2)\omega_{ij}S^{ij}q , \qquad \delta(\partial_k q) = (i/2)\omega_{ij}S^{ij}(\partial_k q) + \omega_k{}^j(\partial_j q) , \tag{2.2b}$$

where S^{ij} are the Lorentz generators, and c^k and ω_{ij} are independent, constant 10 parameters. Here q collectively denotes quarks and leptons, and the Minkowski metric η_{ij} is given by diag(-1,+1,+1,+1).

We now extend translations to <u>extended translations</u> (namely, to general coordinate transformations) for which the parameters c^k are arbitrary functions of space-time points, and demand that the action integral should be invariant under general coordinate transformations and under global Lorentz transformations,

$$\delta x^\mu = \xi^\mu(x) , \qquad \delta q = (i/2)\omega_{ij}S^{ij}q , \tag{2.3}$$

where $\xi^\mu(x)$ are arbitrary four functions and ω_{ij} are <u>constant</u> 6 parameters as before. Since we are now treating general coordinate transformations, we use Greek letters for coordinate indices and distinguish them from Lorentz indices denoted by Latin letters.

To meet with the invariance requirement, we must define those quantities $D_k q$ which change under (2.3) in the same manner as $\partial_k q$ of (2.2b).[9] We define $D_k q$ by

$$D_k q = (\delta_k^\mu + c_k{}^\mu)\partial_\mu q , \tag{2.4}$$

then we get the following transformation rule for $c_k{}^\mu$;

$$\delta c_k{}^\mu = \partial_\nu \xi^\mu c_k{}^\nu + \omega_k{}^j c_j{}^\mu + \delta_k{}^\nu \partial_\nu \xi^\mu + \delta_j{}^\mu \omega_k{}^j . \tag{2.5}$$

The field $c_k{}^\mu$ is the gauge field associated with the translation gauge group. It transforms inhomogeneously under extended translations. The special relativistic limit is obtained by putting $c_k{}^\mu = 0$: When $c_k{}^\mu = 0$, Greek indices are equivalent to Latin ones, and the transformations (2.3) compatible with (2.5) are restricted by

$$\partial_j \xi^k + \omega_j{}^k = 0 , \tag{2.6}$$

from which we get (2.2a).

The transformation law of the translation gauge field given by (2.5) is rather complicated. The field $b_k{}^\mu$ defined by

$$b_k{}^\mu = \delta_k{}^\mu + c_k{}^\mu \tag{2.7}$$

obeys much simpler transformation law,

$$\delta b_k{}^\mu = \partial_\nu \xi^\mu b_k{}^\nu + \omega_k{}^j b_j{}^\mu . \tag{2.8}$$

Also, we define $b^k{}_\mu$ by

$$b^k{}_\mu b_k{}^\nu = \delta_\mu{}^\nu , \qquad b^k{}_\mu b_j{}^\mu = \delta_j{}^k . \tag{2.9}$$

The invariant action integral is now given by

$$S_M = \int b \, d^4x \, L_M(q, D_k q) \tag{2.10}$$

with

$$b = \det(b^k{}_\mu) , \tag{2.11}$$

because b changes like $\delta b = - b\partial_\mu \xi^\mu$, and bd^4x gives invariant volume element.

3. Gravitational field equation

We shall construct a gravitational Lagrangian density in vacuum

$$S_G = \int b \, d^4x \, L_G \tag{3.1}$$

by the following basic postulates: (1) Invariance under the group of extended translations and global Lorentz transformations, (2) L_G be quadratic in the translation gauge field strength,

$$T_{ijk} = b_j{}^\mu b_k{}^\nu (\partial_\nu b_{i\mu} - \partial_\mu b_{i\nu}) , \tag{3.2}$$

and (3) L_G be invariant under the parity operation. The most general gravitational Lagrangian density L_G can then be represented as

$$L_G = \alpha(t_{ijk} t^{ijk}) + \beta(v_i v^i) + \gamma(a_i a^i) \tag{3.3}$$

with α, β and γ three unknown parameters with dimension of $(\text{mass})^2$, where t_{ijk}, v_i and a_i are the three irreducible parts of T_{ijk};

$$t_{ijk} = (1/2)(T_{ijk} + T_{jik}) + (1/6)(\eta_{ki} v_j + \eta_{kj} v_i) - (1/3)\eta_{ij} v_k , \tag{3.4a}$$

$$v_i = T^k{}_{ki} , \tag{3.4b}$$

$$a_i = (1/6)\varepsilon_{ijkm} T^{jkm} . \tag{3.4c}$$

Taking the variation of the total action integral,

$$S = S_G + S_M , \tag{3.5}$$

with respect to $b^k{}_\mu$, we get the following gravitational field equation,

$$2D^k F_{ijk} + 2v^k F_{ijk} + 2H_{ij} - \eta_{ij} L_G = T_{ij} , \tag{3.6}$$

where

$$\tilde{F}_{ijk} = \alpha(t_{ijk} - t_{ikj}) + \beta(\eta_{ij}v_k - \eta_{ik}v_j) - (\gamma/3)\varepsilon_{ijkm}a^m , \quad (3.7a)$$

$$\tilde{H}_{ij} = T_{mni}\tilde{F}^{mn}{}_j - (1/2)T_{jmn}\tilde{F}_i{}^{mn} . \quad (3.7b)$$

Here T_{ij} is the energy-momentum tensor defined by

$$T_{ij} = (1/b)b_{j\nu}[\delta(bL_M/\delta b^i{}_\nu] , \quad (3.8)$$

which reduces to the canonical energy-momentum tensor

$$T_{ij} = -[\partial L_M/\partial(\partial^j q)]\partial_i q + \eta_{ij}L_M \quad (3.9)$$

in the special relativistic limit, $c_k{}^\mu = 0$.

As a simple example, let us consider a spherically symmetric case where the field $b_k{}^\mu$ takes a diagonal form,

$$\begin{aligned} b_{(0)}{}^0 &= A(r,t) , & b_{(0)}{}^\alpha &= 0 = b_{(a)}{}^0 , \\ b_{(a)}{}^\alpha &= B(r,t)\delta_a{}^\alpha , & (a, \alpha &= 1, 2, 3) \end{aligned} \quad (3.10)$$

with A and B two unknown functions of t and $r = (x^\alpha x^\alpha)^{1/2}$, where we have enclosed Latin indices by parentheses. In this case the gravitational field equation <u>in vacuum</u> (namely, $T_{ij} = 0$) can be solved exactly: The functions A and B should be <u>time-independent</u>, and are given by

$$\begin{aligned} A(r) &= (1 - \frac{GM}{pr})^{-p/2}(1 + \frac{GM}{qr})^{q/2} , \\ B(r) &= (1 - \frac{GM}{pr})^{(p-2)/2}(1 + \frac{GM}{qr})^{-(q+2)/2} , \end{aligned} \quad (3.11)$$

where p and q are defined by

$$p = \frac{2}{1-5\varepsilon}\{[(1-\varepsilon)(1-4\varepsilon)]^{1/2} - 2\varepsilon\} , \qquad q = \frac{2}{1-5\varepsilon}\{[(1-\varepsilon)(1-4\varepsilon)]^{1/2} + 2\varepsilon\} \quad (3.12)$$

with

$$\varepsilon = (\alpha+\beta)/(\alpha+4\beta) . \quad (3.13)$$

Here GM is an integration constant: G is Newton's gravitational constant and M can be interpreted as the gravitational mass of the central gravitating body. We notice that when the parameter ε is vanishing, this solution gives the Schwarzschild metric written in the isotropic coordinates with the metric tensor $g_{\mu\nu}$ defined from $b_k{}^\mu$ according to (4.1) below.

4. Geometry of space-time structure

The set of four vectors $\underline{b} = \{b_k{}^\mu\}$ given by (2.7) defines a global set of orthonormal frame with respect to the metric $\underline{g}$ with the metric tensor $g_{\mu\nu}$ given by

$$g_{\mu\nu} = b^k{}_\mu b_{k\nu} \ . \tag{4.1}$$

The spinor fields q of the fundamental particles of spin 1/2 are defined by referring to this orthonormal frame. The operator $D_k = b_k{}^\mu \partial_\mu$ of (2.4) is the covariant derivative with respect to the absolute parallelism which takes b as the parallel vector fields. This absolute parallelism defines the nonsymmetric affine connection Γ,

$$\Gamma^\lambda_{\mu\nu} = b_k{}^\lambda \partial_\nu b^k{}_\mu \ . \tag{4.2}$$

The torsion tensor of this connection is given by

$$T^\lambda{}_{\mu\nu} = \Gamma^\lambda_{\mu\nu} - \Gamma^\lambda_{\nu\mu} = b_k{}^\lambda(\partial_\nu b^k{}_\mu - \partial_\mu b^k{}_\nu) \ , \tag{4.3}$$

and coincides with the translation gauge field strength of (3.2).

Accordingly, the translation gauge theory can be interpreted as a gravitational theory based on the Weitzenböck space-time with absolute parallelism. The translation gauge field defines the parallel vector fields by (2.7), and the translation gauge field strength represents the torsion tensor. In the Weitzenböck space-time, the curvature tensor defined by the connection (4.2) is identically vanishing, and the torsion tensor describes the non-Minkowskian structure of space-time. This situation can be contrasted with that in the Riemann space-time characterized by the curvature tensor alone.

The Riemann-Cartan space-time has the curvature tensor and the torsion tensor, and it is the underlying space-time of Poincaré gauge theory. From this space-time follow two interesting space-time models: One is the Riemann space-time with the curvature tensor alone and the other is the Weitzenböck space-time with the torsion tensor alone. General relativity is a gravitational theory based on the Riemann space-time, while the translation gauge theory is based on the Weitzenböck space-time. Both theories can indeed be formulated as special limiting cases of Poincaré gauge theory.

The translation and the Lorentz gauge field strengths of Poincaré gauge theory represent the torsion and the curvature of the underlying Riemann-Cartan space-time, respectively. We assume that the gravitational Lagrangian density be linear and quadratic in the field strengths. The curvature tensor has one linear invariant and 6 quadratic invariants, while the torsion tensor has 3 quadratic invariants. The most general gravitational Lagrangian density of Poincaré gauge theory is then given by[10)]

$$L_G = a(\text{lenear invariant}) + (\alpha,\beta,\gamma)(3 \text{ quadratic invariants of the torsion tensor}) + (a_1,\ldots,a_6)(6 \text{ quadratic invariants of the curvature tensor}),$$

where the four parameters, a, α, β, and γ, are of dimension $(\text{mass})^2$, and the remaining six parameters, $a_1,\ldots, a_6$, are dimensionless. It has been shown that Poincaré gauge theory reduces to general relativity and to new general relativity (namely, to the translation gauge theory) in the limits, (i) $\alpha\to\infty$, $\beta\to\infty$, $\gamma\to\infty$, and (ii) $a_i\to\infty$, respectively.[11] In the first limit (i), the torsion tensor is vanishing, and the above Lagrangian density reduces to the following quadratic Lagrangian density in general relativity,

$$L_G = aR(\{\}) - bR_{\mu\nu}(\{\})R^{\mu\nu}(\{\}) + cR(\{\})^2 \tag{4.4}$$

with two dimensionless constants b and c related linearly to a_i's. In the second limit (ii), the curvature is vanishing, and the Lagrangian density reduces to (3.3).

5. Comparison with experiments

(A) The equivalence principle

The world line of the fundamental particles of spin 1/2 and of light rays propagating in vacuum is the geodesics of the metric $\underline{g}$ of (4.1), as can be shown by taking the short wave-length limit of the Dirac equation and the Maxwell equation. A macroscopic system such as a planet or a test particle can be described to a good approximation by the macroscopic energy-momentum tensor, which is obtained from the microscopic energy-momentum tensor by averaging in space and in time. When the spin direction of constituent particles is randomly distributed, the antisymmetric part of the microscopic energy-momentum tensor cancels out in the averaging process because of the Tetrode formula,

$$T^{[ij]} = (1/2b)\partial_\nu(bS^{ij\nu}) \tag{5.1}$$

with $S^{ij\nu}$ the spin tensor. It can then be shown from the gravitational field equation (3.6) that the macroscopic, symmetric energy-momentum tensor obeys the conservation law,

$$\nabla_\nu T^{\mu\nu} = 0 \ , \tag{5.2}$$

where ∇_ν is the covariant derivative with respect to the Christoffel symbol. The world line of macroscopic bodies such as planets and test bodies is then the geodesics of the metric $\underline{g}$.

Thus, as far as the effects due to intrinsic spin are negligibly small, the equivalence principle is valid in new general relativity. Violation of the equivalence principle occurs only in the microscopic world: For example, the precession of spin in the torsion field.

(B) Comparison with solar-system experiments

We demand that the gravitational field equation reproduces the correct Newtonian limit: This gives the following condition of the parameters,

$$\alpha\varkappa + 4(\beta\varkappa) + 9(\alpha\varkappa)(\beta\varkappa) = 0 \ , \tag{5.3}$$

where $\varkappa$ is Einstein's gravitational constant, $\varkappa = 8\pi G$. Notice that $\alpha\varkappa$ and $\beta\varkappa$ are dimensionless, and by virtue of (3.13) and (5.3) they are expressed by

$$\alpha\kappa = -1/3(1-\varepsilon)\ , \qquad \beta\kappa = 1/3(1-4\varepsilon)\ . \tag{5.4}$$

From the exact solution (3.11) we see that Eddington-Robertson's post-Newtonian parameters defined by

$$ds^2 = (1 - 2(\frac{GM}{r}) + 2c(\frac{GM}{r})^2 + \ldots)dt^2 - (1 + 2d(\frac{GM}{r}) + \ldots)dx^\alpha dx^\alpha \tag{5.5}$$

are given by

$$c = 1 - \varepsilon/2\ , \qquad d = 1 - 2\varepsilon\ . \tag{5.6}$$

(We notice that Eddington-Robertson's parameters c and d defined above are usually denoted by β and γ, respectively.)

According to the solar-system experiments, the parameter ε should be severely restricted;

$$\varepsilon = -\ 0.004 \pm 0.004\ , \tag{5.7}$$

and therefore, we have for $\alpha\kappa$ and $\beta\kappa$ the following experimental values,

$$\alpha\kappa = -\ 1/3 + (0.001 \pm 0.001)\ , \qquad \beta\kappa = 1/3 + (-\ 0.005 \pm 0.005)\ . \tag{5.8}$$

6. Model with $\alpha+\beta=0$

As we have seen in the previous section, the parameter ε is severely restricted by solar-system experiments. In view of this, we shall now assume that the parameter ε is exactly vanishing. The parameters α and β are then given by

$$\alpha = -\ 1/3\kappa\ , \qquad \beta = 1/3\kappa\ . \tag{6.1}$$

The gravitational Lagrangian density L_G of (3.3) becomes

$$\begin{aligned} L_G &= -(1/3\kappa)(t_{ijk}t^{ijk} - v_i v^i) + \gamma(a_i a^i) \\ &= (1/2\kappa)R(\{\}) + (9/4\lambda)(a_i a^i) + (\text{a total derivative})\ , \end{aligned} \tag{6.2}$$

where $R(\{\})$ is the Riemann-Christoffel scalar curvature defined by the metric tensor of (4.1), and $\lambda = 9\kappa/(4\gamma\kappa - 3)$.

The gravitational field equation of the present case is still complicated compared with the Einstein equation of general relativity, and very little is known about its solutions. We briefly mention the results for spherically symmetric and axially symmetric solutions in vacuum.

(1) Spherically symmetric solution in vacuum:

Irrespectively of whether the source is time-independent or not, the axial-vector part of the torsion tensor a_i should be vanishing, and $b_k{}^\mu$ obeys the Einstein equation in vacuum. According to the Birkhoff theorem, the metric should then be given by the Schwarzschild solution. Therefore, the parallel vector fields $b_k{}^\mu$ are obtained from the special solution (3.10-11) with $\varepsilon = 0$ by a local Lorentz transformation which preserves the condition $a_i = 0$.

(2) Stationary, axially symmetric solutions in vacuum:

Fukui and Hayashi derived a class of axially symmetric solutions in vacuum starting from axially symmetric solutions of the Einstein equation in vacuum (such as the Kerr solution or the Tomimatsu-Sato solutions).[12] In their solutions in vacuum, the axial-vector part of the torsion tensor, a_i, is non-vanishing.

The Lagrangian density of (6.2) is invariant under a restricted class of local Lorentz transformations, namely under those local Lorentz transformations which leave the axial-vector part of the torsion tensor a^μ invariant. The Dirac equation is also invariant under these local Lorentz transformations. The gravitational field equation derived from (6.2), however, is not covariant under such local Lorentz transformations. This fact casts some doubts on the internal consistentcy of the present model with $\alpha+\beta = 0$.[13)] At present we have not definite answer to this problem.

In the weak field situations with $|c_k{}^\mu| \ll 1$, we can expand the field equation into power series of $c_k{}^\mu$ and can keep only lowest order terms. We need not distinguish Greek indices from Latin indices, and can decompose the translation gauge field into symmetric and antisymmetric parts,

$$c_{ij} = -(1/2)h_{ij} - A_{ij} \tag{6.3}$$

with $h_{ij} = h_{ji}$ and $A_{ij} = -A_{ji}$. Notice that the symmetric field h_{ij} is just the weak field correction to the metric tensor: $g_{\mu\nu} = \eta_{\mu\nu} + h_{\mu\nu}$ with $h_{\mu\nu} = \delta_\mu{}^i\delta_\nu{}^j h_{ij}$. In this approximation the gravitational field equation is decoupled into its symmetric and antisymmetric parts,

$$\Box h_{ij} - \partial^k(\partial_i h_{jk} + \partial_j h_{ik}) + \eta_{ij}\partial_m\partial_n h^{mn} - (\eta_{ij}\Box - \partial_i\partial_j)h^k{}_k = -2\kappa T_{(ij)} \,, \tag{6.4a}$$

$$\Box A_{ij} - \partial^k(\partial_i A_{jk} - \partial_j A_{ik}) = -\lambda T_{[ij]} \,. \tag{6.4b}$$

From (6.4a) and (6.4b) follow the conservation laws,

$$\partial_j T^{(ij)} = 0 \,, \qquad \partial_j T^{[ij]} = 0 \,. \tag{6.5}$$

The symmetric part (6.4a) is just the linearized Einstein equation, and describes the massless graviton field of spin 2. The antisymmetric part, on the other hand, represents a massless field of spin 0 interacting with the intrinsic spin of the fundamental particles. If the parameter λ is positive, the energy of this spinless field is positive-definite. So we shall assume λ is positive.

The symmetric field h_{ij} gives rise to the universal attraction (namely, gravitation) between the fundamental particles. The antisymmetric field A_{ij} induces a universal spin-spin interaction which can be described by the interaction Hamiltonian,[14)]

$$H_{spin-spin} = (\lambda/8\pi)\vec{S}_A\cdot[\vec{\nabla}\times(\vec{\nabla}\times\vec{S}_B)]\frac{1}{r}$$
$$= (\lambda/8\pi)[(8\pi/3)(\vec{S}_A\cdot\vec{S}_B)\delta^3(\vec{x}) - r^{-3}\{(\vec{S}_A\cdot\vec{S}_B) - 3r^{-2}(\vec{S}_A\cdot\vec{x})(\vec{S}_B\cdot\vec{x})\}] , \quad (6.6)$$

where $\vec{S}_A$ and $\vec{S}_B$ are the spin vectors of spin 1/2 particles A and B, respectively. This interaction is of the same form as that between two magnetic moments, and is expected to contribute to the hyperfine splitting of energy levels in atoms and muoniums. The theoretical values for the hyperfine splitting based on Q.E.D. are in very good agreement with the experimental values both for atoms and muoniums. So we get the following upper bound for the parameter λ,

$$\lambda/4\pi \lesssim 3\times10^{-4}\ (\mathrm{GeV})^{-2} . \quad (6.7)$$

7. Conclusion

The translation gauge theory (or new general relativity) is a gravitational theory on the Weitzenböck space-time with absolute parallelism. The Weitzenböck space-time is a special case of the Riemann-Cartan space-time with a curvature and a torsion. Analogously to this, the translation gauge theory follows from Poincaré gauge theory in the limit, $a_i \to \infty$. Roughly speaking, the parameters a_i measure the inverse of the coupling strength of the Lorentz gauge field, so the limit $a_i\to\infty$ is the "zero coupling" limit of the Lorentz gauge field.

The translation gauge theory passes all the experimental tests so far carried out if the parameters α and β are finely tuned so that

$$|(\alpha+\beta)/(\alpha+4\beta) \equiv \varepsilon| \lesssim 0.004 . \quad (7.1)$$

This suggests us to assume

$$\alpha + \beta = 0 .$$

The theoretical basis for this choice has not yet been fully understood, however. We shall compare the main consequences of the present theory with those of general relativity in the following Table I.

Table I

	General relativity	New general relativity
Space-time	Riemann space-time	Weitzenböck space-time
Connection	Levi-Civita connection	Non-symmetric affine connection

(Table I continued)

Basic structure	Metric tensor	Parallel vector fields
Gravitation	Riemann-Christoffel curvature tensor	Torsion tensor
Transformation group	General coordinate transformation group (Local Lorentz group)	General coordinate transformation group Global Lorentz group
The Birkhoff theorem	Yes	Yes
Isotropic, gravitational field	The Schwarzschild solution	The Schwarzschild solution
Newtonian approximation	Yes	Yes
Weak field approximation	Symmetric field; massless and spin 2	Symmetric field; massless and spin 2 Antisymmetric field; massless and spin 0
Theory	Macroscopic	Microscopic
Equivalence principle	Yes	Yes, for macroscopic phenomena No, for microscopic phenomena

References

1) R. Utiyama, Phys. Rev. 101 (1956), 1597.
2) T.W.B. Kibble, J. Math. Phys. 2 (1961), 212. See also D.W. Sciama, in "Recent Developments in General Relativity" (Pergamon, Oxford, 1962), p.415.
3) K. Hayashi and T. Nakano, Prog. Theor. Phys. 38 (1967), 491.
4) A. Einstein, Sitzungsber. Preuss. Akad. Wiss. (1928), 217, 224; (1929), 2, 156; (1930), 18, 401.
5) A. Einstein and W. Mayer, Sitzungsber. Preuss. Akad. Wiss. (1930), 110.
6) C. Møller, K. Dan. Vidensk. Selsk. Mat. Fys. Skr. 1 (1961), No.1; 89 (1978), No.13.
7) C. Pellegrini and J. Plebanski, K. Dan. Vidensk. Selsk. Mat. Fys. Skr. 2 (1962), No. 4.
8) K. Hayashi and T. Shirafuji, Phys. Rev. D19 (1979), 3524. Earlier references can be found therein.
9) Here we follow the procedure given in Refs. 2) and 3). Modern treatment based on the notion of fiber bundles can be found in R.J. Petti, Gen. Rel. Grav. 7 (1976), 869; K.A. Pilch, Lett. Math. Phys. 4 (1980), 49; R. Giachetti, R. Ricci and E. Sorace, Lett. Math. Phys. 5 (1981), 85; C.P.Luehr and P. Rosenbaum, J. Math. Phys. 21 (1980), 1432; J. Hennig and J. Nitsch, Gen. Rel. Grav. 13 (1981), 947.
10) K. Hayashi and T. Shirafuji, Prog. Theor. Phys. 64 (1980), 866, 883, 1435, 2222; 65 (1981), 525; 66 (1981), 318, 2258.
F.W. Hehl, in "Cosmology and Gravitation" (ed. by P.G. Bergmann and V. de Sabbata, Plenum, 1980), p.5.
11) K.Hayashi and T. Shirafuji, the second paper of Ref. 10).
12) M. Fukui and K. Hayashi, Prog. Theor. Phys. 66 (1981), 1500.
13) W. Kopczynski, J. Phys. A15 (1981), 493.
K. Hayashi and T. Shirafuji, Phys. Rev. D24 (1981), 3312.
F. Müller-Hoissen and J. Nitsch, Göttingen preprint (1981).
14) S. Miyamoto and T. Nakano, Prog. Theor. Phys. 45 (1971), 295.

GENERALIZATIONS OF GRAVITATIONAL THEORY BASED ON GROUP COVARIANCE

Leopold Halpern
Research Institute for Fundamental Physics, Kyoto 606
and
Dept. of Physics, Florida State University
Tallahassee, Fla 32306/U.S.A

The justification for the validity of a symmetry group should be established as firmly as possible in the physical conditions and the mode of its breaking in modifications of the latter.

I proceed here in the opposite direction, letting myself be guided by the mathematical beauty of a symmetry group, I develop a formalism to derive the physics based on it. The physical theory is thus dictated by mathematics (not at all however by mathematicians).

One can expect from such a procedure in spite of its mathematical rigour, a physical fairy tale of determined harmony in which things may happen that are strange to physical reality. Fairy tales have however often a significant content of truth and I think it worth while to understand the models as well as possible before dismissing it as unrelated to reality.

In the beginning there was the group, a semisimple Lie group G with a semisimple Lie subgroup H, such that G/H, the factor space is homeomorphic to the unperturbed manifold of space-time. The group manifold G has a natural projection π : $G \to G/H$ so that $P(G, G/H, \pi, H)$ form a principal fibre bundle with typical fibre H.[1-4] Locally G is homeomorphic to $G/H \times H$. for most cases considered this is globally true; P is trivial.

G has a natural metric the Cartan-Killing metric γ given in a local orthogonal frame in terms of the structure constants by:

$$(1) \qquad \gamma_{RS} = C_{R\ V}^{\ U} C_{S\ U}^{\ V}$$

it's projection $g = \pi\gamma$ is the space-time metric of the universe.[5-9] G may for example be chosen as SO(4.1) or SO(3.2), the De Sitter groups and H as SO(3.1), the Lorentz group which yield the De Sitter or anti De Sitter universe as factor spaces.

It is postulated in general that physical quantities in the space-time manifold G/H are obtained as the projection of geometric quantities on G.

The orbits of one dimensional subgroups of G, which are the geodesics of the space with the metric γ, have as their projection on G/H all the geodesics of this space with metric g, but besides this other lines that have the form of trajectories of charged particles in electromagnetic fields. The projection from the higher dimensional space yields in general more than only the corresponding geometrical quantities on the base space - more things may occur in the fairy tale than what can be brought into accord with our (limited) experience - but these features could be the effect of inner degrees of freedom in a classical theory which should be brought in accord with reality by imposing quantum conditions. Especially the spiral motion analog to that of a charged particle in a magnetic field for the De Sitter groups, which unify momentum and angular momentum, may be the classical analog of spin motin.

The metric of the group manifold can only describe a space without inhomogenous mass distribution. This metric for a r-parameter semisimple group fulfills the relation:

$$(2) \qquad R_{UV} - \frac{1}{2}\gamma_{UV} R + \frac{r-2}{8}\gamma_{UV} = 0$$

This can be interpreted as the homogenous Einstein equations in r dimensions with a cosmological member. (The radius of the De Sitter universe is of order unity so that here the cosmological term is of conventional magnitude)

The generalisation to the case of inhomogenous local mass distributions is made by introducing a source term to eq.(2) which generalizes the metric γ in such a way that the global topology of the manifolds and their character of a principal fibre bundle with group and typical fibre H is not altered. We have arrived at peculiar versions of multidimensional Kaluza-Klein theories with non-Abelian, in general non compact gauge groups H. If one believes in the fundamental character of exact invariance groups, one should analyze this mathematically consistent scheme with all suitable semisimple groups for its physical content.

There is no doubt that this scheme is closer adjusted to the invariance group than that of any other theory, even if the latter is formulated on the group manifold.

The effect of the metric in r dimensions with (r-k) Killing vectors, which must exist to meet the requirements of a principal fibre bundle, is equivalent to the effect of the metric projected on the k-dimensional base manifold together with (r-k) Yang-Mills fields. The latter are obtained from the curvature two form Ω of a connection on the principal fibre bundle P. (The generalisation) from that on the group manifold) This connection can be given by a Lie algebra valued one form ω:

$$(3) \qquad \Omega = d\omega + [\omega, \omega]$$

Horizontal vectors B are defined by $\omega(B) = 0$.

Ω has only horizontal components; it can be expressed in a local natural coordinate system where $P = B \times H$ by (r-k) Yang Mills fields:

$$(3a) \qquad F^M_{ik} = B^M_{k,i} - B^M_{i,k} + C^M_{PQ} B^P_i B^Q_k \qquad (M, P, Q \cdots k+1 \cdots r)$$

with (r-k) Yang-Mills potentials B^M_i and the structure constants of H.

The Lagrangian for eq.(2) has then the following k-dimensional form:

$$(4) \qquad \mathcal{L} = \sqrt{\gamma}(R^{(r)} + \lambda) = \sqrt{g}(R^{(k)} + \tfrac{1}{4}F^{ikM}F_{ikM} + \lambda)$$

it does not depend on the coordinates of H. (γ metric on P, g metric on B) We require that the torsion two form vanishes. Horizontal and vertical vector spaces are perpendicular.

The projection of geodesics on P on B fulfills in our coordinates:

$$(5) \qquad \ddot{x}^i + \{{}^{\,i}_{jk}\}\, \dot{x}^j\dot{x}^k = F^{iM}_k \dot{x}^k C_M$$

where the generalized "charges" C_M are given by the vertical components of an initial tangent vector of the geodesic.

The solution of the homogenous field equations (2) which constitutes the group manifold G has a nonflat metric g as well as a nonvanishing curvature form Ω due to the cosmological member. There exist thus "cosmological fields" F^M even in this case which can give rise to the nongeodesic motion. The Maurer-Cartan equations on the group manifold:

$$(6) \qquad dA^R + C^R_{ST}A^SA^T = 0$$

($R,S,T = 1\cdots r$) for the left invariant forms A^R give:

$$(6a) \qquad F^M = dA^M + C^M_{PQ}A^PA^Q = - C^M_{EF}A^EA^F$$

(E, F summed over $1\cdots k$, P, Q over $k+1\cdots r$) for the cosmological fields. Although these fields do not vanish, the total of their energy densities on the De Sitter universes vanishes. The energy of any one field F^M can thus be either positive or negative definite; this results from the non-compactness of the gauge group H. The cosmological fields are not felt except by a particle with suitable charge ($\equiv$ initial condition) C^M. The latter must be chosen so that no energy can be drawn from the fields. This property remains true along the world line. It is a generalisation of the restriction in relativity that a world line must be time like. In the De Sitter case it restricts the motion in a suitable frame to the analog of a spiral motion of a charge in a magnetic field.

The fairy tale could be well brought in accord with reality by letting the charges vanish or making them unobservably small; but this is not in the spirit of the present considerations. We have to see whether not some deep truth is in the fairy tale that properly applied may even give a better insight into reality. The classical equations of motion already in the De Sitter case are much more complicated. The Hamilton-Jacobi equation in a way has an oscillatory wave like character. The nature of the quantum of action in this context has first to be better understood to see whether the quantized equations of motion do not describe also a spin motion in the De Sitter case.

A heuristic considerations which contributed to this development was the following: The De Sitter groups for a large radius of the universe may be regarded as a unification of momentum and angular momentun in a similar way as the group of rotations on a sphere unifies rotation and translations of $\mathbb{E}^2$.

A system of reference which rotates should thus be on an equal footing with a uniformly translated system. Making a passive transformation to such a system we recognize that an observer there really sees the spiral motion described in our fairy tale - not just for one particle, but even for all the macroscopic bodies. From experience we know that he must pay for this by experiencing inertial forces - a result that can not adequately be derived from the general theory of relativity. The present formalism provides in addition to the metric g still the fields F^M which should give a stronger account of Mach's principle if they are interpreted as a spin-spin interaction in the De Sitter cases; it is strongly felt that spin and orbital angular momentum cannot be fully separated. To seek to account for this in the present theory one would have to describe the motion of orbiting bodies in detail including the gravitational fields generated by the fields causing the binding forces.

The only example presented here was that of the De Sitter group. The formalism is applicable to higher dimensional groups G. The conformal group SO(4.2) is 15-dimensional and has a 10-dimensional subgroup H. The metric base space B is thus five dimensional and can be interpreted to unify gravity and electromagnetism in a Kaluza-Klein theory of higher order by using the projective formalism of Veblen and Jordan.[10,11]

The present formalism is not extended to supersymmetry but a presentation by D. Ebner has shown that higher dimensional groups can take account of the antisymmetry of Fermions. Limiting oneself here to this case may provide a desirable criterium to restrict the growing number of possible theories.

Acknowledgements.

This research was supported by D.O.E. and written at the Institute of Fundamental Physics of the University of Kyoto. I thank Prof. H. Sato and the Director of the Institute for their hospitality and Prof. R. Utiyama and the organizers of the symposium on Gauge Theory and Gravitation for support of my participantion.

Literature references

The mathematical formalism required can be found in:

1. L.P. Eisenhart, "Continous Groups of Transformation", Princeton (1933)
2. N. Steenrod, "The Topology of Fibre Bundles", Princeton (1974)
3. K. Nomizu, "Lie Groups and Differential Geometry", Mathemat. Soc. Japan 2 (1956)
4. Y. Choquet, C. De Witt, M. Dillard, "Analysis, Manifolds and Physics" North Holland (1982)

The ideas and the formalism were gradually developed in:

5. L. Halpern, "Gen. Relat & Gravitation", Vol.8 No.8 p623 (1977)
6. L. Halpern, Int. J. Theor. Phys. Vol.18 No.11 p845 (1979)
7. L. Halpern, Int. J. Theor. Phys. Vol 20 No.4 p297 (1981)
8. L. Halpern, Int. J. Theor. Phys. to appear shortly (Proc. Meeting in Honour of the 80th birthday of P. A. M. Dirac)
9. L. Halpern, Proceedings Internat. Meeting on Differential Geometry and Physics Trieste July 1981

10. O. Veblen, "Projektive Relativitätstheorie", Vieweg, Braunschweig 1932
11. P. Jordan, "Schwerkraft und Weltall" Vieweg, Braunschweig (1952)

Algebraic Construction of Static Axially Symmetric Self-Dual Fields*

R. SASAKI

Research Institute for Theoretical Physics
Hiroshima University, Takehara, Hiroshima 725, Japan

In this short Note we will present a short introduction to the recent work by us[1),2)] on the problem of solving self-dual gauge field equations for an arbitrary compact semisimple gauge group G. The discussion is mostly focused on the simple case of the static and axially symmetric configuration, in which the effective space-time dimension is reduced to two and we can apply some of the solution generation techniques of two dimensional soliton theories.[3)] The central result is an algebraic method for the construction of solutions starting from one particular solution and solutions of the associated linear scattering problem. Here we see an interesting interplay of soliton theoretic techniques and group theoretic concepts. A main application is the construction of axially symmetric monopole solutions in gauge theories with an arbitrary group G and a single Higgs field in the adjoint representation.

Ansatz The ansatz for the static axially symmetric gauge potential is

$$A_z,\ A_\rho \in \mathcal{K}, \quad A_t,\ A_\phi \in \mathcal{P}\ , \tag{1}$$

in which $\mathcal{K}$ is some maximal subalgebra of $\mathcal{G}$ and $\mathcal{P}$ is its complementary subspace; $\mathcal{G} = \mathcal{K} + \mathcal{P}$,

$$[\mathcal{K},\mathcal{K}] \leq i\mathcal{K},\ [\mathcal{K},\mathcal{P}] \leq i\mathcal{P},\ [\mathcal{P},\mathcal{P}] \leq i\mathcal{K}\ . \tag{2}$$

The coefficient functions depend only on z and ρ.

---

*) Report on work in collaboration with F. A. Bais.

Generalized Self-Duality Equations with static axial symmetry

$$\partial_y A_{\bar{y}} - \partial_{\bar{y}} A_y - i[A_y, A_{\bar{y}}] + i[\phi_y, \phi_{\bar{y}}] = 0 , \qquad (3.a)$$

$$\partial_y \phi_{\bar{y}} - i[A_y, \phi_{\bar{y}}] + \tfrac{1}{2}(M_y \phi_{\bar{y}} + M_{\bar{y}} \phi_y) = 0 , \qquad (3.b)$$

$$\partial_{\bar{y}} \phi_y - i[A_{\bar{y}}, \phi_y] + \tfrac{1}{2}(M_y \phi_{\bar{y}} + M_{\bar{y}} \phi_y) = 0 , \qquad (3.c)$$

in which we used the coordinates $y \equiv \rho + iz$, $\bar{y} \equiv \rho - iz$ and gauge potentials $A_y = \frac{1}{2}(A_\rho - iA_z)$, $\phi_y = -\frac{1}{2}(A_\phi + iA_t)$, etc. In order to enlarge the symmetry of the system we have introduced the additional variables M_y and $M_{\bar{y}}$ which obey

$$\partial_y M_{\bar{y}} = \partial_{\bar{y}} M_y = -M_y M_{\bar{y}} . \qquad (4)$$

For the choice $M_y = M_{\bar{y}} = 1/2\rho$, eq. (3) reduces to the ordinary static, axially symmetric self-duality equation.

K-invariance The equation (3) as well as the ansatz (1) are invariant under the following gauge transformation,

$$A_\nu \to A'_\nu = \Omega A_\nu \Omega^{-1} - i\partial_\nu \Omega \Omega^{-1}, \quad \phi_\nu \to \phi'_\nu = \Omega \phi_\nu \Omega^{-1} , \qquad (5)$$

in which $\nu = y, \bar{y}$ and $\Omega = \Omega(y,\bar{y}) \varepsilon K$. This is a subgroup K of the original gauge group G which preserves the structure of the ansatz (1).

Pure gauge and Triangularity Two of the three equations (3) imply that the following combination of gauge potentials

$$a_\nu = A_\nu + i\phi_\nu, \quad \nu = y, \bar{y} , \qquad (6)$$

is pure gauge, i.e.,

$$a_\nu = -i(\partial_\nu g) g^{-1}, \quad g \varepsilon G^*, \quad \mathcal{G}^* = \mathcal{K} + i\mathcal{P} . \qquad (7)$$

A theorem of group theory (Iwasawa decomposition[4]) states that G^* factorizes into two parts,

$$G^* = KT , \qquad (8)$$

in which the Lie algebra of T consists of abelian and nilpotent parts only. By an appropriate K-transformation one can always make a_ν as

$$a_\nu = -i(\partial_\nu \tau)\tau^{-1} , \quad \tau \varepsilon T . \tag{9}$$

This gauge will be called "triangular".

<u>Σ-invariance</u> From a solution of (3) in the triangular gauge another solution in the triangular gauge is obtained in terms of the following <u>discrete</u> transformation

$$\left.\begin{aligned} A_y \rightarrow \tilde{A}_y = RA_yR^{-1}, \ \phi_y \rightarrow \tilde{\phi}_y = S\phi_yS^{-1} + M_y\eta, \ M_y \rightarrow M_y , \\ A_{\bar{y}} \rightarrow \tilde{A}_y = SA_{\bar{y}}S^{-1}, \ \phi_{\bar{y}} \rightarrow \tilde{\phi}_y = R\phi_{\bar{y}}R^{-1} + M_{\bar{y}}\eta, \ M_{\bar{y}} \rightarrow M_{\bar{y}} , \end{aligned}\right\} \tag{10}$$

in which R, S and η are constant matrices. Their explicit forms are found by considering the isomorphisms of the extended Dynkin diagrams of each algebra $\mathcal{G}$, which also characterizes the Kac-Moody algebra[4).]

<u>Γ-invariance</u> The generalized self-dual equations (3) are invariant under

$$A'_\nu = A_\nu, \ \phi'_y = \gamma^{1/2}\phi_y, \ \phi_{\bar{y}}' = \gamma^{-1/2}\phi_{\bar{y}}, \ M'_y = \gamma M_y, \ M_{\bar{y}}' = \gamma^{-1}M_{\bar{y}}, \tag{11}$$

in which a scalar function γ should satisfy the following completely integrable Riccati equation

$$d\gamma = (\gamma-1)(\gamma M_y dy + M_y d\bar{y}) . \tag{12}$$

The Γ, Σ and K are three important symmetry transformations in terms of which new solutions are generated successively.

<u>Associated Linear Problem</u> By combining two facts that for the solution of (3), a_ν (6) is pure gauge and that the Γ-transformation maps a solution to another, we get the following linear problem,

$$\partial_y\mathcal{R} = i(A_y + i\gamma^{1/2}\phi_y)\mathcal{R}, \ \partial_{\bar{y}}\mathcal{R} = i(A_{\bar{y}} + i\gamma^{-1/2}\phi_{\bar{y}})\mathcal{R}. \tag{13}$$

The integrability condition is (3) and (4).

Triangularity Restoration. If a Γ-transformation is applied to a triangular solution, the result is not triangular any more. But the triangularity can be restored by a K-transformation in terms of Ω, which can be obtained from a solution $\mathcal{R}$ of the linear problem (13) in terms of the Iwasawa decomposition,

$$\mathcal{R} = \Omega^{-1}\tau, \quad \mathcal{R} \in G^*, \quad \Omega \in K, \quad \tau \in T \, . \tag{14}$$

Solution Generation By combining the Σ, Γ and K-transformations we can generate a host of new solutions from a given one $\mathring{A} \equiv (\mathring{A}_\nu, \mathring{\phi}_\nu, \mathring{M}_\nu)$ of eq. (3). A simple scheme is, for example,

$$\underset{(\text{Tr.})}{\mathring{A}} \to \underset{(\text{Tr.})}{\Sigma\mathring{A}} \to \underset{(\text{Non Tr.})}{\Gamma\Sigma\mathring{A}} \to \underset{(\text{Tr.})}{K\Gamma\Sigma\mathring{A}} \equiv \overset{1}{A} \to \underset{(\text{Tr.})}{\Sigma\overset{1}{A}} \to \cdots \, , \tag{15}$$

in which Tr. stands for triangular. This procedure can be repeated an arbitrary number of times. New parameters are introduced through constants of integration for γ and $\mathcal{R}$ at each stage.

Algebraic Construction The main point of our work is that we can construct a hierarchy of solutions $\overset{1}{A}$, $\overset{2}{A}, \cdots$, algebraically. The only analytical work needed is to solve the associated linear problem (12) and (13) for the initial solution A. We denote the solutions as

$$\mathring{A} \; ; \; (\mathring{\mathcal{R}}_1, \mathring{\gamma}_1) \, , \, \cdots \, , \, (\mathring{\mathcal{R}}_m, \mathring{\gamma}_m) \, , \, \cdots \, , \tag{16}$$

in which the suffix indicates a particular choice of constants of integration. The γ-functions and the triangularity restoring Ω functions at each stage can be constructed from (16) algebraically and step by step. For an explicit construction, a close relationship between the Σ-transformation and the associated linear problem must be revealed. We refer to our papers[1),2)] for further details.

Summary and Comments Applied to the simplest case, i.e., $G = SU(2)$, the above procedure reproduces all the solution generation techniques of Neugebauer[5)] for the Ernst equation[6)] in general relativity, which describes the stationary axially symmetric vacuum gravitational field.

The algebraic solution generation method makes use of some specific combinations of elements of an infinite dimensional group of symmetries for eq. (3). This group is a natural generalization of Geroch-Kinnersley-Chitre[7] group for the Ernst equation. We believe that further investigation of the infinite dimensional group is very important for deeper understanding of gauge theories.

References

1) F. A. Bais and R. Sasaki, Nucl. Phys. B195 (1982) 522, Phys. Lett. 113B (1982) 35, 39.
2) F. A. Bais and R. Sasaki, in preparation.
3) See for example; R. K. Bullough and P. J. Caudrey eds., Solitons, Topics in current physics (Springer, Berlin 1980).
4) S. Helgason, Differential Geometry, Lie Groups and Symmetric Spaces, Academic Press, New York, 1978.
5) G. Neugebauer, J. Phys. A12 (1979), L67.
6) F. J. Ernst, Phys. Rev. 167 (1968) 1175.
7) R. Geroch, J. Math. Phys. 12 (1971) 918, W. Kinnersley and D. M. Chitre, J. Math. Phys. 18 (1977), 1538.

SPACE-TIME STRUCTURE OF GRAVITATIONAL SOLITONS

Akira Tomimatsu

Research Institute for Theoretical Physics
Hiroshima University, Takehara, Hiroshima 725, Japan

The axially symmetric and stationary soliton solution of Einstein's equations in a vacuum has been found by means of the inverse scattering problem technique. This solution has metric of the form

$$ds^2 = g_{ab}dx^a dx^b + h(d\rho^2 + dz^2) , \tag{1}$$

where $a,b = t,\phi$. For the N-soliton metric on a flat background for which $g^o_{ab} = \mathrm{diag}(-1,\rho^2)$ and $h^o = 1$, we obtain the following expressions[1]:

$$\begin{aligned} g_{ab} &= \prod_{k=1}^{N} |\mu_k/\rho| (\chi(\lambda=0;\rho,z))_{ac} g^o_{cb} \\ &= \prod_{k=1}^{N} |\mu_k/\rho| \{g^o_{ab} - \sum_{k,\ell=1}^{N} (\Gamma^{-1})_{k\ell} N_{ka} N_{\ell b}\} , \end{aligned} \tag{2}$$

$$h = c_o \rho^{N^2/2-N} (\prod_{k=1}^{N} |\mu_k/\rho|)^{N-1} \prod_{k>\ell}^{N} (\mu_k-\mu_\ell)^{-2} \det(\Gamma_{k\ell}) , \tag{3}$$

$$N_{kt} = -1,\ N_{k\phi} = c_k \mu_k^{-1} \rho^2 ,$$

$$\Gamma_{k\ell} = (\rho^2 c_k c_\ell - \mu_k \mu_\ell)/(\rho^2 + \mu_k \mu_\ell) ,$$

where c_o and c_k are any constants. The matrix function χ depends on a complex spectral parameter λ, and has N poles at the points $\lambda = \mu_k(\rho,z) = w_k - z + \varepsilon_k\{\rho^2 + (w_k-z)^2\}^{1/2}$, where w_k are any constants and $\varepsilon_k = \pm 1$. From the conditions of asymptotic flatness and $\det(g_{ab}) < 0$, we require that N is even and $\sum_{k=1}^{N} \varepsilon_k = 0$. Then, we subdivide all the μ_k into pairs with opposite signs ε_k, and rewrite the parameters w_k and c_k as follows,

$$\begin{aligned} &\mu^{\pm}_\gamma = w^{\pm}_\gamma - z \pm \{\rho^2 + (w^{\pm}_\gamma - z)^2\}^{1/2} , \quad \gamma = 1,2,\ldots,N/2, \\ &w^{\pm}_\gamma = z_\gamma \mp m_\gamma \cos\lambda_\gamma , \quad c^{\pm}_\gamma = \cot((\alpha_\gamma \pm \lambda_\gamma)/2) . \end{aligned} \tag{4}$$

The N-soliton metric turns out to be a nonlinear superposition of N/2 Kerr-NUT metrics aligned along the common rotational axis (z-axis). Each soliton pair describes the Kerr-NUT metric with the Kerr and NUT rotation parameters λ_γ, α_γ and the mass m_γ located at the point $z = z_\gamma$. This stationary configuration of several Kerr-NUT masses will be realized as a result of the presence of any singularities in the space-time or a dynamical balance between the gravitational attraction and the rotational repulsion. If there exists no singular structure, great interest from the physical point of view attaches to the N-soliton metric, because it can represent a dynamical equilibrium of many interacting black holes. In this paper we investigate the space-time structure of the multi-soliton metric, by taking the two Kerr-NUT case (N=4) as a typical example.

The metric components can be derived from the Ernst potential ξ, which is given by[2)]

$$\xi \equiv \frac{N}{D} = \begin{vmatrix} S_1^- & S_1^+ & S_2^- & S_2^+ \\ 1 & 1 & 1 & 1 \\ w_1^- & w_1^+ & w_2^- & w_2^+ \\ (w_1^-)^2 & (w_1^+)^2 & (w_2^-)^2 & (w_2^+)^2 \end{vmatrix} \Bigg/ \begin{vmatrix} S_1^- & S_1^+ & S_2^- & S_2^+ \\ 1 & 1 & 1 & 1 \\ w_1^- & w_1^+ & w_2^- & w_2^+ \\ w_1^-S_1^- & w_1^+S_1^+ & w_2^-S_2^- & w_2^+S_2^+ \end{vmatrix} , \qquad (5)$$

$$S_\gamma^\pm = \mp e^{-i\beta_\gamma^\pm}\{\rho^2 + (w_\gamma^\pm - z)^2\}^{1/2} , \quad \beta_\gamma^\pm = \alpha_\gamma \pm \lambda_\gamma \ (\gamma = 1,2) ,$$

for the 4-soliton metric. We can set

$$m_\gamma > 0 , \quad \cos\lambda_\gamma > 0 , \quad z_1 = -z_2 \equiv z_0 > 0 .$$

If $w_1^+ = w_2^-$ ($2z_0 = \sum_{\gamma=1}^{2} m_\gamma \cos\lambda_\gamma$), this metric reduces to the Kerr-NUT metric. For a double coincidence of the poles, i.e., $w_1^\pm = w_2^\pm$ ($z_0 = 0$)[2),3)], we obtain the $\delta = 2$ Tomimatsu-Sato and Kinnersley-Chitre metrics. Therefore, we consider only the case that any $w_\gamma^\pm$ do not coincide. Then, we can classify the 4-soliton metric into two types;

(I) separated type of two Kerr-NUT metrics, i.e., $2z_0 > \sum_{\gamma=1}^{2} m_\gamma \cos\lambda_\gamma$,

(II) overlapped type, i.e., $2z_0 < \sum_{\gamma=1}^{2} m_\gamma \cos\lambda_\gamma$.

We construct diagrams (Fig. 1) that represent clearly such different types.[4)] We place on the horizontal axis (z-axis) four points $z = w_\gamma^\pm$. The axis is divided into five different regions (a,...,e). The vertical bars drawn from these points are directed up at the

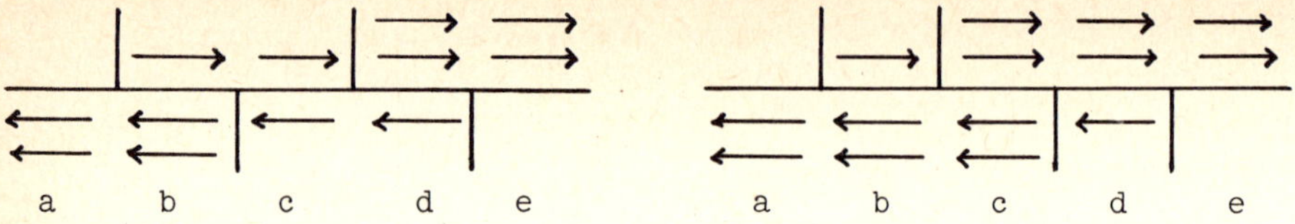

(I) (II)

Fig. 1

points $z = w_\gamma^+$ and down at the points $z = w_\gamma^-$. An arrow along the axis means that one of the $\mu_\gamma^\pm$ at this region of the axis is of the order of $O(\rho^2)$.

For the static field[4], i.e., $\alpha_\gamma = \lambda_\gamma = 0$, the metric coefficient h becomes of the order of $O(\rho^{2(s-2)(s-3)})$ near each region of the axis (s is the number of arrows). Then, for the metric of type II, the invariant of the Riemann curvature tensor $\Lambda \equiv R_{\mu\nu\lambda\sigma}R^{\mu\nu\lambda\sigma}$ $(\mu,\nu,\ldots = t,\phi,\rho,z)$ becomes infinite at the middle region (c).

For the rotational field, the invariant Λ does not increase without limit at the z-axis. In general, such a singularity appears at the point $\xi = -1$. In order to look for this point, we consider a simple case such as $m_1 = m_2 \equiv m$, $\lambda_1 = \lambda_2 \equiv \lambda$ $(\cos\lambda \equiv p)$, $\alpha_1 = \alpha_2 = 0$. Then, we have $c_1^+ = -c_1^- = c_2^+ = -c_2^- \equiv c$, $c^2 = (1+p)/(1-p) > 1$. The relation $\xi = -1$ means that $\mathrm{Re}(N+D) = \mathrm{Im}(N+D) = 0$. We see that $\mathrm{Im}(N+D)$ becomes an odd function of z, so the singular point will exist at the plane $z = 0$, where the metric has the form

$$g_{tt} = -B^{-1}[(\eta_+-\eta_-)^2(c^2\eta_+\eta_-+1)^2 - c^2(1-\eta_+^2\eta_-^2)^2] ,$$

$$g_{t\phi} = -\rho c(\eta_+-\eta_-)(1-\eta_+^2\eta_-^2)(\eta_+\eta_-B)^{-1}$$

$$\times [(\eta_+^2+\eta_-^2-\eta_+\eta_-)(c^2\eta_+\eta_-+1) - (c^2+\eta_+^3\eta_-^3)] , \qquad (6)$$

$$g_{\phi\phi} = \rho^2(\eta_+^2\eta_-^2B)^{-1}[(\eta_+-\eta_-)^2(c^2+\eta_+^3\eta_-^3)^2 - c^2(1-\eta_+^2\eta_-^2)^2(\eta_+^2+\eta_-^2-\eta_+\eta_-)^2],$$

$$B \equiv (\eta_+-\eta_-)^2(c^2+\eta_+\eta_-)^2 - c^2(1-\eta_+^2\eta_-^2)^2 ,$$

$$\eta_\pm \equiv \rho^{-1}\mu_1^\pm(\rho,z=0) = \rho^{-1}[z_o \mp mp \pm \{(z_o \mp mp)^2 + \rho^2\}^{1/2}] .$$

All the metric components g_{ab} become infinite at the point $B = 0$, while the metric coefficient h vanishes at the same point ($B \propto$ det

($\Gamma_{k\ell}$)).

For the metric of type II, i.e., $z_o < mp$, $\eta_+(\rho)$ is monotonously increasing, and

$$(\eta_+-\eta_-)(c^2+\eta_+\eta_-) - |c|(1-\eta_+^2\eta_-^2) = \begin{cases} -|c| < 0 & \text{as } \rho \to 0, \\ 2(c^2-1) > 0 & \text{as } \rho \to \infty. \end{cases}$$

It can be proved that the algebraic equation $B = 0$ has only one root ρ_o (For example, when $z_o \to mp$, we obtain $\rho_o \simeq (mp-z_o)|\tan\lambda|$). This means that the metric of type II includes one ring-singularity around the region (c). On the other hand, for the metric of type I, $\eta_+(\rho)$ is monotonously decreasing. If the equation $B = 0$ holds, we have

$$|c| = [1-\eta_+^2\eta_-^2+\{(1-\eta_+^2\eta_-^2)^2 - 4\eta_+\eta_-(\eta_+-\eta_-)^2\}^{1/2}]/2(\eta_+-\eta_-) .$$

This relation turns out to be incompatible with the constraint $|c| > 1$. For any other choice of the parameters we expect that the presence of this kind of singularity is a crucial difference between the two types of the 4-soliton metric.

Next, we examine some properties of the z-axis of the 4-soliton metric of type I.[5),6)] We rewrite the metric as follows,

$$ds^2 = -f(dt-\omega d\phi)^2 + f^{-1}[\rho^2 d\phi^2 + e^{2\mu}(d\rho^2 + dz^2)] . \qquad (7)$$

On the z-axis we find that $\partial\omega/\partial z = \partial\mu/\partial z = 0$ except at the points $z = w_\gamma^\pm$, and $f \propto \prod_{\gamma=1}^{2}(z-w_\gamma^+)(z-w_\gamma^-)$. We denote the constant values of ω and μ at each region of the axis by ω_A and μ_A ($A = a,\ldots,e$). If the z-axis has a local Euclidean property of a line, for any infinitesimal spacelike circle around the z-axis the ratio of circumference to radius should be 2π, i.e., $g_{\phi\phi}/\rho^2 g_{\rho\rho} \to 1$ as $\rho \to 0$. This requires that $\omega_A = \mu_A = 0$ and gives some restrictions to a choice of the parameters. By a suitable choice of the coordinates (t,ϕ) and c_o (see Eq. (3)), we can always set $\omega_e = \mu_e = 0$. Then, the conditions $\omega_a = \mu_a = 0$ lead to

$$4z_o^2(m_1v_1 + m_2v_2) + 4m_1m_2z_o(q_2u_1 - q_1u_2) - (m_1^2p_1^2 - m_2^2p_2^2)(m_1v_1 - m_2v_2) = 0 , \qquad (8)$$

where $e^{i\alpha_\gamma} \equiv u_\gamma + iv_\gamma$ and $e^{i\lambda_\gamma} \equiv p_\gamma + iq_\gamma$.

At the regions (b) and (d), we cannot find any choice of the parameters satisfying the conditions $\omega_A = \mu_A = 0$ $(A = b,d)$. Using the coordinate θ_γ ($\gamma = 1$ at the region (d) and $\gamma = 2$ at the region (b)) defined by $z - z_\gamma = m_\gamma p_\gamma \cos\theta_\gamma$ $(0 \leq \theta_\gamma \leq \pi)$, we can prove that each region has the structure of a closed 2-sphere. Furthermore, there exists a Killing vector $\xi^\nu_{(A)} \equiv \omega_A \xi^\nu_{(t)} + \xi^\nu_{(\phi)}$ which becomes null on this closed surface Σ_A.[7] Here, $\xi^\nu_{(t)}$ and $\xi^\nu_{(\phi)}$ are the Killing vectors associated with the stationarity and axial symmetry respectively. We verify that $\xi^\nu_{(A)}$ lies in Σ_A*) since it is orthogonal to the vector $n^{(A)}_\nu \equiv \xi_{(A)\sigma;\nu}\xi^\sigma_{(A)}$ which is normal to Σ_A. The vector $n^{(A)}_\nu$ becomes null when $\xi^\nu_{(A)}$ does, so the regions (b) and (d) can be regarded as two horizons separated by the region (c).

The structure of the middle region (c) between two black holes is important. If there is no artificial "support" between the masses, this region should have a regular structure of a line. Hence, the conditions $\omega_c = \mu_c = 0$ seem to assure a dynamical balance between the gravitational attraction and the rotational repulsion. These conditions lead to

$$\cos(\alpha_1 - \alpha_2) = 0\ , \tag{9}$$

$$(4z_o^2 - m_1^2p_1^2 + m_2^2p_2^2)v_1 + 4m_2z_o(q_2u_1 + \sin(\alpha_1 - \alpha_2)) = 0\ . \tag{10}$$

It is remarkable that due to Eqs. (8) $\sim$ (10) the distance between two masses, $2z_o$, is fixed by the masses and angular momenta. If $\omega_c \neq 0$, the metric has a pathological structure of causality violation (i.e., the existence of a closed timelike curve), since $g_{\phi\phi} \simeq -f\omega_c^2 < 0$ near the region (c). We conclude that only the 4-soliton metric of type I with the parameters satisfying the conditions (8) $\sim$ (10) will avoid any singular structures in the space-time.

REFERENCES

1) V.A. Belinsky and V.E. Zakharov, Sov. Phys.-JETP 50 (1979), 1.
   G.A. Alekseev, Sov. Phys. Dokl. 26 (1981), 158.
2) D. Kramer and G. Neugebauer, Phys. Letters 75A (1980), 259.
3) A. Tomimatsu, Prog. Theor. Phys. 63 (1980), 1054.
4) G.A. Alekseev and V.A. Belinsky, Sov. Phys.-JETP 51 (1980), 655.
5) M. Kihara and A. Tomimatsu, Prog. Theor. Phys. 67 (1982), 349.
6) A. Tomimatsu and M. Kihara, Prog. Theor. Phys. 67 (1982), 1406.
7) C.V. Vishveshwara, J. Math. Phys. 9 (1968), 1319.
   A. Tomimatsu and H. Sato, Lett. Nuovo Cim. 8 (1973), 740.

*) This surface should include the t-direction too.

DYON BLACK HOLE IN THE TOMIMATSU-SATO-YAMAZAKI SPACE-TIME

Masahiro Kasuya
Institute for Nuclear Study, University of Tokyo,
Midori-cho, Tanashi, Tokyo 188, Japan

The symmetry between electric and magnetic charge which is inherent in Maxwell's equations does not seem to be realized in nature. Dirac pointed out, however, that quantum mechanics does not preclude the existence of magnetic monopoles [1], and Schwinger proposed the dyon (a pole possessing both electric and magnetic charges) [2]. This dyon lies in the Abelian theory U(1).

On the other hand, in non-Abelian theory, 't Hooft and Polyakov obtained spherically symmetric classical monopole solution of the SO(3) Yang-Mills theory coupled with a triplet Higgs field [3]. Shortly afterwards, Julia and Zee showed that the same theory also admitted dyon [4]. The magnetic monopoles and the dyons may play an important role in grand unified theories.

A solution of the Einstein-Maxwell equation in the Kerr space-time was obtained by Newman et al. [5]. This solution represents a rotating mass and electric charge. Tomimatsu-Sato [6] discovered the series of solutions for the gravitational field of a rotating mass, following Ernst's formulation of axisymmetric stationary fields [7]. Furthermore Yamazaki obtained the charged Kerr-Tomimatsu-Sato family of solutions with arbitrary integer distortion parameter for gravitational fields of rotating masses [8].

We present an exact stationary rotating dyon solution in the Tomimatsu-Sato-Yamazaki space-time [9]. Our solution is characterized by five parameters (mass M, angular momentum J, electric charge Q, magnetic charge Φ, and distortion parameter δ). Let us start with the following Lagrangian density, which describes the electromagnetic field induced by an Abelian dyon in curved space-time ($\hbar$=c=1)

$$L = \sqrt{-g}\,\left(-\frac{1}{16\pi G}R - \frac{1}{4}g^{\mu\rho}g^{\nu\sigma}F_{\mu\nu}F_{\rho\sigma}\right) , \tag{1}$$

where $F_{\mu\nu} = \partial_\mu A_\nu - \partial_\nu A_\mu + {}^*G_{\mu\nu}$ and ${}^*G_{\mu\nu}$ is the Dirac string term. We can express the line element of stationary and axisymmetric space-time in the form

$$ds^2 = f^{-1}[e^{2\gamma}(d\rho^2+dz^2)+\rho^2 d\varphi^2]-f(dt-\omega d\varphi)^2 , \tag{2}$$

where f, ω and γ are functions of ρ and z only.

Our exact dyon solution for (1) and (2) can be expressed as follows: The metric functions are

$$f = \mathcal{A}_\delta/\mathcal{B}_\delta ,$$
$$\omega = 2GMq(1-y^2)\mathcal{C}_\delta/\mathcal{A}_\delta , \tag{3}$$
$$e^{2\gamma} = \mathcal{A}_\delta/[p^{2\delta}(x^2-y^2)^{\delta^2}] .$$

We shall follow the notation of Ref. [9].
It is worth while noting that γ has the same form as in vacuum field. Here the relations among the Weyl coordinates (t,ρ,z,φ), the Boyer-Lindquist coordinates (t,r,θ,φ), and the prolate spheroidal coordinates (t,x,y,φ) are

$$\rho = K(x^2-1)^{1/2}(1-y^2)^{1/2} , \quad z = Kxy ,$$
$$r = Kx + GM , \quad \cos\theta = y ,$$
$$(GMp\sigma)^2 = (K\delta)^2 = (GM)^2-a^2-\frac{G}{4\pi}(Q^2+\Phi^2) , \tag{4}$$
$$a = J/M = GMq\sigma ,$$
$$p^2+q^2 = 1 , \quad \sigma^2+|\tau|^2 = 1 , \quad \tau = (4\pi G)^{1/2}\tau_0 .$$

Next the electromagnetic potentials are

$$4\pi GM\sigma^2\mathcal{B}_\delta A_0 = Q(\sigma\eta+\zeta)+\Phi\sigma\kappa , \quad A_1=A_2=0 ,$$
$$A_3 = \omega A_0 + \frac{K}{8\pi GM\sigma^2}[Q(\sigma\hat{U}+\hat{V}-\frac{\delta}{pq})-\Phi\sigma(\hat{w}-\frac{2\delta}{p})] , \tag{5}$$

with the dyon solution ξ of the Ernst equation given by

$$\sigma^{-1}\xi = \frac{\eta+i\kappa}{\zeta} = \frac{\mu+\nu}{\mu-\nu} , \tag{6}$$

ζ, η and κ being real functions, μ and ν being complex functions of x, y, p and q, and taking $4\pi GM\tau_0 = Q+i\Phi$. The presence of a Dirac string in our solution may be seen from A_3 in (5). In fact the monopole term in A_3 does not vanish on the negative semi-infinite line of the symmetry axis. Our dyon solution reduces to the magnetic monopole solution in the case $Q = 0$.

On the other hand, the non-Abelian dyon solution can be obtained from the Abelian dyon solution and vice versa using the extended Arafune-Freund-Goebel singular gauge transformation [9],[10],[11].

Let us discuss the space-time structure of our dyon solution. The metric functions $\mathcal{A}_\delta$ and $\mathcal{B}_\delta$ in (3) are written in the form, using (6), [12]

$$\begin{aligned} \mathcal{A}_\delta &= (\mu\nu^*+\mu^*\nu)/2 , \\ \mathcal{B}_\delta &= [(\sigma+1)^2\mu\mu^*+(\sigma^2-1)(\mu\nu^*+\mu^*\nu)+(\sigma-1)^2\nu\nu^*]/(2\sigma)^2 . \end{aligned} \tag{7}$$

The ergosurfaces are obtained by taking $\mathcal{A}_\delta = 0$. On the equatorial plane (y=0) the complex functions μ and ν in (6) become real $\mathcal{M}$ and $\mathcal{N}$. The metric functions $\mathcal{A}_\delta$ and $\mathcal{B}_\delta$ are then of the form

$$\begin{aligned} \mathcal{A}_\delta(y=0) &= \mathcal{M}(x,p)\mathcal{N}(x,p) , \\ \mathcal{B}_\delta(y=0) &= [\{(\sigma+1)\mathcal{M}(x,p)+(\sigma-1)\mathcal{N}(x,p)\}/(2\sigma)]^2 , \\ \mathcal{N}(x,p) &= (-1)^\delta\mathcal{M}(-x,p) . \end{aligned}$$

The position of ergosurfaces on the equatorial plane is determined by $\mathcal{A}_\delta(y=0) = 0$, and there exist ring singularities determined by $\mathcal{B}_\delta(y=0) =0$, i.e., $(\sigma+1)\mathcal{M}+(\sigma-1)\mathcal{N} = 0$, on the equatorial plane. The metric g_{11} becomes infinity at $x = \pm 1$. We find that the number of ergosurfaces is δ for $x>1$ and also δ for $x<-1$, and that the number of ring singularities is $[\delta/2]$ for $x>1$ and $\delta-[\delta/2]$ for $x<-1$.

We obtain the proper area A of the surface x=1 in the Tomimatsu-Sato-Yamazaki space-time for our dyon

$$A = \frac{4\pi}{\delta}(GM)^2(1+2p\sigma+\sigma^2)\left(\frac{q}{p}\right)^{\delta-1}\sqrt{(-1)^{\delta^2-1}} . \tag{8}$$

Therefore our dyon solution has an event horizon for arbitrary odd integer δ; there exists an event horizon at x=1, i.e.,

$$r = GM + \frac{1}{\delta}\,[(GM)^2-a^2-\frac{G}{4\pi}(Q^2+\Phi^2)]^{1/2} .$$

Thus, for arbitrary odd integer δ, our dyon solution represents a black hole with four hairs provided that

$$(GM)^2 \geqq a^2 + \frac{G}{4\pi}(Q^2+\Phi^2) .$$

The special case $\delta=1$ is the Kerr-Newman case of our dyon. Furthermore it can be easily seen that the series of Weyl solutions (a=q=0) has no event horizons except for the case $\delta=1$, i.e. the Schwarzschild and Reissner-Nordström solutions.

We have presented an exact dyon solution for which the space-time metric takes the Tomimatsu-Sato-Yamazaki form. Our solution has five parameters, i.e., the mass M of the dyon, the angular momentum J, the electric charge Q, the magnetic charge Φ, and the distortion parameter δ, and reduces to the rotating monopole solution in the case Q=0. Using our extended Arafune-Freund-Goebel singular gauge transformation, the non-Abelian dyon solution may be obtained from the Abelian dyon solution, and vice versa. Therefore an observer at large distances cannot distinguish between the Abelian dyon and the non-Abelian dyon. In the Abelian dyon we can treat Q and Φ as independent. However, this is not the case for the non-Abelian dyon, and the finite energy Prasad-Sommerfield solution [13] in flat space-time is known. Finally there is the problem of naked ring singularities of the fields with $\delta = 3,5,7,\cdots$. This remains an open problem since Einstein's general relativity is not applicable to the ring singularities.

References

[1] P.A.M. Dirac, Proc. Roy. Soc. Lond. A133 (1931) 60; Phys. Rev. 74 (1948) 817.
[2] J. Schwinger, Phys. Rev. 144 (1966) 1087.
[3] G. 't Hooft, Nucl. Phys. B79 (1974) 276; A.M. Polyakov, JETP Lett. 20 (1974) 194.
[4] B. Julia and A. Zee, Phys. Rev. D11 (1975) 2227.
[5] E.T. Newman, E. Couch, K. Chinnapared, A. Exton, A. Prakash and R. Torrence, J. Math. Phys. 6 (1965) 918.
[6] A. Tomimatsu and H. Sato, Phys. Rev. Lett. 29 (1972) 1344; Prog. Theor. Phys. 50 (1973) 95.
[7] F.J. Ernst, Phys. Rev. 167 (1968) 1175; Phys. Rev. 168 (1968) 1415; Phys. Rev. D7 (1973) 2520.
[8] M. Yamazaki, J. Math. Phys. 18 (1977) 2502; J. Math. Phys. 19 (1978) 1376.
[9] M. Kasuya, Phys. Rev. D25 (1982) 995; Phys. Lett. 109B (1982) 380.
[10] J. Arafune, P.G.O. Freund and C.J. Goebel, J. Math. Phys. 16 (1975) 433.
[11] M. Kasuya, Phys. Lett. 103B (1981) 353; Prog. Theor. Phys. 67 (1982) 499.
[12] M. Kasuya, Phys. Lett. B (to be published).
[13] M.K. Prasad and C.M. Sommerfield, Phys. Rev. Lett. 35 (1975) 760.

MASSIVE GAUGE THEORIES IN THREE DIMENSIONS (= AT HIGH TEMPERATURE)

R. Jackiw
Center for Theoretical Physics
Laboratory for Nuclear Science and Department of Physics
Massachusetts Institute of Technology
Cambridge, Massachusetts 02139

I. INTRODUCTION

Gauge theories in three dimensions provide interesting case studies of gauge invariant quantum field theoretic phenomena. Also they are physically relevant: a field theory on a 3-dimensional [Euclidean] space summarizes the high-temperature behavior of a theory in four-dimensional space-time.

Recently it has been established that the special topological properties of odd-dimensional space allow the construction of a gauge invariant mass term in three dimensions.[1] Moreover, for non-Abelian gauge groups, the configuration space of the quantum theory is not simply connected, and as a consequence the mass must be quantized, somewhat analogously to Dirac's monopole quantization condition. While all this appears to be a peculiar feature of 3-dimensional theories, it is an interesting phenomenon, whose certain aspects have 4-dimensional analogs. Also there may be a direct physical [high temperature] significance to the mass. For these reasons it is profitable to study the subject, and I shall report here on this research.[2]

II. ABELIAN THEORY

Consider the following Lagrange density in 3-dimensional space-time.

$$\mathcal{L} = -\frac{1}{4} F^{\mu\nu} F_{\mu\nu} + \frac{\mu}{4} \varepsilon^{\mu\nu\alpha} F_{\mu\nu} A_\alpha$$

$$F_{\mu\nu} = \partial_\mu A_\nu - \partial_\nu A_\mu \qquad (2.1)$$

Dimensional arguments show μ has dimension of mass. Although the Lagrange density is not gauge invariant, the equation of motion which follows from (2.1) is

$$\partial_\mu F^{\mu\nu} + \mu {*F}^\nu = 0 \tag{2.2}$$

Here we have defined the dual field, which in three dimensions is a vector.

$$*F^\mu = \frac{1}{2} \varepsilon^{\mu\alpha\beta} F_{\alpha\beta}$$

$$F^{\alpha\beta} = \varepsilon^{\alpha\beta\mu} *F_\mu \tag{2.3}$$

Note that the dual field is identically conserved.

$$\partial_\mu *F^\mu = 0 \tag{2.4}$$

This Bianchi identity is a consequence of the definitions of $F^{\mu\nu}$ and $*F^\mu$; alternatively, it follows from the equation of motion (2.2), owing to the antisymmetry of $F^{\mu\nu}$. Under a gauge transformation the Lagrange density changes by a total derivative.

$$A_\mu \to A_\mu + \partial_\mu \theta \tag{2.5a}$$

$$\mathcal{L} \to \mathcal{L} + \frac{\mu}{2} \partial_\mu (*F^\mu \theta) \tag{2.5b}$$

This is why the equation of motion is gauge invariant.

While it is clear that μ has dimension of mass, it still remains to be established that it is indeed a mass term for the field. This is most easily done by writing the field equation (2.2) in terms of the dual tensor (2.3). Eq. (2.2) is equivalent to

$$(\mu g^{\mu\alpha} + \varepsilon^{\mu\alpha\beta} \partial_\beta) *F_\alpha = 0 \tag{2.6a}$$

Multiplying this with the differential operator $(\mu g^{\nu\mu} - \varepsilon^{\nu\mu\gamma}\partial_\gamma)$ yields

$$(\Box + \mu^2)\, {}^*F^\nu = 0 \tag{2.6b}$$

which demonstrates clearly that the gauge field excitations are massive.

An analysis of the kinematics shows that the massive vector meson carries non-vanishing spin $\mu/|\mu| = \pm 1$. The existence of a single excitation with only one value of the spin -- as opposed to two, each differing in sign from the other -- signals reflection non-invariance. Of course, the lack of this symmetry is already evident from the Lagrangian, which contains the reflection non-invariant structure $\varepsilon^{\alpha\beta\gamma}$. One may regain a P and T conserving system by working with a doublet of models, one with mass μ, the other with $-\mu$, and defining parity and time inversion to include a field interchange.

It is gratifying that μ^2 occurs in (2.6b) with the correct sign for a propagating particle. Although we have no <u>a priori</u> control over this sign [$\mathcal{L}$ is linear in μ], we may understand that it must emerge the way it does by considering the energy-momentum tensor $\theta^{\mu\nu}$. When coupling our theory to an external metric, $(\mu/4)\int d^3x\, \varepsilon^{\mu\nu\alpha} F_{\mu\nu} A_\alpha$ is already a coordinate invariant world scalar, without additional metric factors. Hence the variation of action $I = \int d^3x$ with respect to the metric [this variation defines the energy-momentum tensor] does not see the mass term. Consequently $\theta^{\mu\nu}$ has its conventional Maxwell form.

$$\theta^{\mu\nu} = -F^{\mu\alpha}F^\nu{}_\alpha + \frac{1}{4} g^{\mu\nu} F^{\alpha\beta}F_{\alpha\beta} \tag{2.7}$$

In particular the energy $\mathcal{E}$ is a positive definite quantity,

$$\mathcal{E} = \frac{1}{2}\int d^2\vec{x}\,(\vec{E}^2 + B^2) \tag{2.8a}$$

$$\vec{E} = -\vec{\nabla}A^0 - \dot{\vec{A}}, \qquad B = -\frac{1}{2}\varepsilon^{ij}F_{ij} = \vec{\nabla}\times\vec{A} \tag{2.8b}$$

and the system's excitations cannot be tachyonic. Of course $\theta^{\mu\nu}$ remains conserved in our theory, as a consequence of the field equation (2.2).

The fact that the action associated with our mass term is a world scalar is evidence for its topological nature. This will have profound implication for the quantum theory of the non-Abelian generalization, which we shall discuss later. Here I want to record another curious topological property. Consider the time component of the field equation (2.2), in the presence of an external charge density ρ. This is the analog of Gauss' law; in our theory it reads

$$\vec{\nabla} \cdot \vec{E} - \mu B = \rho \tag{2.9a}$$

Upon integrating (2.9a) over all space, the first term vanishes, since the fields, being massive, decrease exponentially at large distances. One is left with

$$-\mu \int d^2\vec{x} B = \int d^2\vec{x} \rho = Q \tag{2.9b}$$

The magnetic flux passing out of our 2-dimensional space is proportional to the external charge Q. Correspondingly, the magnetic potential is long range, even though the magnetic field is short range.

$$\vec{A} \underset{r\to\infty}{\longrightarrow} -\vec{\nabla} \frac{Q}{2\pi\mu} \tan^{-1} y/x \tag{2.10}$$

This is similar to the electromagnetic configuration supported by vortices in the Higgs model.

Let us note that apart from a total derivative, $\mathcal{L}$ may be written in a gauge invariant form, which, however, is spatially non-local, and Lorentz non-invariant. This follows from the identity

$$\frac{\mu}{4}\varepsilon^{\mu\nu\alpha}F_{\mu\nu}A_\alpha = \frac{\mu}{2}B\frac{\vec{\nabla}}{\nabla^2}\cdot\vec{E} - \frac{\mu}{2}\vec{E}\cdot\frac{\vec{\nabla}}{\nabla^2}B - \frac{\mu}{4}\varepsilon^{\mu\nu\alpha}\,\partial_\alpha(F_{\mu\nu}\frac{\vec{\nabla}}{\nabla^2}\cdot\vec{A})$$

Such an explicitly gauge invariant formulation is not available for the non-Abelian generalization.

An interesting interacting generalization involves coupling fermions to (2.1). Their gauge invariant Lagrange density is

$$\mathcal{L}_F = i\,\bar{\psi}\gamma^\mu(\partial_\mu - ieA_\mu)\psi - m\bar{\psi}\psi \tag{2.11}$$

In three space-time dimensions, the Dirac algebra may be realized by 2x2 [Pauli] matrices, and the fermion field is a 2-component spinor, describing a particle [and an anti-particle] with spin $\frac{1}{2}\frac{m}{|m|} = \pm\frac{1}{2}$.

Correspondingly, the mass term violates P and T symmetries and belongs in a Lagrange density which contains the reflection non-invariant vector meson mass term. Indeed in a theory with only one mass term, the other will be induced by radiative corrections. [Reflection invariance is regained by supplementing (2.11) with a Lagrange density for another fermion field, with a mass term of opposite sign to (2.11). Reflection transformations are again defined to include field exchange, and the model becomes equivalent to a 4-component Dirac theory.]

Feynman-Dyson perturbation theory is straight-forwardly carried out. It is both infrared and ultraviolet finite in the Landau gauge, where the free boson and fermion propagators read, respectively

$$D_{\mu\nu}(p) = \frac{-i}{p^2-\mu^2+i\varepsilon}\,[P_{\mu\nu}(p)-i\mu\varepsilon_{\mu\nu\alpha}P^{\alpha}/P^2] \tag{2.12a}$$

$$S(p) = \frac{i}{\not p - m} \tag{2.12b}$$

Consequently, this is the only non-trivial field theory which is known to possess a perturbation expansion entirely free of divergences -- not even normal ordering need be performed, provided Lorentz and gauge invariance are maintained.

III. NON-ABELIAN THEORY

The 3-dimensional mass term can be generalized to a non-Abelian gauge theory. The gauge field Lagrange density is

$$\mathcal{L} = \frac{1}{2g^2}\,\mathrm{tr}\,F^{\mu\nu}F_{\mu\nu} - \frac{\mu}{2g^2}\,\varepsilon^{\mu\nu\alpha}\,\mathrm{tr}(F_{\mu\nu}A_\alpha - \tfrac{2}{3}A_\mu A_\nu A_\alpha) \tag{3.1}$$

We use a matrix notation

$$A_\mu = gT^aA^a_\mu$$

$$F_{\mu\nu} = gT^aF^a_{\mu\nu} = \partial_\mu A_\nu - \partial_\nu A_\mu + [A_\mu, A_\nu] \tag{3.2}$$

which employs the representation matrices of the group.

$$[T^a, T^b] = f^{abc} T^c \qquad (3.3)$$

The coupling constant is g, while μ/g^2 is dimensionless. The field equations which follow from (3.1) are gauge covariant,

$$\mathcal{D}_\mu F^{\mu\nu} + \frac{\mu}{2} *F^\nu = 0 \qquad (3.4a)$$

$$\mathcal{D}_\mu = \partial_\mu + [A_\mu, \ ] \qquad (3.4b)$$

and from our previous consideration of the non-interacting limit (g=0), we know that μ indeed provides a mass for the field. The dual field

$$*F^\mu = \frac{1}{2} \varepsilon^{\mu\alpha\beta} F_{\alpha\beta}$$

$$F^{\alpha\beta} = \varepsilon^{\alpha\beta\mu} *F_\mu \qquad (3.5)$$

satisfied the Bianchi identity

$$\mathcal{D}_\mu *F^\mu = 0 \qquad (3.6)$$

as a consequence of the definitions (3.2) and (3.5), or alternatively as a consequence of the field equation (3.4a). The dual of (3.4a) is

$$\mathcal{D}_\alpha *F_\beta - \mathcal{D}_\beta *F_\alpha - \mu F_{\alpha\beta} = 0 \qquad (3.7a)$$

and another covariant divergence converts this, with the help of (3.14) and the Ricci identity $[\mathcal{D}_\alpha, \mathcal{D}_\beta] = [F_{\alpha\beta}, \ ]$ into

$$(\mathcal{D}_\alpha \mathcal{D}^\alpha + \mu^2) *F^\mu = \varepsilon^{\mu\alpha\beta} [*F_\alpha, *F_\beta] \qquad (3.7b)$$

which is the non-Abelian analogue of (2.6b).

$\mathcal{L}$ is not invariant against gauge transformations; rather it changes by a total derivative. Consider a finite transformation

$$A_\mu \to U^{-1} A_\mu U + U^{-1} \partial_\mu U \tag{3.8a}$$

The response of the action to the gauge transformation (3.8a) is

$$\begin{aligned} \int d^3x \mathcal{L} \to \int d^3x \mathcal{L} + \frac{\mu}{g^2} \int d^3x \, \varepsilon^{\alpha\beta\mu} \, \partial_\alpha \, \mathrm{tr} \, [\partial_\beta U U^{-1} A_\mu] \\ + \frac{\mu}{3g^2} \int d^3x \, \varepsilon^{\alpha\beta\gamma} \, \mathrm{tr} \, (\partial_\alpha U U^{-1} \, \partial_\beta U U^{-1} \, \partial_\gamma U U^{-1}) \end{aligned} \tag{3.8b}$$

The second term on the right-hand side, which is manifestly a total divergence, is the analogue of the Abelian term (2.5b). We shall only consider gauge transformations which tend to the identity at temporal and spatial infinity.

$$U(x) \xrightarrow[x\to\infty]{} \pm I \tag{3.9}$$

This restriction is made to avoid convergence problems in (3.8). Also, it reflects our assumption of asymptotic space-time uniformity. With Eq. (3.9), we may conclude that the A-dependent surface integral in (3.8b) vanishes. The last term in (3.8b), which has no Abelian analog, can also be converted to a surface integral once the integrand is re-written as a total derivative. This can be made manifest by introducing an explicit parametrization for U. We choose the gauge group to be SU(2) [more generally, we consider a SU(2) subgroup of the gauge group] and make use of the exponential parametrization.

$$\begin{aligned} U(x) &= \exp T^a \, \theta^a(x) \\ T^a &= \sigma^a/2i \\ \theta^a &= \hat{\theta}^a |\theta| \end{aligned} \tag{3.10}$$

It follows that

$$\int d^3x \mathcal{L} \to \int d^3x \mathcal{L} + \mu \frac{8\pi^2}{g^2} \, w(U) \tag{3.11}$$

where we have introduced the "winding number" of the gauge transformation U.

$$w(U) = \frac{1}{24\pi^2} \int d^3x \, \varepsilon^{\alpha\beta\gamma} \, \mathrm{tr}[\partial_\alpha UU^{-1} \partial_\beta UU^{-1} \partial_\gamma UU^{-1}]$$
$$= \frac{-1}{16\pi^2} \int d^3x \, \varepsilon^{\alpha\beta\gamma} \varepsilon^{abc} \partial_\alpha [\hat{\theta}^a \partial_\beta \hat{\theta}^b \partial_\gamma \hat{\theta}^c \, x(|\theta| - \sin|\theta|)] \tag{3.12}$$

This quantity, though given by a surface integral, is not in general zero. It vanishes only for homotopically trivial U's -- those continuously deformable to I. However, as a consequence of the fact that

$$\Pi_3(SU(2) = \Pi_3(S_3) = \text{group of all integers} \tag{3.13}$$

there are U's which are not continuously deformable to the identity. Indeed, the gauge functions U can be arranged into homotopically inequivalent classes, labelled by the integers, and $w(U)$ equals precisely that integer.[3] These considerations are, of course, familiar from the analysis of topological structure in 4-dimensional Yang-Mills theory.[3] That they should reappear in the 3-dimensional theory is not surprising, in view of the further mathematical/topological connections which we shall draw later.

We conclude that the action is not gauge invariant, but changes by $\mu(8\pi^2/g^2)\, W(U) = \mu(8\pi^2/g^2) \times$ integer. While the action is of no particular consequence in classical mechanics, in quantum mechanics the exponential of the action, $\exp i\int d^3x\mathcal{L}$, determines probability amplitudes and must be gauge invariant (see also below). Hence a change in the action, can be tolerated only if it is an integral multiple of 2π. Consequently the requirement of gauge invariance gives a quantization condition on the dimensionless ratio $4\pi\mu/g^2$.

$$4\pi \frac{\mu}{g^2} = n \qquad n = 0, \pm 1, \ldots \tag{3.14}$$

A Euclidean formulation leads to the same conclusion. The functional integral requires $\exp - \int d^3x\mathcal{L}$ to be gauge invariant, but the mass term's contribution to the action is purely imaginary; a factor of i appears when the continuation to imaginary time [Euclidean space] is performed. The winding number is a world scalar; hence it takes the same integer value regardless of the space's signature. The quantization condition (3.14) follows as before; it is entirely due to the internal group.

The way that gauge transformations act on the mass term may also be appreciated once it is recognized that its action is a well-known mathematical entity -- the "Chern-Simons secondary class characteristic."[4]

This is defined in the following way. In even number of dimensions one can construct a gauge invariant quantity -- the "Pontryagin density" P_{2n} -- whose integral over the even-dimensional space is an invariant that characterizes the topological class to which the gauge field belongs. Examples in two and four dimensions are

$$P_2 = \frac{1}{2\pi} {}^*F = \frac{1}{4\pi} \varepsilon^{\mu\nu} F_{\mu\nu} \tag{3.15a}$$

$$P_4 = -\frac{1}{16\pi^2} \mathrm{tr}\ {}^*F^{\mu\nu} F_{\mu\nu} \tag{3.15b}$$

[The 2-dimensional Pontryagin density arises in 2-dimensional massless QED as the anomalous divergence of the axial vector current, and is responsible for mass generation in that model. The 4-dimensional Pontryagin density is associated with θ-vacua in 4-dimensional Yang-Mills theories.] Since the integral is a topological invariant, this gauge invariant Pontryagin density can also be written as a total derivative of a gauge variant quantity.

$$P_{2n} = \partial_\mu X^\mu_{2n} \tag{3.16}$$

The formulas corresponding to (3.15) are

$$X^\mu_2 = \frac{1}{2\pi} \varepsilon^{\mu\nu} A_\nu \tag{3.17a}$$

$$X^\mu_4 = -\frac{1}{16\pi^2} \varepsilon^{\mu\alpha\beta\gamma} \mathrm{tr}(F_{\alpha\beta}A_\gamma - \tfrac{2}{3} A_\alpha A_\beta A_\gamma) \tag{3.17b}$$

On odd-dimensional spaces Pontryagin classes do not exist. But one may construct another topological quantity, the Chern-Simons secondary characteristic class. This is gotten by integrating one component of X^μ_{2n} over the 2n - 1 dimensional space which does not include that component. The integral is gauge invariant against homotopically trivial gauge transformations; otherwise, it changes by the winding number of the transformation.

We recognize that our mass term action is proportional to the 3-dimensional Chern-Simons structure.

This observation about the mass term in Yang-Mills theory has an immediate parallel in the construction of a topological mass for 3-dimensional gravity from the 4-dimensional *RR Hirzebruch-Pontryagin density. But this subject is outside the scope of my lectures, hence those interested are referred to the literature.[5]

Let us explore further the need for the mass quantization condition, and show that the theory is inconsistent without it. The functional integral for a 3-dimensional, massive gauge theory [in Euclidean space] is given by

$$Z = \int \mathcal{D}A^\mu \exp -\{\int d^3x\, (-\tfrac{1}{2}\,\mathrm{tr}\, F^{\mu\nu}F_{\mu\nu}) - i\mu\frac{8\pi^2}{g^2}\, W(A)\}$$

$$W(A) = -\frac{1}{16\pi^2}\int d^3x\, \varepsilon^{\mu\nu\alpha}\,\mathrm{tr}\,(F_{\mu\nu}A_\alpha - \tfrac{2}{3}A_\mu A_\nu A_\alpha) \qquad (3.18)$$

As the functional integration ranges over all gauge potentials, for any given A^μ it also encounters its gauge copies A'^μ.

$$A'_\mu = U^{-1}A_\mu U + U^{-1}\partial_\mu U \qquad (3.19)$$

where the gauge functions U fall into homotopically distinct classes labelled by the integers n. Now the usual gauge fixing prescriptions, which are [implicitly] contained in $\mathcal{D}A^\mu$, remove gauge copies arising from the homotopically trivial gauge transformations. [Recall that Faddeev-Popov procedures are formulated in infinitesimal terms.] However, there seems to be no way of removing gauge copies in (3.18) arising from non-trivial, large gauge transformations. Thus we may write Z as

$$Z = \sum_{n=-\infty}^{\infty} Z_n \qquad (3.20)$$

Here Z_0 results by performing an integration over vector potentials, with no gauge copies; Z_1 results from integrating over vector potentials related by a gauge transformation in the first homotopy class to those occurring in the integral for Z_0; etc. But it is clear that once we have determined Z_0, Z_n for $n \neq 0$ may be evaluated by changing variables in the functional integral from A^μ to A'^μ, which is defined to be the gauge transform of A^μ with a large gauge function of the nth homotopy class. Such a change of variables does not affect the measure nor the usual action, since both are gauge invariant. In the mass term, W(A) changes by an integer, so we get

$$Z = Z_0 \sum_{n=-\infty}^{\infty} e^{in\mu \frac{8\pi^2}{g^2}} \qquad (3.21)$$

Now we see that if the mass term is not quantized, the infinite sum vanishes, by destructive interference. On the other hand, when the quantization condition holds, the sum becomes $\sum_{n=-\infty}^{\infty} 1$, which, though infinite, may be harmlessly cancelled by a normalizing denominator in the definition of Z.

Thus the result: the massive gauge theory vanishes in the absence of mass quantization. The same conclusion may be established by canonical reasoning.[6] Lack of time and space prevent giving details, but the essential steps are two in number. First, one realizes that the homotopy formula (3.13) implies that at fixed time the canonical configuration space consisting of spatial components of the vector potentials, is not simply connected. Second, one finds that Gauss' law, which in the canonical formalism is a constraint on physical states, i.e. a [functional] differential equation that physical wave functionals satisfy, cannot be globally integrated unless the mass term is quantized.[7]

In conclusion, let me speculate concerning the physical significance of the mass term. As I have already remarked, finite temperature perturbation theory for simple models suggests that in the high temperature limit a 3-dimensional version of that same model comes into play. Of course such a dimensional reduction makes no reference to non-perturbative phenomena. Moreover, for a non-Abelian gauge theory, which we know to be rich in non-perturbative effects, neither the 4-dimensional finite temperature perturbation theory, nor zero temperature perturbation theory in three dimensions makes sense owing to infrared divergences. The discovery of the 3-dimensional mass term allows the conjecture that the high temperature limit of a non-Abelian, 4-dimensional gauge theory is governed by a 3-dimensional, massive yet gauge invariant Yang-Mills theory, of the type here described.

There is no derivation of this fact; but neither can it be falsified, since naive perturbation theory does not exist and we do not have sufficient control over the formalism to extract non-perturbative behavior. Confronting such a hiatus, we invoke the principle of "naturalness" to aid in constructing the effective Lagrangian. The 3-dimensional effective Lagrangian should possess all terms with quantum numbers of the 4-dimensional theory, whose high temperature asymptote is under discussion. According to this criterion, the gauge invariant mass should be present, since its reflection non-invariance mirrors the reflection non-invariance of the 4-dimensional θ-vacua. Indeed, the topological setting of our mass term puts into evidence an intimate mathematical connection between it and the quantity responsible for the θ vacua. However, it is not known at present whether this mathematical relationship can be the basis for a physical derivation.

If we accept the gauge invariant mass as a proper term in the effective Lagrangian which summarizes high temperature behavior of physical non-Abelian gauge theories, we get another bonus, beyond infrared regularity. Owing to the quantization condition (3.14), the mass becomes evaluated in terms of the coupling constant. Recalling that in a high-temperature effective Lagrangian, the dimensionful 3-dimensional coupling constant g is related by a power of the temperature T to the dimensionless 4-dimensional coupling e, we find

$$\mu = \frac{g^2}{4\pi} n = \frac{e^2 T}{4\pi} n = \alpha T n \tag{3.22}$$

The integer structure to μ is most fascinating. A non-vanishing mass presumably arises from a non-vanishing θ, and discontinuities in the former for the 3-dimensional model are suggestive of different phases in the latter for the 4-dimensional theory. That different values of θ correspond to different phases has been occasionally suggested. Clearly it would be most satisfying if more understanding of these speculative ideas could be obtained.

REFERENCES

1. W. Siegel, Nucl. Phys. B156, 135 (1979); R. Jackiw and S. Templeton, Phys. Rev. D23, 2291 (1981); J. Schonfeld, Nucl. Phys. B185, 157 (1981); S. Deser, R. Jackiw and S. Templeton, Phys. Rev. Lett. 48, 975 (1982) and Ann. Phys. (NY) 140, 372 (1982); H. Nielsen and H. Woo (unpublished).
2. Other reviews are R. Jackiw in "Asymptotic Realms of Physics" (A. Guth, K. Huang, and R. Jaffe, editors), MIT Press, Cambridge, MA, 1983 and Arctic Summer School Proceedings (1982); S. Deser, DeWitt Festschrift, to appear.
3. R. Jackiw and C. Rebbi, Phys. Rev. Lett. 37, 172 (1976); R. Jackiw, Rev. Mod. Phys. 52, 661 (1980).
4. S. Chern, "Complex Manifolds without Potential Theory", 2 ed. Springer Verlag, Berlin, 1979.
5. See Deser, Jackiw and Templeton, Ref. 1; Deser, Ref. 2.
6. The canonical description is due to J. Goldstone and E. Witten unpublished; for details see Jackiw, Ref. 2 (second cited work).
7. An analogous quantization condition has been obtained by E. Witten in a 4-dimensional SU(2) gauge theory, Princeton University preprint (unpublished). One begins with the observation that $\Pi_4(SU(2)$ $= \Pi_4(S_3)$ = cyclic group of two integers, to conclude that the 4-dimensional gauge functions $U(t,\vec{x})$ fall into two homotopically distinct classes. Next one finds that when N species of left-handed

Weyl fermions in the fundamental [doublet] representation are coupled to the SU(2) gauge field, their functional [fermionic] determinant is not invariant against homotopically non-trivial gauge transformations. Rather it changes by the factor $(-1)^N$; hence N must be even.

GLUON CONDENSATION AND CONFINEMENT OF QUARKS

R. Fukuda

Research Institute for Fundamental Physics
Kyoto University, Kyoto 606, Japan

The purpose of my talk is to argue that if gluon condenses in the form of color singlet, then the dual variable becomes a good coordinate in the low energy region and acquires a mass gap.

The vacuum of Quantum Chromodynamics (QCD, quarks are neglected for the moment) will be filled with nontrivial configuration of gluonic fields $A_\mu^a(x)$: they condense in the vacuum in the form of color singlet and of Lorentz scalar. We have to take into account the effect of the gluon condensation when we study the vacuum state and the excitation spectra.

Any gluonic operator can be used for the discussion of the gluon condensation as long as it has correct quantum numbers: color singlet and $J^{PC}=0^{++}$. We can use $G_{\mu\nu}^2(x)$, $A_\mu^a(x)A^{a\mu}(x)$ etc. Here $G_{\mu\nu}^a$ is the covariant field strength. The gauge has to be fixed of course in the case $A_\mu^a(x)A^{a\mu}(x)$. The most convenient operator should be picked up according to the following criteria, 1) The condensation of the selected operator is easy to discuss. 2) In case it condenses the effect on the excitation spectra is clearly seen. The operator $G_{\mu\nu}^2$ has been studied but it lacks the second property. The operator chosen here is the zero momentum mode $A_\mu^{a(0)}$ of $A_\mu^a(x)$ in the axial gauge $A_3^a(x)=0$,

$$A_\mu^a(x) = A_\mu^{a(0)} + A_\mu^{a'}(x) \ , \qquad \int A_\mu^{a'}(x)d^4x = 0 \ . \tag{1}$$

Although $\langle A_\mu^{a(0)}\rangle=0$ we can assume $\langle A_\mu^{a(0)}A^{a(0)\mu}\rangle\neq 0$. The residual gauge symmetry in $A_3^a(x)=0$ gauge is fixed by specifying the prescription to avoid the singularity of the gluon propagator $\langle A_\mu^a A_\nu^b\rangle_p$ at $p_3=0$ in momentum space. The conventional one is the principal part prescription. We assume for the moment the condensation of $A_\mu^{a(0)}$ in the above sense. At the end of the talk the condensation is discussed.

In case $A_\mu^{a(0)}$ condenses, the excitation spectra are determined by substituting (1) in the Lagrangian,

$$\int \mathcal{L}(A_\mu^a)d^4x = -\frac{1}{4}\, G_{\mu\nu}^{(0)2}\Omega$$

$$- \frac{1}{2} \iint A^{\mu a'}(x) M^{ab}_{\mu\nu}(x-y) A^{\nu b'}(y) d^4x d^4y$$

$$+ (A')^3 \text{ term} + (A')^4 \text{ term} , \qquad (2)$$

where $\Omega = \int d^4x$. The matrix $M^{ab}_{\mu\nu}(p)$ is <u>not</u> a positive definite matrix so that the squared mass matrix $M^{ab}_{\mu\nu}(p=0)$ may have negative (i.e. tachyonic) eigenvalues. We have seen that this is indeed the case for SU(2) color group. For SU(3) the same phenomenon occurs since SU(2) is a subgroup of SU(3). Therefore $A^{a'}_{\mu}$ is not a stable coordinate: we have to condense above tachyonic unstable modes to reach the really stable vacuum. The situation is similar to that discovered by Nielsen and Oleson[1]. Instead of condensing unstable modes we make a dual transformation and find that the unstable modes are absent if the theory is written in terms of the dual potential. Moreover the condensation of $A^{a(0)}_{\mu}$ yields a positive definite squared mass matrix to the dual potential.

The dual formalism in the axial gauge has been given by Halpern[2] with the result

$$\int [dA_\mu] \exp\{- \frac{i}{4} \int (\partial_\mu A^a_\nu - \partial_\nu A^a_\mu + g f^{abc} A^b_\mu A^c_\nu)^2 d^4x\} \qquad (3)$$

$$= \int [dG_{\mu\nu}] \delta(D_\mu \tilde{G}^{\mu\alpha}) \exp\{- \frac{i}{4} \int G^2_{\mu\nu} d^4x\} \qquad (4)$$

$$= \int [dG_{\mu\nu}] [dB^\nu] \delta(n_\mu B^\mu) \exp i \int (B^a_\nu D_\mu \tilde{G}^{\mu\nu} - \frac{1}{4} G^2_{\mu\nu}) d^4x \qquad (5)$$

$$= \int [dB^\alpha] \exp i \int \mathcal{L}(B) d^4x . \qquad (6)$$

In (4) $D^{ab}_\mu = \delta^{ab}\partial_\mu + g f^{acb} A^c_\mu$ with $A^c_\mu(x) = \frac{1}{\partial_3} G^c_{3\mu}$ and $n_\mu = (0,0,0,1)$. In the following indices $\alpha, \beta, \gamma \cdots$ take 0,1,2, while $\mu, \nu, \rho, \sigma \cdots$ take 0,1,2,3. The Lagrangian for the dual potential B^α is written symbolically as[2]

$$\mathcal{L}(B) = - \frac{1}{4} (\widetilde{\partial_\mu B^a_\nu - \partial_\nu B^a_\mu}) N^{-1\,ab}_{\mu\nu,\rho\sigma} (\widetilde{\partial_\rho B^b_\sigma - \partial_\sigma B^b_\rho}) \qquad (7)$$

where the tilde indicates the dual tensor as $\tilde{F}_{\mu\nu} \equiv \frac{1}{2}\varepsilon_{\mu\nu\rho\sigma} F^{\rho\sigma}$, $B^a_3 = 0$ and $N^{ab}_{\mu\nu,\rho\sigma} = [1 + g \int^{x_3} dx'_3 B]^{ab}_{\mu\nu,\rho\sigma}$.

In the representation (5) we see that the theory is invariant under $G^a_{\mu\nu} \to G^a_{\mu\nu}$ and

$$B^a_\mu \to B^a_\mu + D^{ab}_\mu \Lambda^b + g f^{abc} \int^{x_3} G^b_{3\mu} \Lambda^c \, dx'_3$$

with $\Lambda^a(x)$ arbitrary function. This is due to the following identity

$$G^a_{\mu\nu} = \partial_\mu A^a_\nu - \partial_\nu A^a_\mu + gf^{abc}A^b_\mu A^c_\nu + \varepsilon_{3\gamma\mu\nu}J^{a\gamma} ,$$

where $A^a_\mu = \frac{1}{\partial_3}G^a_{3\mu}$ and $J^{a\gamma} = \int^{x_3} D^{ab}_\mu \tilde{G}^{b\mu\gamma} dx'_3$. The Ward identities are derived from the above transformation and we see, by the similar procedure as in the conventional case, that the dual potential is massless to any finite order of perturbation: writing the propagator as

$$i\langle B^a_\mu B^b_\nu \rangle_p = \frac{\delta^{ab} g_{\mu\nu}}{p^2_\mu + \Pi(p^2_\mu, (n_\mu p^\mu)^2)} + \text{gauge term} ,$$

we can show that Π behaves near $p^2_\mu = 0$ as Ap^2_μ with some constant A. We have also checked explicitly in the form of (6) that the one loop correction produces the correct amount of logarithmic divergence: $A \sim \frac{11}{3} g^2 C_2(G) \ln\Lambda$ which corresponds to the β-function behaving as $\beta(g) \underset{g\sim 0}{\sim} -\frac{11}{3} g^3 C_2(G)$ where $f^{abc}f^{dbc} = \delta^{ad} C_2(G)$.

Now in the presense of the condensation $A^{a(0)}_\mu$, we separate it as in (1) and from the variable $A^{a'}_\mu$ we switch to the dual variable. Since $G^a_{3\mu} = \partial_3 A^a_\mu = \partial_3 A^{a'}_\mu$, $A^{a'}_\mu(p) = iP \frac{1}{p_3} G^a_{3\mu}(p)$ where P denotes the principal part. Writing $[dA_\mu] = [dA^{(0)}_\mu][dA'_\mu]$ and following the same procedure as above we arrive at

$$(3) = \int [dA^{(0)}][dB^\alpha] \exp i \int \mathcal{L}(A^{(0)}_\mu, B_\alpha)$$

where

$$\mathcal{L}(A^{(0)}_\mu, B_\alpha) = -\frac{1}{4} \widehat{(D^{(0)}_\mu B_\nu - D^{(0)}_\nu B_\mu)}^a N^{-1\ ab}_{\mu\nu,\rho\sigma} \widehat{(D^{(0)}_\rho B_\sigma - D^{(0)}_\sigma B_\rho)}^b , \tag{8}$$

$$D^{(0)ab}_\mu \equiv \delta^{ab}\partial_\mu + gf^{acb}A^{c(0)}_\mu .$$

Expanding $\int \mathcal{L}(A^{(0)}_\mu, B^\alpha) d^4x$ in the form

$$\frac{1}{2}\int B^{a\alpha}(p) \mathcal{M}^{ab}_{\alpha\beta}(A^{(0)}, p) B^{b\beta}(-p) \frac{d^4p}{(2\pi)^4} + B^3 \text{ term} + \cdots , \tag{9}$$

we define the mass of B by the term

$$\frac{1}{2} B^{a\alpha}(0) \mathcal{M}^{ab}_{\alpha\beta}(A^{(0)}, 0) B^{b\beta}(0) = \frac{1}{2}(\varepsilon^{\alpha\beta\gamma} gf^{abc} A^{c(0)}_\beta B^b_\gamma(0))^2 .$$

Since it is written as a square, we have a positive semi-definite squared mass matrix. The dual potential is a stable coordinate: unstable modes which are present in the spectrum of $A^{a'}_\mu$ field are eliminated by the dual transformation.

Once the dual potential aquires mass gap, the system is in the magnetic Higgs phase — dual to the usual Higgs phase. The dual loop introduced by 't Hooft[3] shows the perimeter law and we expect the area law for the Wilson loop and the color electric flux tube will be formed as a dual Meissner effect. We do not discuss this scenario here.

The solution to the U(1) problem is also provided by the above mechanism. From (5) we see that B^{ai} and $\varepsilon_{ii'}G^a_{3i'}$ are canonically conjugate pairs (Here i,i'=1 or 2, $\varepsilon_{12}=-\varepsilon_{21}=1$, $\varepsilon_{11}=\varepsilon_{22}=0$) and by (9), propagators take the form in the low energy region

$$i\langle B^a_\alpha B^b_\beta\rangle_p = i(1/m^2)\langle G^a_{3\alpha}G^b_{3\beta}\rangle_p = \delta^{ab}g_{\alpha\beta}(1/p_\mu^2-m^2) \ . \qquad (10)$$

Consider $K^\mu=(g^2/16\pi^2)\varepsilon^{\mu\nu\rho\sigma}A^a_\nu(\partial_\rho A^a_\sigma+\frac{g}{3}f^{abc}A^b_\rho A^c_\sigma)$. We have to explain why $\partial_\mu K^\mu$ does not vanish at zero momentum. If we substitute (1) in $K_3(x)$ there appears a term proportional to $A^{a'}_\alpha(x)A^{b(0)}_\beta A^{c(0)}_\gamma$ which gives a non-zero contribution to

$$\int d^4x\langle\partial_\mu K^\mu(x)\partial_\nu K^\nu(0)\rangle = \int d^4x\langle\partial_3 K_3(x),\partial_3 K_3(0)\rangle$$

because of (10). Recall here $\partial_3 A^{a'}_\mu=G^a_{3\mu}$. The above quantity gives a non-zero mass to η' meson as is well known.

Finally we discuss the condensation of $A^{a(0)}_\alpha$. The effective potential $V(A^{a(0)}_\alpha)$ of $A^{a(0)}_\alpha$ can be calculated in loop expansion. We know that V has a non-trivial minimum at one loop level. Our conclusion is that we can develop the series expansion of V in such a way that the position of the minimum found at one loop level is shifted slightly for small coupling if the effect of the higher order terms of the series are taken into account. These shifts are in accordance with the renormalization group equations.

Details are found in Ref.4 where we also discuss the whole subjects in this talk.

References

1 N.K. Nielsen and P. Olesen, Nucl. Phys. B144, 376 (1978).

2 M.B. Halpern, Phys. Rev. D19, (1979); Phys. Letters 81B, 245 (1979).

3 G. 't Hooft, Nucl. Phys. B138, 1 (1978).

4 R. Fukuda, "Gluon Condensation and the Field Strength Formulation in Quantum Chromodynamics" preprint RIFP-469 (to appear in Prog. Theor. Phys. August 1982).

GENERALIZED STRING AMPLITUDE AND WAVE EQUATION FOR HADRONS

H. Suura
School of Physics and Astronomy
University of Minnesota
Minneapolis, Minnesota 55455
USA

A gauge invariant hamiltonian formulation of hadron dynamics involving a generalized string amplitude is proposed. Resulting partial integro-differential wave equations reflect the shielding of the long range potential and determine a universal logarithmic derivative of the hadronic wave functions. Because of this boundary condition, the new wave equation allows confined solutions without an explicit confinements potential in it.

Instead of the conventional string amplitude $\exp\{ ig \int \vec{\underset{\sim}{A}} \cdot d\vec{x}\}$ used in gauge invariant formulations of hadronic systems, I propose to study a modified string operator on a t plane

$$U(2,1) = \exp\{ g \int_1^2 [i\, \vec{\underset{\sim}{A}}(x) + \lambda \vec{\underset{\sim}{E}}(x)] \cdot d\vec{x} \}, \qquad (1)$$

($\underset{\sim}{A} = A^a \lambda^a/2$). In spite of its explicitly non-covariant definition, the operator is useful in taking into account an infinite number of soft gluons created on the string as well as the shielding of the long range force due to splittings of the string. Using (1), I construct a gauge invariant quark operator

$$q_{\alpha\beta}(1,2) = \mathrm{Tr}^c\, [\psi_\alpha(1)\, U^T(2,1)\, \psi_\beta^{\,+}(2)], \qquad (2)$$

where α and β are Dirac indices, and flavor indices have been suppressed. Tr^c is the trace over color spin. Time development of the operator (2) can be derived in the same way as in my previous paper,[1] which was for the case $\lambda = 0$. Two new features arise because of the $\underset{\sim}{E}$ term inserted in (1). Time derivative of A(x) is equivalent to a c-number operator $\frac{1}{i}\frac{\partial}{\partial\lambda}$, since $\dot{A} = -E$. Time derivative of E is equal to insertion of the quark current -j, which, after use of a Fiertz identity, gives a product of two-string operators. In the following I consider q(1,2) at a large distance and neglect all the derivatives of the field operators (B, $\nabla \times E$ and $\nabla \times B$). Strictly speaking, transverse (to the string) momentum of gluons should be kept under the large distance approximation. However, it would give rise to the transverse oscillation of the string, which will be studied elsewhere. Thus

*Supported in part by the U. S. Department of Energy Contract No. DE-AC02-82ER-40051.

$$i\,\dot{q}\,(1,2) = H_D\, q(1,2) + \frac{\partial}{\partial\lambda} q\,(1,2)$$
$$- \frac{i}{2} g^2\lambda \int_1^2 d\vec{x} \cdot \sum q^{(a)}(1,x)\vec{\alpha} \quad q^{(a)}(x,2)$$
$$+ \frac{i}{6} g^2\lambda \int_1^2 d\vec{x} \cdot \vec{j}_s(x) \quad q(1,2) \cdots \tag{3}$$

where

$$H_D\, q(1,2) = [-i\alpha \cdot \nabla_1 + \beta m]\, q - q\, [i\alpha \cdot \overleftarrow{\nabla}_2 + m] \,. \tag{4}$$

$\sum_a$ in the third term means a summation over all possible flavors. $\vec{j}_s$ is the color- and flavor-singlet quark current, and represents an emission of a singlet vector meson from the string. For vacuum-vacuum matrix element of (3), I may neglect this term, and also keep only the vacuum intermediate state neglecting all the hadronic ones, since the latter tend to give short range forces. Defining

$$S(1,2) = \langle 0 \mid q(1,2) \mid 0 \rangle , \tag{5}$$

I obtain an equation

$$H_D\, S(1,2) + \frac{\partial}{\partial\lambda} S(1,2) - \frac{i}{2} g^2\lambda \int_1^2 d\vec{x} \cdot S(1x)\vec{\alpha}\, S(x,2) = 0 \tag{6}$$

S(1,2) can be expanded in Dirac matrices like

$$S(1,2) = -i\, \alpha\cdot\hat{r}\; S_1(r) + \beta\; S_2(r) + i\beta\; \alpha\cdot\hat{r}\; S_3(r) \quad . \tag{7}$$

If S(1,2) were the equal-time limit of a quark propagator for which a spectral representation holds, then we would have $S_3=0$. Since no spectral representation holds for S(1,2) which has no color-singlet hadronic intermediate states we may include S_3 term. In fact as shown below S_3 term is necessary in order to obtain the spontaneous breaking of the chiral symmetry which says

$$\langle\, \bar{\psi}(0)\; \psi\,(0)\, \rangle_0 = -S_2(0) \neq 0. \tag{8}$$

Introducting (7) into (6) I obtain

$$\frac{\partial S_1(r)}{\partial\lambda} - \frac{1}{2} g^2\lambda \int_0^r dz\, [S_1(r-z)S_1(z) + S_2(r-z)S_2(z) + S_3(r-z)S_3(z)] = 0,$$

$$-2\left(\frac{\partial S_3(r)}{\partial r} + \frac{2}{r} S_3(r)\right) + \frac{\partial S_2(r)}{\partial\lambda} - g^2\lambda \int_0^r dz\; S_1(r-z)\; S_2(z) = 0,$$

$$2\; \frac{\partial S_2(r)}{\partial r} + \frac{\partial S_3}{\partial\lambda} - g^2\lambda \int_0^r dz\; S_1(r-z)\; S_3(z) = 0 \,. \tag{9}$$

If $S_3 = 0$, then the third equation tells that S_2 = const. and hence must vanish. The solution $S_2=S_3=0$ represents a chirally symmetric solution. The time reversal requires that the amplitudes S_1, S_2 and S_3 in (7) are all real, if the parameter λ is taken to be real. The solution of eq. (9) is not well defined without a precise knowledge of the boundary conditions in variable λ . Nevertheless we may obtain a physically reasonable solution in the following way. Neglecting quark kinetic energy terms in (9) I obtain an asymptotic solution for large r,

$$S_1(r) = A(\lambda)\, e^{-D(\lambda)r}; \; A = -\frac{dD}{d\lambda}/g^2\lambda$$

$$S_2(r) = b_2 S_1 \;, \quad S_3(r) = b_3 S_1 \;, \tag{10}$$

where b_2 and b_3 are constants independent of satisfying $b_2^2 + b_3^2 = 1$. Thus, I have

$$S_1'/S_1 = S_2'/S_2 = S_3'/S_3 = -D \tag{11}$$

To obtain a picture of the overall solution, I consider three regions in r. In the inner region I, the system is free (short range forces are being neglected), so that $S_2 = 1$ and $S_3 = 0$. In the outer region III, (10) gives the leading asymptotic terms. In the intermediate region II, I may extrapolate from outside and set

$$-\frac{\partial S_{2,3}}{\partial\lambda} = (W-Kr)\, S_{2,3} \tag{12}$$

where

$$W = -A^{-1}\frac{dA}{d\lambda} \;, \quad K = -\frac{dD}{d\lambda} \;.$$

The Kr term should cancel the integral terms in (9) in region III. However, in region II, the integral involves $S_{2,3}$ in region I, so that cancellation will not be complete. Thus, in region II, neglecting the integral terms, the second and third equations give the Breit-type equation with an effective eigenvalue W and a linear potential Kr. The equation must be solved with a boundary condition (11) imposed at a certain radius r = R inside region II. In this way the Klein paradox associated with a linear potential[2] is completely avoided. Details of the solution, as well as its relation to the pion wave function will be discussed elsewhere.

References

1. H. Suura, Phys. Rev. D20, 1412 (1979).
2. D. A. Geffen and H. Suura, Phys. Rev. D16, 3305 (1977).

STOCHASTIC QUANTIZATION AND LARGE N REDUCTION

J. Alfaro and B. Sakita
Department of Physics, City College
of City University of New York, N.Y. 10031

The stochastic quantization method of Parisi and Wu is used in order to understand the quenched momentum prescription for large N theories. The Hermitian matrix field theory model is studied first and then the same method is applied to SU(N) gauge theory.

I. Introduction

After a work of Eguchi and Kawai[1], some very intensive and active studies on the reduction of degrees of freedom at large N were done recently by various groups[2,3] to reach the quenched momentum prescription[3] for large N theories. These authors[3] have based their discussion on a detailed analysis of planar Feynman graphs for all order of perturbation theory. We believe that the large N reduction phenomenon is so general that it is likely to have more transparent explanations.

In this report we present our study on this problem using the stochastic quantization method of Parisi and Wu[4]. Since in this method the average over the random variables is taken at the end of calculation of correlation functions (Green's functions), it is possible to derive the quenched models by viewing a part of the random average as a quenched average.

Since the stochastic quantization method is relatively new and since in this method we believe there are subjects that require further studies especially for gauge theories, we first review the method in the next section. In section III we discuss the reduction of degrees of freedom for large N by taking the Hermitian matrix field model as an example[5]. In IV, we discuss the same problem for SU(N) gauge theory. In this case there exists some confusions whether we really obtained the quenched Eguchi-Kawai model or not. This confusion is mainly due to our lack of knowledge on the stochastic quantization. We suggest our resolution.

II. Stochastic Quantization

In this section, we review the stochastic quantization method of Parisi and Wu[3].

Let us consider an Euclidean field theory of a system of Bose field $\phi_\ell(x)$. x denotes a d-dimensional space time point while ℓ represents a set of internal indices. The correlation functions (Green's functions) of the theory are given by a functional average defined by

$$\langle \phi_{\ell_1}(x_1)\phi_{\ell_2}(x_2)\cdots\phi_{\ell_n}(x_n)\rangle \equiv \frac{\int \mathcal{D}\phi\,\phi_{\ell_1}(x_1)\phi_{\ell_2}(x_2)\cdots\phi_{\ell_n}(x_n)e^{-S[\phi]}}{\int \mathcal{D}\phi\; e^{-S[\phi]}} \tag{2.1}$$

where $S[\phi]$ is the action of the system.

One interprets (2.1) as a statistical average of dynamical variables $\phi_\ell(x)$ with Boltzman statistical weight. The basic idea of stochastic quantization of Parisi and Wu is that one regards the average (2.1) as the large (fictitious) time equilibrium limit of a statistical average in an inequilibrium system. The time evolution of this statistical system can be described by Fokker-Planck equation:

$$-\frac{1}{2}\frac{\partial\psi}{\partial t} = \hat{H}_{HF}\psi \tag{2.2}$$

$$\hat{H}_{HP} = \int dx\left[-\frac{1}{2}\sum_\ell \frac{\delta^2}{\delta\phi_\ell^2(x)} + \frac{1}{4}\left\{\frac{1}{2}\sum_\ell\left(\frac{\delta S}{\delta\phi_\ell(x)}\right)^2 - \sum_\ell \frac{\delta^2 S}{\delta\phi_\ell^2(x)}\right\}\right] . \tag{2.3}$$

The probability distribution function at time t, $P[\phi,t]$, is related to ψ by

$$\psi[\phi,t] = e^{\frac{1}{2}S[\phi]}\quad P[\phi,t] . \tag{2.4}$$

It is easy to see that $\hat{H}_{HF}\psi=0$ when $P=e^{-S[\phi]}$, so that ψ is stationary. One assumes that at large t the system reaches to this stationary state. If this is the case, the average can also be calculated by using Langevin equation

$$\frac{\partial\phi_\ell(x,t)}{\partial t} = -\frac{\delta S}{\delta\phi_\ell(x,t)} + \eta_\ell(x,t) \tag{2.5}$$

where t is a fictitious time. $\eta_\ell(x,t)$ is a random source function with

Gaussian distribution, namely it possesses the following property of averages:

$$\langle \eta_{\ell_1}(x_1,t_1)\eta_{\ell_2}(x_2,t_2)\cdots\eta_{\ell_n}(x_n,t_n)\rangle$$

$$= \sum_{\substack{\text{possible}\\ \text{combination}}} \prod_{\text{pairs}} \langle \eta_{\ell_i}(x_i,t_i)\eta_{\ell_j}(x_j,t_j)\rangle_\eta \tag{2.6}$$

$$\langle \eta_\ell(x,t)\eta_{\ell'}(x',t')\rangle_\eta = 2\delta_{\ell\ell'}\delta(x-x')\delta(t-t') \ . \tag{2.7}$$

In (2.6), when n is odd the average is zero. The connection between Langevin and Fokker-Planck is given by

$$\langle F[\phi^\eta(\cdot,t)]\rangle_\eta = \int \mathcal{D}\phi F[\phi(\cdot)]P[\phi,t] \tag{2.8}$$

where $\phi^\eta(x,t)$ is a solution of Langevin equation with initial condition $\phi^\eta(x,0)=\phi^0(x)$ while $P[\phi,t]$ is a solution of Fokker-Planck equation with initial condition $P[\phi,0]=\delta(\phi-\phi^0)$. Accordingly, the stochastic quantization prescription of Parisi and Wu is simply expressed by

$$\langle \phi_{\ell_1}(x_1)\phi_{\ell_2}(x_2)\cdots\phi_{\ell_n}(x_n)\rangle$$

$$= \lim_{t\to\infty} \langle \phi^\eta_{\ell_1}(x_1,t)\phi^\eta_{\ell_2}(x_2,t)\cdots\phi^\eta_{\ell_n}(x_n,t)\rangle_\eta \ . \tag{2.9}$$

Since $\phi^\eta_\ell(x,t)$ is a solution of Langevin equation (2.5) with a certain initial condition, in general the expression (2.9) depends on the initial field configuration ϕ^0. An implicit assumption made in the stochastic quantization is that the final result is independent of the initial condition. But it seems to us this is a point requires further investigations and a source of confusions. We shall come back to this point in Section IV.

III. Reduction of Degrees of Freedom for Large N

In this section we apply the stochastic quantization method of Parisi and Wu to derive the quenched momentum prescription for large N theories. We illustrate it for the Hermitian matrix model defined by the action

$$S[\phi] = \int dx \ \mathrm{tr}\{\tfrac{1}{2}(\partial_\mu\phi)^2 + \frac{m^2}{2}\phi^2 + \frac{g}{4!N}\phi^4\}$$

where $\phi(x)$ is an N×N Hermitian matrix field $(\phi(x)=\phi^+(x))$.

We note the action has an SU(N) symmetry and a reflection symmetry:

$$\left.\begin{array}{lll} \phi(x) \longrightarrow u\phi(x)u^+ & & u \in SU(N) \\ \phi(x) \longrightarrow -\phi(x) & & \end{array}\right\} . \tag{3.2}$$

Therefore, assuming SU(N) symmetry is unbroken we consider only the following Green's functions

$$\langle tr(\phi(x_1)\phi(x_2)\cdots\phi(x_n))\rangle \tag{3.3}$$

which is invariant by the transformations.

The corresponding Langevin equation to (3.1) is given by

$$\frac{\partial\phi_{ij}(x,t)}{\partial t} = (\Box - m^2)\phi_{ij}(x,t) - \frac{g}{3!N}(\phi^3(x,t))_{ij} + \eta_{ij}(x,t) \tag{3.4}$$

where the random source matrix function $\eta(x,t)$ is assumed to have the Gaussian distribution:

$$\langle\eta_{ij}(x,t)\eta_{i'j'}(x',t')\rangle_\eta = 2\delta_{ij'}\delta_{ji'}\delta(x-x')\delta(t-t') . \tag{3.5}$$

One may formally solve Langevin equation (3.4) by iteration, and since each term of the solution is classified by a tree diagram the solution can be expressed as

$$\phi^\eta_{ij}(x,t) = \sum\cdots(\eta\eta\cdots\eta)_{ij} . \tag{3.6}$$

Inserting (3.6) into (3.3) we can express Green's function (3.3) in terms of the η averages as following:

$$\begin{aligned} &\langle tr(\phi(x_1)\cdots\phi(x_n))\rangle \\ &= \lim_{t\to\infty}\sum_{m=0}^{\infty}\left(\frac{g}{N}\right)^m \int\cdots\int dy_1dt_1dy_2dt_2\cdots dy_{2m+n}dt_{2m+n} \\ &\quad K_m(x_1\cdots x_n,t\ ;\ y_1t_1,y_2t_2,\cdots,y_{2m+n}t_{2m+n}) \\ &\quad \langle tr(\eta(y_1,t_1)\eta(y_2,t_2)\cdots\eta(y_{2m+n},t_{2m+n}))\rangle_\eta . \end{aligned} \tag{3.7}$$

If we insert the Gaussian distribution property of η given by (2.6) and (3.5) into (3.7), we should be able to obtain the standard perturbation expansion as shown by Parisi and Wu. A nice point of (3.7) is that in (3.7) the SU(N) indices appear only through η so that we can discuss the large N limit by examining only the η averages.

Next we show that the reduced form of η defined by

$$\eta_{ij}(x,t) = \left(\frac{\Lambda}{2\pi}\right)^{d/2} e^{i(p_i-p_j)x}\ \bar{\eta}_{ij}(t) \tag{3.8}$$

has in a sense the same Gaussian distribution property in the large N limit. In this expression Λ is a momentum cutoff. In a discrete lattice version of the theory, Λ is related to the inverse of lattice distance a; $\Lambda \sim \frac{1}{a}$.

In this report we describe the proof only for $\langle tr(\eta(x,t)\eta(x',t'))\rangle_\eta$, which should have the property

$$\langle tr(\eta(x,t)\eta(x',t'))\rangle_\eta = 2N^2\delta(x-x')\delta(t-t') . \tag{3.9}$$

In the reduced case, using (3.8) we obtain

$$\sum_{ij} \left(\frac{\Lambda}{2\pi}\right)^d \langle e^{i(p_i-p_j)(x-x')}\ \bar{\eta}_{ij}(t)\bar{\eta}_{ji}(t')\rangle_{p,\bar{\eta}} . \tag{3.10}$$

As a reduced average over p_i and $\bar{\eta}$ we choose the integration over p_i in a hypercube $[\frac{\Lambda}{2},-\frac{\Lambda}{2}]^d$ $\left(\int \prod_{\alpha,i=1}^{N} \left(\frac{dp_i^\alpha}{\Lambda}\right)\right)$ and the Gaussian distribution for $\bar{\eta}$:

$$\langle\bar{\eta}_{ij}(t)\bar{\eta}_{i'j'}(t')\rangle = 2\delta_{ij'}\delta_{i'j}\delta(t-t') . \tag{3.11}$$

Then, $i\neq j$ contribution of (3.10) is given by

$$\left(\frac{\Lambda}{2\pi}\right)^d 2(N^2-N)\delta(t-t')\int \prod_{k,\alpha} \frac{dp_k^\alpha}{\Lambda} e^{i(p_i-p_j)(x-x')}$$

$$\approx 2N^2\delta(t-t')\delta(x-x') \sim O(N^2)$$

while $i=j$ contribution is given by

$$\left(\frac{\Lambda}{2\pi}\right)^d 2N\delta(t-t')\int \sum_{k,\alpha} \frac{dp_k^\alpha}{\Lambda} = \left(\frac{\Lambda}{2\pi}\right)^d 2N\delta(t-t') \sim O(N) .$$

In the large N limit we neglect $i=j$ contribution against $i\neq j$. Then we obtain the same expression as (3.9).

In order to see the degree of largeness of N we have used we compare these two contributions by integrating over x. We obtain the following criterion:

$$N >> (\frac{\Lambda}{2\pi})^d L^d . \qquad (3.12)$$

Since Λ is the cutoff momentum and it is ~1/a, the criterion (3.12) is equivalent to

N >> number of space-time points .

For a large N which satisfies the condition (3.12) it is possible to prove that the reduced η (3.8) has Wick decomposition property (2.6), provided N>>n.

Next we look for a solution of Langevin equation when η is given by the reduced form (3.8). We first make an ansatz for ϕ,

$$\phi^{\eta}_{ij}(x,t) = e^{i(p_i-p_j)x} \bar{\phi}^{\eta}_{ij}(t) \qquad (3.13)$$

and insert it into (3.4). Then, we obtain

$$\frac{\partial\bar{\phi}^{\eta}_{ij}(t)}{\partial t} = -[(p_i-p_j)^2+m^2]\bar{\phi}^{\eta}_{ij}(t) - \frac{g}{3!N} (\bar{\phi}^{\eta^3})_{ij} + \bar{\eta}_{ij}(t)(\frac{\Lambda}{2\pi})^d , \qquad (3.14)$$

which can be considered as a reduced Langevin equation for $\bar{\phi}$.

Combining these equations together we obtain finally

$$\langle tr(\phi(x_1)\phi(x_2)\cdots\phi(x_n))\rangle$$

$$= \lim_{t\to\infty} \langle tr(\phi^{\eta}(x_1,t)\cdots\phi^{\eta}(x_n,t))\rangle_{\eta} \qquad \text{(stochastic quantization)}$$

$$= \lim_{t\to\infty} \int \Pi \frac{dp}{\Lambda} \sum e^{i(p_i-p_j)x_1\cdots} \langle\bar{\phi}^{\eta}_{ij}(t_1)\bar{\phi}^{\eta}_{jk}(t_2) \cdots \rangle_{\bar{\eta}}$$

(Reduction of η and ϕ)

$$= \int \Pi \frac{dp}{\Lambda} \sum e^{i(p_i-p_j)x_1\cdots} \langle\bar{\phi}_{ij}\bar{\phi}_{jk} \cdots \rangle_{\bar{S}}$$

(stochastic quantization)

(3.15)

where

$$\langle\cdots\rangle_{\bar{S}} \equiv \frac{\int d\bar{\phi}(\cdots)e^{-\bar{S}[\bar{\phi}]}}{\int d\bar{\phi}e^{-\bar{S}[\bar{\phi}]}} \qquad (3.16)$$

and

$$\bar{S}[\bar{\phi}] = \left(\frac{2\pi}{\Lambda}\right)^d \left[\sum_{ij} \frac{1}{2}((p_i - p_j)^2 + m^2)\bar{\phi}_{ij}\bar{\phi}_{ji} + \frac{g}{4!N}\,\mathrm{tr}(\bar{\phi}^4)\right] . \tag{3.17}$$

The expression (3.15) is nothing but the quenched momentum prescription proposed by the authors of reference 2.

IV. SU(N) Gauge Theory

The action of SU(N) gauge theory is given by

$$S = \int dx \, \frac{1}{4e^2} \, \mathrm{tr} \, F_{\mu\nu}F_{\mu\nu} \tag{4.1}$$

where e is a coupling constant. The field strength $F_{\mu\nu}$ is expressed in terms of N×N Hermitian matrix vector potential A_μ as

$$F_{\mu\nu} = \partial_\mu A_\nu - \partial_\nu A_\mu - i[A_\mu, A_\nu] . \tag{4.2}$$

The Langevin equation of this theory is given by

$$\dot{A}_\mu(x,t) = \frac{1}{e^2}\{\partial_\nu F_{\mu\nu}(x,t) - i[A_\nu, F_{\nu\mu}]\} + \eta_\mu(x,t) , \tag{4.3}$$

where $\eta_\mu(x,t)$ is a random source function with the same Gaussian distribution property as η in III. According to Parisi and Wu, if one solves this Langevin equation by perturbation with the initial condition

$$A_\mu(x,0) = 0 \tag{4.4}$$

and calculate the gauge invariant correlation functions by the stochastic method one obtains the ordinary perturbation results in Landau gauge. We note that Langevin eq.(4.3) and the η-average property given by (2.6) and (3.5) are invariant by the following (fictitious) time independent gauge transformations:

$$\begin{aligned} A_\mu(x,t) &\longrightarrow u(x)A_\mu(x,t)u^+(x) + iu(x)\partial_\mu u^+(x) \\ \eta_\mu(x,t) &\longrightarrow u(x)\eta_\mu(x,t)u^+(x) \\ u(x) &\in SU(N) . \end{aligned} \tag{4.5}$$

Since the SU(N) gauge theory described by the stochastic quantization in this way is a slight generalization of the Hermitian model of the previous section, it is not difficult to see that we arrive at the following large N reduced model:

$$\left.\begin{aligned} \eta_\mu(x,t) &= \left(\frac{\Lambda}{2\pi}\right)^{d/2} e^{iP\cdot x}\, \bar{\eta}_\mu(t)\, e^{-iP\cdot x} \\ A_\mu(x,t) &= e^{iP\cdot x}\, \bar{A}_\mu(t)\, e^{-iP\cdot x} \end{aligned}\right\} \qquad (4.6)$$

$$\dot{\bar{A}}(t) = \frac{1}{e^2}\,[P_\nu-\bar{A}_\nu(t),[P_\nu-\bar{A}_\nu(t),P_\mu-\bar{A}_\mu(t)]] + \left(\frac{\Lambda}{2\pi}\right)^{d/2} \bar{\eta}_\mu(t) \qquad (4.7)$$

$$\bar{A}_\mu(0) = 0\ , \qquad (4.8)$$

where P is a diagonal matrix with momenta p_i's in its diagonal elements.

Now it is straightforward to calculate the gauge invariant correlation functions in the large N limit by using the perturbative stochastic method to obtain the results of the quenched momentum prescription. However, there is an unexpected complication in the formal level. If we change reduced variables from $\bar{A}_\mu$ to $\tilde{A}_\mu$ by

$$\tilde{A}_\mu = \bar{A}_\mu - P_\mu\ , \qquad (4.8)$$

the Langevin equation in terms of new variables becomes

$$\dot{\tilde{A}}(t) = -\frac{1}{e^2}\,[\tilde{A}_\nu(t)[\tilde{A}_\nu(t),\tilde{A}_\mu(t)]] + \left(\frac{\Lambda}{2\pi}\right)^{d/2} \bar{\eta}_\mu(t)\ , \qquad (4.9)$$

which is independent of P_μ. On the other hand the initial condition becomes

$$\tilde{A}_\mu(0) = -P_\mu \qquad (4.10)$$

which depends on P_μ. Therefore, if the results of the stochastic quantization are independent of the initial condition the momenta are trivially integrated out and we obtain a matrix model (Eguchi-Kawai model) as a large N reduced model rather than the quenched momentum form.

We believe that this is due to the assumption that the stochastic quantization does not depend on the initial condition chosen to solve the Langevin equation. In gauge theories, due to the gauge symmetry there are no drift forces along the direction of gauge orbit. Thus, the system does not reach to the equilibrium distribution along the gauge orbit and the final distribution depends on the initial configuration. This is in a sense a gauge fixing (weighted average). The criterion that this phenomenon occurs is the existence of a continuous energy spectrum above the ground state of Fokker-Planck Hamiltonian because of the flatness of the potential in a certain direction in configuration space. In the reduced Langevin equation (4.9), the situation is more complicated. We note first that the Langevin equation (4.9) is invarinat by the following time independent reduced gauge transformation:

$$\left.\begin{array}{ll} \tilde{A}_\mu(t) \to u\tilde{A}_\mu(t)u^+ & \\ & u \in SU(N) \\ \eta_\mu(t) \to u\eta_\mu(t)u^+ & \end{array}\right\} . \tag{4.11}$$

We also note that $\tilde{A}_\mu(0)=-P_\mu$ cannot be obtained by the gauge transformation from $\tilde{A}_\mu(0)=0$. Thus, if there remains the P_μ dependence it must be due to the continuous spectrum which does not related to the known symmetry. We found an indication that this is the case. We like to devote the rest of this section for our preliminary study[5].

The action and the Fokker-Planck equation corresponding to reduced Langevin equation (4.9) are given by

$$S[\tilde{A}] = -\frac{\alpha^{-1}}{4} \operatorname{tr}[\tilde{A}_\mu,\tilde{A}_\nu]^2 \tag{4.12}$$

$$\tilde{H}_R = -\frac{1}{2}\sum_{\mu,ij} \frac{\delta^2}{\delta\tilde{A}^{ij}_\mu \delta\tilde{A}^{ji}_\mu} + \frac{1}{4}V[\tilde{A}] \tag{4.13}$$

$$V[\tilde{A}] = \frac{1}{2}\alpha^{-2}\operatorname{tr}[\tilde{A}_\nu,[\tilde{A}_\nu,\tilde{A}_\mu]]^2 - 2\alpha^{-1}(d-1)N\operatorname{tr}(\tilde{A}_\mu - \frac{1}{N}\operatorname{tr}\tilde{A}_\mu)^2 \tag{4.14}$$

$$\alpha = e^2 \left(\frac{\Lambda}{2\pi}\right)^d . \tag{4.15}$$

We notice that when the configuration of $\tilde{A}_\mu$ is diagonal the first term in Fokker-Planck potential V is zero while the second term negative and bottomless. Thus, at least for the weak coupling (small α) the diagonal components of $\tilde{A}_\mu$ should be treated non-perturbatively. We therefore separate $\tilde{A}_\mu$ into the diagonal component b_μ and the off diagonal component B_μ:

$$\tilde{A}^{ij}_\mu = (b_\mu+B_\mu)^{ij} \equiv b^i_\mu\delta_{ij} + B^{ij}_\mu .$$

We insert it into V and then expand V in the powers of B_μ. To leading order in α we obtain

$$V = \frac{1}{2}\alpha^{-2} \sum_{i\neq j} (b_\sigma^i - b_\sigma^j)^2 \{\delta_{\mu\nu} - \frac{(b^i - b^j)_\mu (b^i - b_j)_\nu}{(b_\sigma^i - b_\sigma^j)^2}\} B_\mu^{ij} B_\nu^{ji}$$

$$- 2\alpha^{-1}(d-1)N \sum_i (b^i - \frac{1}{N}\sum_k b_\mu^k)^2 \ . \qquad (4.16)$$

We then calculate the zero point energy due to B_μ fluctuation keeping b_μ fixed (Born-Oppenheimer). We obtain

$$\frac{1}{4}(d-1)\alpha^{-1} \sum_{ij} (b^i - b^j)^2$$

which is precisely cancelled by the second term in V. Therefore at least in the lowest order in α b_μ becomes effectively a cyclic variable near the ground state due to the off diagonal fluctuations, and there is no effective drift force along the direction of diagonal A .

Acknowledgement

This research is supported by a National Science Foundation grant and CUNY-PSC BHE Faculty Research Award. One of us (B.S.) acknowledges the warm hospitality at RIFP Kyoto when this report was written and typed. He thanks Mrs. K. Honda for an excellent typing.

References

1. T. Eguchi and H. Kawai, Phys. Rev. Letters 48, 1063 (1982).
2. G. Bhanot, U. Heller and H. Neuberger, Phys. Letters 113B, 47 (1982).
   G. Parisi, Phys. Letters 112B, 463 (1982)
3. D.J. Gross and Y. Kitazawa, Princeton Univ. preprint (1982).
   S. Das and S. Wadia, Univ. of Chicago preprint (1982).
   T. Eguchi and H. Kawai, Tokyo Univ. preprint (1982).
   G. Parisi and Zheng Yi-Cheng, Frascati preprint (1982).
   A.A. Migdal, Landau Institute preprint (1982).
3. G. Parisi and Wu Yong-Shi, Scientia Sinica, 24, 483 (1981).
4. J. Alfaro and B. Sakita, Phys. Letters B (to be published).
5. More detailed study on this subject with a possible supersymmetry discussion will be published elsewhere.

VARIATIONAL METHODS IN THE MASTER FIELD FORMULATION FOR QCD_{3+1}

Masa-aki Sato
Department of Physics, New York University
4 Washington Place, New York, NY 10003, U. S. A.

ABSTRACT

Master fields are formulated for finite N-QCD_{3+1}. They satisfy classical Yang-Mills equations with an infinite number of internal indices and an infinite number of constraints. Master fields and constraints on them in the large N limit are derived from the finite N master fields and constraints using vacuum dominance among color singlet states. The large N constraints can be explicitly solved and the solutions involve arbitrary functions which are used as trial functions in variational calculations.

I. FINITE N MASTER FIELDS

In this letter we will report the outline of our previous works[1,2] on master field methods for QCD in 4-dimensional Minkowski space (QCD_{3+1}).

We choose the axial gauge ($A_3 = 0$) and SU(N) for the color group. Then the dynamical variables are the 1 and 2 components of matrix operator gluon fields, $(\hat{A}_\alpha)_{ab}(x)$ ($\alpha = 1, 2$; $a, b = 1, \cdots, N$), and the conjugate momenta, $(\hat{\Pi}_\alpha)_{ab}(x)$, which satisfy usual canonical commutation relations. The operator Hamiltonian, $\hat{H}$, is given by

$$\hat{H} = \int d^3x \; 1/2 \; \mathrm{Tr}[\hat{\Pi}_\alpha^2(x) + (\partial_3 \hat{A}_\alpha(x))^2 + \hat{F}_{12}^2(x)]$$
$$+ 1/4 \int d^3x \, d^3y |x_3 - y_3| \delta^2(x_\alpha - y_\alpha) \; \mathrm{Tr}[(\hat{\nabla}_\alpha \hat{\Pi}_\alpha)(x)(\hat{\nabla}_\beta \hat{\Pi}_\beta)(y)], \tag{1}$$

where Tr represents a trace over color indices and all operator products are color ordered products (COP): adjoint operator products which are made from $\hat{A}_\alpha$ and $\hat{\Pi}_\alpha$ such that the order of the operators and the order of the matrix products in the color space coincides. An example of COP is given by

$$(\hat{A}_1 \hat{\Pi}_2 \hat{\Pi}_3 \hat{A}_4)_{ab} \equiv (\hat{A}_{\alpha_1})_{ac_1}(x_1)(\hat{\Pi}_{\alpha_2})_{c_1c_2}(x_2)$$
$$\times (\hat{\Pi}_{\alpha_3})_{c_2c_3}(x_3)(\hat{A}_{\alpha_4})_{c_3b}(x_4). \tag{2}$$

Since the Hamiltonian (1) and the total momentum operator are invariant under global color SU(N) transformations, the energy-momentum eigenstates can be classified by the irreducible representations of the color SU(N) group.

Finite N master fields, $A_\alpha(x)$ and $\Pi_\alpha(x)$, are defined as reduced matrix elements of $\hat{A}_\alpha(x)$ and $\hat{\Pi}_\alpha(x)$ between all singlet and adjoint energy-momentum eigenstates such as [1,2,3]

$$< s|(\hat{A}_\alpha)_{ab}(x)|g, cd> = \sqrt{N/(N^2-1)}$$
$$\times (\delta_{ad}\delta_{bc} - 1/N \, \delta_{ab}\delta_{cd})(A_\alpha(x))(s\,;g), \tag{3}$$

where s(g) represents the quantum number of a singlet (adjoint) energy-momentum eigenstates, $|s>$ ($|g, ab>$). The master fields, $(A_\alpha(x))(n;n')$ and $(\Pi_\alpha(x))(n;n')$, can be treated as C-number matrix fields with matrix indices n (= s or g). Since COP creates only singlet or adjoint states from singlet and adjoint states (when COP is applied to an adjoint state one of the indices of the adjoint state and the neighboring index of COP should be contracted), any matrix element of COP between singlet and adjoint states can be expressed as a matrix product of the master fields such as

$$< s|(\hat{A}_1 \hat{\Pi}_2 \hat{\Pi}_3 \hat{A}_4)_{ab}|g, cd> = \sqrt{N/(N^2-1)}$$
$$\times (\delta_{ad}\delta_{bc} - 1/N \, \delta_{ab}\delta_{cd})(A_1 \Pi_2 \Pi_3 A_4)(s\,;g), \tag{4}$$

where a matrix notation is used for products of the master fields;

$$(A_1 \Pi_2)(s\,;\,g) = \sum_{n=s',g'} (A_1)(s\,;\,n)(\Pi_2)(n\,;\,g). \tag{5}$$

Taking matrix elements of the operator equations of motion between the singlet and adjoint states, it can be shown that the finite N master fields satisfy the classical Yang-Mills equations with an infinite number of internal indices, n (= s or g). The canonical commutation relations can be inverted into an infinite number of constraints. They are derived by considering the difference between two traces of any COP and the induced COP by the cyclic permutation. An example of the constraints is given by

$$\begin{aligned}(A_1 A_2 \Pi_3 A_4 A_5 A_6)(s_1\,;\,s_2) &- (A_6 A_1 A_2 \Pi_3 A_4 A_5)(s_1\,;\,s_2)\\ = -iN\delta(3,\,6)\{\sum_{s_3} &(A_1 A_2)(s_1\,;\,s_3)(A_4 A_5)(s_3\,;\,s_2)\\ &- 1/N^2\,(A_1 A_2 A_4 A_5)(s_1\,;\,s_2)\}.\end{aligned} \tag{6}$$

There is another set of constraints which fix an integer value of N and correspond to the Mandelstam constraints[4] for the Wilson loops.

II. THE LARGE N LIMIT

In the large N limit contributions from the vacuum dominate over those from other excited singlet states[4,5]. Then the large N master fields can be obtained from the finite N master fields by reducing an infinite number of the singlet indices to a single index which corresponds to the vacuum. They also satisfy the classical Yang-Mills equations. Large N constraints are obtained from the constraints corresponding to the commutation relations by replacing the sum over intermidiate singlet states with the vacuum index. In the large N limit Mandelstam-like constraints can be neglected.

The large N constraints can be explicitly solved and the solutions involve arbitrary functions which should be determined by using dynamical equations. To solve the large N-QCD_{3+1} exactly, the solutions should involve an infinite number of arbitrary functions, which gives us great difficulty. Therefore we will use solutions with a finite number of arbitrary functions and determine them by minimizing the vacuum energy. The simplest solution involves a 1-particle wave function as arbitrary functions. The variational calculation, in which the simplest solution is used as a trial function,[1] gives a logalismic potential between quarks and anti-quarks along the Z-axis under straightforward cut off. More realistic trial functions were given in Ref. 2. The Lorentz invariance and the gauge invariance of our method will be discussed within the master field formulation in covariant gauges in a forthcoming paper.

REFERENCES

1. M. Sato, Phys. Lett. 113B (1982) 315.
2. M. Sato, New York University preprint, NYU/TR7/82 (to be published in Phys. Rev. D).
3. K. Bardakci, Nucl. Phys. B178 (1981) 263; Nucl. Phys. B193 (1982) 245; M. B. Halpern, Nucl. Phys. B188 (1981) 61.
4. S. Mandelstam, Phys. Rev. D19 (1979) 2391.
5. G. 'tHooft, Nucl. Phys. B72 (1974) 461; E. Witten, Nucl. Phys. B160 (1979) 57.

CONFINEMENT BY THICK MAGNETIC VORTICES

Tamiaki Yoneya
Institute of Physics, University of Tokyo
Komaba, Tokyo 153, Japan

Abstract: An argument is presented showing that the pure lattice gauge theories which are infrared unstable at the origin of the coupling constant are always in confining phase, provided that (i) the gauge group is compact and contains nontrivial center;(ii) Lagrangian does not contain long range interaction. The steps for an attempt at a proof of confinement are suggested.

1. Introduction

Although Monte Calro simulations on the lattice have provided several convincing evidences for believing in confinement in QCD, we are yet far from microscopic understanding of the mechanism of confinement in the continuum limit. In the lattice-regularized theory, the continuum theory corresponds to the critical region where the physical correlation length is much larger than the lattice spacing a.

In this talk, I wish to report some simple observations on the general structure of the large distance behavior of lattice gauge theories. I will first present a rigorous inequality [1] for magnetic flux free energies which holds in any lattice gauge models with local action and with compact gauge group containing nontrivial center. Then, I will show that if this inequality is supplemented by a simple assumption, which is apparently of kinetical nature, it leads to an interesting consequence for confinement in the large distance critical region.

2. Center transformation and magnetic vortices

The inequality is based on the center invariance of the gauge field Lagrangian. For definiteness, we take the standard Wilson action of pure gauge theory.

$$S = \frac{1}{g^2}\sum_P \mathrm{Tr}\,(U_P + U_P^+)\ , \qquad U_P = U_{x\mu} U_{x+\hat{\mu}\nu} U^+_{x+\hat{\nu}\mu} U^+_{x\nu} \tag{1}$$

where $U_{x\mu}$ is the parallel transporter from the point x to a nearest neighbough point $x+\hat{\mu}$. The center transformation is defined by $U_{x\mu} \to \gamma_\mu U_{x\mu}$, γ_μ being an arbitrary element of the gauge group center C(G), under which the action is invariant. The center invariance is preserved even if the transformation is restricted to an infinitely extended d-1 dimensional plane-layer of parallel links. Furthermore, if a center transformation is performed in a finitely extended layer of parallel links, the action density changes its value only at the (d-2 dimensional) boundary of the layer, as in Fig. 1. The Wilson loop which winds around the boundary changes by the group center. It is appropriate to call the energetic object, which is produced at the boundary of the layer, the magnetic vortex (or simply vortex).

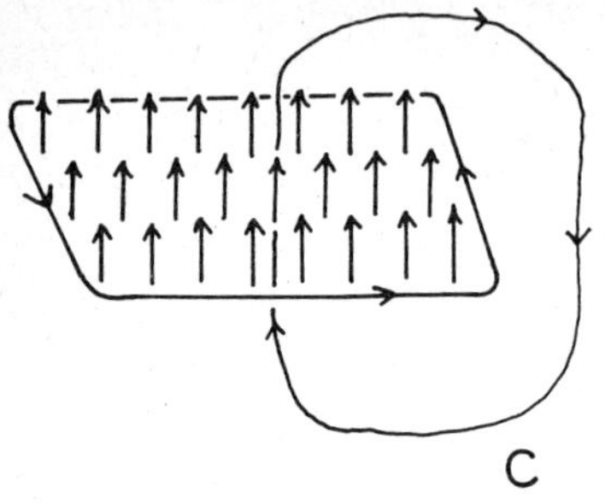

Fig. 1. The layer of parallel links and the vortex at its boundary. C is the Wilson loop winding around the vortex.

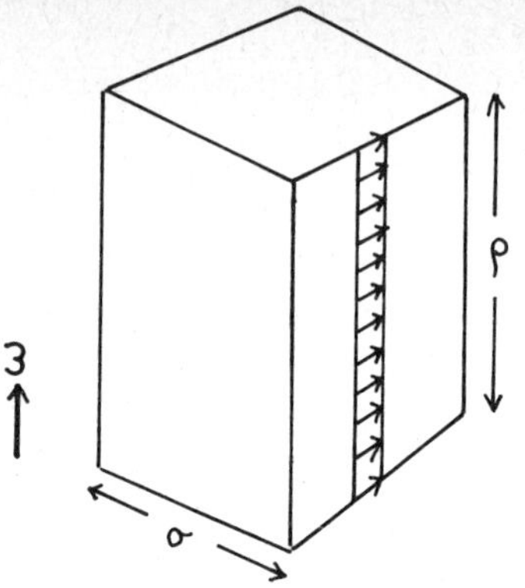

Fig. 2. The 3 dimensional rectangular box with periodic boundary condition in the 3rd direction. The strip of parallel links on the surface is the intersection with the layer of the center transformation.

To discuss the possible role of the vortices in the large distance behaviour, we have to consider the vortices with larger and larger thickness. For this purpose, we extract a finite hypertube which completely encloses the boundary of the layer of the center transformation. The center transformation then induces the change of the boundary condition on the surface of the hypertube. Instead of a hypertube, for simplicity, let us consider d-dimensional rectangular box in which the periodic boundary condition is imposed in its last d-2 directions. (See Fig. 2.) Let the sides of the box, which is denoted by $\Lambda(\rho,\sigma)$,be $(\sigma,\sigma,\rho,\cdots,\rho)$. The center transformation for the box is equivalent to a change of the boundary condition on the strip which is the intersection of the surface and the layer of the center transformation. The partition function for $\Lambda(\rho,\sigma)$ is defined

$$Z_{\Lambda(\rho,\sigma)}\{u\} = \int \prod_{U_{x\mu}\subset\Lambda(\rho,\sigma)} dU_{x\mu}\ \exp \frac{1}{g^2} \sum_{P\subset\Lambda(\rho,\sigma)} \mathrm{Tr}\,(U_P+U_P^+) \tag{2}$$

where $\{u\}$ represents the boundary value of the parallel transporter matrices on the surface $\partial\Lambda(\rho,\sigma)$. In (2), the summation and the product are extended only over plaquettes and links which are completely contained inside $\Lambda(\rho,\sigma)$. Then, the vortex free energy corresponding to a center element γ is given by

$$F^{(\gamma)}_{\Lambda(\rho,\sigma)}\{u\} = -\ln\left[Z_{\Lambda(\rho,\sigma)}\{u_\gamma\} / Z_{\Lambda(\rho,\sigma)}\{u\}\right] \tag{3}$$

where $\{u_\gamma\}$ denotes the center-transformed boundary value. We now define [2]

$$<\hat{\gamma}> = \underset{\{u\}}{\mathrm{Max}}\left[\sum_{\gamma\in C(G)} \gamma\ \exp - F^{(\gamma)}_{\Lambda(\rho,\sigma)}\{u\} \Big/ \sum_{\gamma\in C(G)} \exp - F^{(\gamma)}_{\Lambda(\rho,\sigma)}\{u\}\right] \tag{4}$$

If the gauge group is compact and the Lagrangian does not contain long range interaction, the following inequality [1] can be proven

$$\langle \hat{\gamma} \rangle_{\Lambda(\rho.n\sigma)} \leq \left[\langle \hat{\gamma} \rangle_{\Lambda(\rho.\sigma)} \right]^{n^2} . \qquad (5)$$

This interrelates the vortices with different thickness. Let us next discuss the possible implications of this simple inequality to confinement problem. For this purpose, the properties of $\langle \hat{\gamma} \rangle_{\Lambda(\rho.\sigma)}$ in the large scale limit is crucial.

3. Large scale behaviour of vortex free energy and confinement

If the length of the vortex is much larger than its thickness, we expect the following behaviour for the free energy

$$\hat{F}^{(\gamma)}_{\Lambda(\rho.\sigma)} \equiv \operatorname*{Max}_{\{u\}} F^{(\gamma)}_{\Lambda(\rho.\sigma)}\{u\} \simeq \rho^{d-2} f(\sigma) , \quad r\,(\equiv \sigma/\rho) \ll 1 . \qquad (6)$$

The expected behaviour of the free energy density $f(\sigma)$ for large σ is the following [3], [4]:

(i) confinement phase $\qquad f(\sigma) \sim$ const. $\exp -\alpha\sigma^2$, (7a)

(ii) Higgs phase $\qquad f(\sigma) \sim$ const. , (7b)

(iii) massless phase (d=4) $\qquad f(\sigma) \sim$ const.$/\sigma^2$. (7c)

There is no rigorous justification for these properties ($d \geq 3$). But any reasonable approximation schemes appropriate to each phase predict such behaviours. For instance, the strong coupling expansion predicts (7a) with α being the string tension. Among these properties, that the maximum vortex free energy is proportional to ρ^{d-2} for large ρ seems generally valid independently of different phases. Unfortunately, however, we have no rigorous control over how large ρ must be for the uniform validity of (7). To take into this plausible property account, we adopt the following working <u>hypothesis</u> (or <u>conjecture</u>): In compact pure lattice gauge theories, there exists a constant $k = k(\rho_0.\sigma_0.r)$ for any ρ, σ and r satisfying

$$\rho \gg \rho_0 , \quad \sigma \gg \sigma_0 , \quad r = \sigma/\rho \ll 1$$

such that

$$\langle \hat{\gamma} \rangle_{\Lambda(2\rho.\sigma)} \leq k \langle \hat{\gamma} \rangle_{\Lambda(\rho.\sigma)} . \qquad (8)$$

When C(G) = Z(2), $\gamma = \pm 1$ and $\langle \hat{\gamma} \rangle_{\Lambda(\rho.\sigma)} = \tanh \hat{F}^{(-1)}_{\Lambda(\rho.\sigma)}$.

Hence, if (6) is valid, we have (8) with $k \sim 2^{d-2}$. Compactness of gauge group is required because only for compact gauge group one can prove that $\lim_{\rho\to\infty} \rho^{-(d-2)} \hat{F}_{\Lambda(\rho.\sigma)}$

have finite limit. Under this hypothesis, we prove

Lemma : Under the hypothesis (8), if there exists an allowed set of ρ, σ such that

$$\langle \hat{\gamma} \rangle_{\Lambda(\rho,\sigma)} < \kappa^{-1} \tag{9}$$

then,

$$\langle \hat{\gamma} \rangle_{\Lambda(\lambda\rho,\lambda\sigma)} \leq \text{const.} \exp - \alpha (\lambda\sigma)^2$$

for sufficiently large λ , with

$$\alpha \geqq - \sigma^{-2} \ln \kappa \langle \hat{\gamma} \rangle_{\Lambda(\rho,\sigma)} . \tag{10}$$

Proof : Combining the inequality (7) and our hypothesis (8) we have

$$\langle \hat{\gamma} \rangle_{\Lambda(2\rho,2\sigma)} \leq \kappa \left(\langle \hat{\gamma} \rangle_{\Lambda(\rho,\sigma)} \right)^4 .$$

Repeating this inequality n times (see Fig. 3), we have for large n,

$$\langle \hat{\gamma} \rangle_{\Lambda(2^n\rho,2^n\sigma)} \leq \kappa \left[\kappa \langle \hat{\gamma} \rangle_{\Lambda(\rho,\sigma)} \right]^{4^n} .$$

By setting $\lambda = 2^n$, one arrives at our claim.

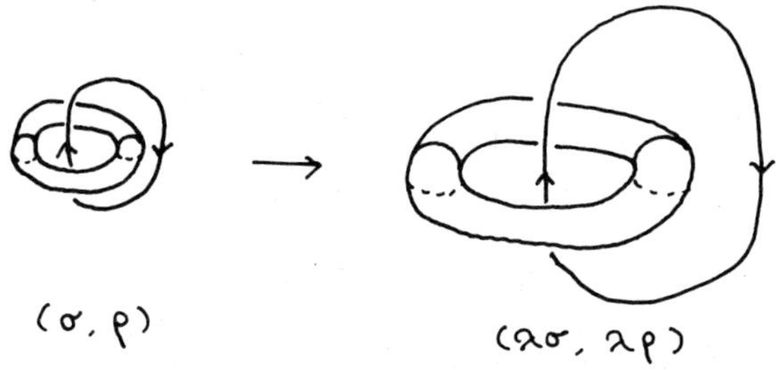

Fig. 3. Uniform dilatation of the vortex and the Wilson loop.

This lemma means that under the assumption (6), the vortex free energy vanishes exponentially with λ^2 provided that the vortex free energy is sufficiently small so that (9) is satisfied for some ρ and σ. Hence, (9) is sufficient for confinement. Clearly, this condition is always satisfied when the bare coupling constant is sufficiently large. Furthermore, even if the bare coupling constant is arbitrarily small, it is possible that (9) is satisfied at sufficiently large scale because of the renormalization effect. We also note that the area-law decay is a rather universal property of gauge systems in the sense that if $f(\sigma)$ is known to decay faster than $\sigma^{-(d-2)}$, it automatically decays exponentially with the cross section. The Mack-Petkova inequality [2] then implies that the tension given by (10) is a lower bound for the exact string tension.

In theories which are not scale invariant, $\hat{F}_{\Lambda(\lambda\rho,\lambda\sigma)}$ would either vanishes or diverges in the limit $\lambda \to \infty$. (We can show [1] that $f(\sigma)$, if it exists, is monotonically decreasing function of σ .) On the other hand, at the scale of cutoff, the inverse free energy (which is dimensionless) is proportional to the bare coupling constant. Hence, at general scale λ , the inverse vortex free energy can serve as an effective coupling constant. For the effective coupling constant in the U(N) model, we can indeed prove the following inequality [1] as suggested from the perturbation theory,

$$g^2_{eff}(\lambda) \geqslant \lambda^{4-d} g^2 \quad . \tag{11}$$

In asymptotically free theories, or more generally, in theories which are infrared unstable at the origin of the coupling constant, the effective coupling constant should increase at large distances. ((11) shows that this happens at least if $d < 4$ in the U(N) model.) Thus the only possibility seems to be that the maximum vortex free energy vanishes as $\lambda \to \infty$. In summary, our conclusion is that the pure lattice gauge theories which are infrared unstable at the origin of the coupling constant are always in confinig phase with linear confinig potential, provided that (i) Lagrangian is local; (ii)the center of gauge group is nontrivial. The crucial assumption in our argument was that the proportionality of the vortex free energy to its length is valid uniformly for any ρ, σ if r is sufficiently small.

Clearly, this conclusion is almost equivalent to the old infrared slavery conjecture. An exact proof, if any, of confinement is now reduced to establish this fundamental assumption and the infrared instablity in nonperturbative fashion. Finally, it should be remarked that our discussion is not affected by the presence of matter fields if the matter fields are in the adjoint representation of the gauge group.

References

1 T. Yoneya, Nucl. Phys. B205 [FS](1982)130.

2 G. Mack and V. B. Petkova, Ann. Phys. 125 (1980)117.

3 See G. Mack, "Properties of lattice gauge theory models at low temperature", in Progress in Gauge Theories, eds. G. 'tHooft et al (Plenum Press, New York, 1980).

4 G. 'tHooft, Nucl. Phys. B153 (1979)141.

RENORMALIZABILITY OF MASSIVE YANG-MILLS THEORY

Takashi Fukuda, Hiroaki Matsuda, Yoshinori Seki
and Kan-ichi Yokoyama

Research Institute for Theoretical Physics
Hiroshima University, Takehara, Hiroshima 725, Japan

In this note, we aim at revealing the renormalizable structure of massive Yang-Mills (YM) theories on the basis of the formalism in our previous work,[1)] quoted as [I], though there had so far been found several counter-observations.[2)] Our formalism consists of four kinds of auxiliary scalar fields other than a massive gauge field A_μ^a; a massless scalar fields ξ^a, in addition to the usual Lagrange multiplier field η^a and a pair of Faddeev-Popov ghost fields C^a and $\overline{C}^a$. In terms of these fields, the Lagrangian density in [I], takes the form in the Landau gauge

$$\left.\begin{aligned} L &= -\tfrac{1}{4}(G_{\mu\nu}^a)^2 - A_\mu^a\partial_\mu\eta^a + i\partial_\mu\overline{C}^a D_\mu^{ab}C^b - \tfrac{1}{2}m^2(K_\mu^a - A_\mu^a)^2, \\ K_\mu^a &= K^{ab}\partial_\mu\xi^b \;, \quad K^{ab} = \frac{\ell}{g}[1 + \sum_{n=1}^{\ell} \frac{1}{(n+1)!}(i\ell\xi)^n]_{ab}, \end{aligned}\right\} \qquad (1)$$

where $G_{\mu\nu}^a = \partial_\mu A_\nu^a - \partial_\nu A_\mu^a + gf^{abc}A_\mu^b A_\nu^c$, $D_\mu^{ab} = \delta^{ab}\partial_\mu + gf^{acb}A_\mu^c$, $\ell = g/m$ and $\xi = T^a\xi^a$. As has been shown in [I], (1) is invariant under a Becchi-Rouet-Stora (BRS) transformation[3)] with a nilpotent character. As is elucidated by Kugo and Ojima (KO),[4)] such an exact BRS symmetry enables us to impose a KO-type of supplementary conditions on physical states, which guarantees our physical subspace to be of positive semi-definite due to the quartet mechanism confining the members of the quartet (ξ^a, η^a, C^a, $\overline{C}^a$) into zero-norm subspace. Therefore our physical S-matrix elements, which are taken between any states containing spin-triplet physical gauge-bosons (the Proca particles) alone, is manifestly unitary.

In the perturbation approach, the interaction Lagrangian is obtained by subtracting the free part from (1). Here we should note: 1) The Feynman propagator of gauge bosons takes the form

$$D_{\mu\nu}(k) \sim [\delta_{\mu\nu} - k_\mu k_\nu D(k)]\Delta(k;m^2) \qquad (2)$$

instead of the usual Proca type, and has the same asymptotic behavior

($k \to \infty$) as that in massless YM theories; namely, (2) takes a renormalizable form. 2) A nonpolynomial character of interactions emerges with respect to the field ξ^a. Therefore, at the first sight, one may guess the theory is unrenormalizable when the ordinary perturbation expansion in the coupling constant or loop expansion is carried out.[2] It is, however, not the case, bacause the exponential-type interaction provides a promising way of constructing renormalizable massive YM theories, as pointed out by several authors.[5),6)]

We attack the renormalizability problem on the basis of nonpolynomial Lagrangian theories. Hereafter, we consider the case of SU(2), for simplicity. In this case, nonpolynomial parts of the interaction Lagrangian can be expressed as follows:

$$L^{\xi}_{int} = P^{ab}_{\Delta}\partial_\mu\xi^a\partial_\mu\xi^b \exp[it\ell(\lambda\xi)] , \qquad (3)$$

$$L^{A}_{int} = mQ^{ab}_{\Delta}A^a_\mu\partial_\mu\xi^b \exp[it\ell(\lambda\xi)] . \qquad (4)$$

Here, $(\lambda\xi) = \lambda^a\xi^a$ and the operator P^{ab}_{Δ} and Q^{ab}_{Δ} are defined by

$$P^{ab}_{\Delta} = \int_0^1 dt t^{\Delta}\int d\Omega \, \frac{(t-1)^2}{2t} \left(\frac{\partial^2}{\partial\lambda^a\partial\lambda^b} - \delta^{ab}\frac{\partial^2}{\partial\lambda^n\partial\lambda^n}\right) , \qquad (5)$$

$$Q^{ab}_{\Delta} = \int_0^1 dt t^{\Delta}\int d\Omega\left[\frac{(t-1)}{2t}\left(\frac{\partial^2}{\partial\lambda^a\partial\lambda^b} - \delta^{ab}\frac{\partial^2}{\partial\lambda^n\partial\lambda^n}\right) - i\varepsilon^{abc}\frac{\partial}{\partial\lambda^c}\right], \qquad (6)$$

where $d\Omega = \sin\theta d\theta d\phi/4\pi$, $\lambda^a = (\sin\theta\cos\phi, \sin\theta\sin\phi, \cos\theta)$ and Δ is a positive number, which is introduced in order to guarantee commutativity of the t- and the other integrations in the S-matrix calculation and is taken to be zero finally.

We shall deal with the S-matrix elements by introducing so called superpropagators. Now, we consider only the case of L^{ξ}_{int} alone, for simplicity. The S-matrix is given by

$$S = 1 + \sum_{N\geq 1}\frac{i^N}{N!}\int dx^N S(x^N), \quad x^N = (x_1,\ldots,x_N) , \qquad (7)$$

where $S(x^N)$ is expressed by using the Hori's formula[7)] as follows:

$$\begin{aligned} S(x^N) = {} & \left[\prod_{i=1}^N P^{a_ib_i}_{\Delta_i}\right] \times \exp\left[-\frac{1}{2}\sum_{i,j}\partial_\mu\partial_\nu D_{ij}\frac{\partial^2}{\partial(\partial_\mu\xi^a_i)\partial(\partial_\nu\xi^a_j)}\right] \\ & \times \prod_{i=1}^N\left[\partial_\mu\xi^{a_i}_i + \sum_j(i\ell t_j\lambda_j^{a_i})\partial_\mu D_{ij}\right]\left[\partial_\mu\xi^{b_i}_i + \sum_j(i\ell t_j\lambda_j^{b_i})\partial_\mu D_{ij}\right] \\ & \times \exp\left[-\frac{1}{2}\sum_{i,j}\{\ell^2 t_i t_j(\lambda_i\lambda_j)D_{ij}\}\right] \times :\prod_{i=1}^N \exp[i\ell t_i(\lambda_i\xi_i)]:, \end{aligned} \qquad (8)$$

where $\xi_i^a = \xi^a(x_i)$ and $\delta^{ab}D_{ij} = \delta^{ab} < 0|T\xi^a(x_i)\xi^b(x_j)|0>$. This is the sum of various kinds of N-th order (with respect to L_{int}^{ξ}) diagrams which are specified by the way how to contract $\partial_\mu \xi^a$ with the other fields. Each N-th order diagram has the following structure: Every two vertex-points are connected with a propagator of the forms $e^{\kappa D}$, $\partial_\mu D e^{\kappa D}$, etc. called superpropagator (SP), in contrast with the usual Feynman propagator D. Thus an N-th order diagram is a supergraph expressed as the N(N-1)/2 products of SPs.

We are now in a position to construct all SPs appearing in general supergraphs. As is well known, $e^{\kappa D}$ is the typical SP in the case for an exponential interaction of scalar fields, and the exponential is expressed by the Sommerfeld-Watson transform:

$$e^{\kappa D} = \frac{i}{2}\int_\Gamma dz \frac{\kappa^z D^z}{\tan\pi z\Gamma(z+1)} + 1 , \qquad (9)$$

where the contour Γ encloses the positive real axis in the z plane. The contour is then opened out to lie parallel to the imaginary axis in $0 < \mathrm{Re}z < 1$; namely, $\Gamma \to C$ ($0 < \mathrm{Re}z < 1$; $z \in C$). The Fourier transform (FT) of D^z is defined for $0 < \mathrm{Re}z < 2$ by[8]

$$FT[D^z] = \int d^4x e^{ikx}[\frac{1}{4\pi^2(x^2+i\varepsilon)}]^z = \frac{i\pi(4\pi)^{2-2z}(k^2-i\varepsilon)^{z-2}}{\sin\pi z\Gamma(z)\Gamma(z-1)}. \qquad (10)$$

Substituting (10) into (9) with $\Gamma \to C$, we obtain the FT of $e^{\kappa D}$.

Due to the fact that (3) contains the derivatives $\partial_\mu \xi^a$, we need some more SPs. We express them symbolically as

$$f(\partial_\mu D, \partial_\mu\partial_\nu D)e^{\kappa D} = \frac{i}{2}\int_{C_f} dz \frac{\kappa^z f(\partial_\mu D, \partial_\mu\partial_\nu D)D^z}{\tan\pi z\Gamma(z+1)} , \qquad (11)$$

where $f(\partial_\mu D, \partial_\mu\partial_\nu D)$ stands for all possible factors, $\partial_\mu D$, $\partial_\mu\partial_\nu D$, $(\partial_\mu D)(\partial_\nu D)$, $(\partial_\mu\partial_\nu D)(\partial_\nu D)$, $(\partial_\mu D)(\partial_\nu D)^2$, $(\partial_\mu\partial_\nu D)^2$, $(\partial_\mu\partial_\nu D)(\partial_\mu D)(\partial_\nu D)$ and $(\partial_\mu D)^2(\partial_\nu D)^2$, and the contour C_f ($\mathrm{Re}z < 0$) is taken in a range where the FT of $f(\partial_\mu D, \partial_\mu\partial_\nu D)D^z$ can be well defined.

The generalized function $f(\partial_\mu D, \partial_\mu\partial_\nu D)D^z$ is dealt with as follows: Consider $(\partial_\mu\partial_\nu D)e^{\kappa D}$, for example. Supposing D is an ordinary classical function, $D = (x^2)^{-1}$, we obtain the identity

$$(\partial_\mu\partial_\nu D)D^z = \frac{2}{(z+1)(z+2)}\{\partial_\mu\partial_\nu - \frac{\delta_{\mu\nu}}{4}\Box\}D^{z+1}, \quad (\Box = \partial_\mu\partial_\mu) \qquad (12)$$

which holds also for the generalized function D except in the neighborhood of $x_\mu = 0$. Evidently, (12) fails at $z = 0$, since the left hand side then becomes $\partial_\mu\partial_\nu D$ while the right hand side is not so because of the fact $\Box D = i\delta^4(x)$ [$\neq 0$] for the generalized function D. Information in the neighborhood of $x_\mu = 0$ lacks this point. Hence, (12) is to be elaborated so that the second term of the right hand side should vanish at $z = 0$, and then we are to have a valid formula for all x_μ in place of (12). Noting that $(\partial_\mu\partial_\nu D)D^z$ always appears as the integrand like (11), we adopt

$$(\partial_\mu\partial_\nu D)D^z = \lim_{\delta\to o} \frac{2}{(z+1)(z+2)}\{\partial_\mu\partial_\nu - \frac{\delta_{\mu\nu}}{4}(\frac{z}{z+\delta})^N \Box\}D^{z+1} \qquad (13)$$

as the valid formula,[6] where N is an integer (≥ 1) and δ is such a positive number that the singularity at $z = -\delta$ lies to the left of the z-contour, and the limit $\delta \to 0$ is taken after the z-integration. Similar arguments lead to formulae for the other remaining $f(\partial_\mu D, \partial_\mu\partial_\nu D)D^z$, from which their FT together with the position of C_f are obtained by use of (10). Next, we note that the operator $P^{a_i b_i}_{\Delta_i}$ defined by (5) may yield extra singularities in (8) in the complex z_{ij} plane of several variables; in fact, the t_i-integration of $P^{a_i b_i}_{\Delta_i}$ produces factors like $(\sum z_{ij} + n + \Delta_i)^{-1}$ with n any positive integer. Such a singularity, however, can be avoided: Take the positive number Δ_i to be so large that the contours C_{ij} guarantee $(\sum \mathrm{Re} z_{ij} + n + \Delta_i) > 0$, then we can safely bend the contour C_{ij} back to Γ_{ij} in (8), since we can assure that the inequality $(\mathrm{Re} z_{ij} + n) \geq 1$ always holds at any pole-point with respect to z_{ij} when we count the residues of the z_{ij}-integrals. Hence, the limit $\Delta_i \to 0$ brings no singular result.

In this stage, we can grasp a power-counting scheme for supergraphs, which estimates the degree of ultraviolet divergences in their momentum integrations. In (11), we transform z, the real part of which is already in a certain range, into z' such that $z \to z'$ $(0 < \mathrm{Re} z' < 1)$. We then find that the FT of $f(\partial_\mu D, \partial_\mu\partial_\nu D)e^{\kappa D}$ becomes proportional to $(k^2)^{\alpha(n)}$ with $\alpha(n) = z'-2+(n/2)$ according to n (= 0,1,2,3,4) being the number of derivatives in f. Hence, with respect to asymptotic behaviors as $k \to \infty$, the FTs of $fe^{\kappa D}$ can be classified into five types by n; denote each of them by S(k;n). Our power-counting scheme is as follows: Consider an N-th order supergraph, in which the numbers of S(k;n) are I_n. Then, it holds

$$I \equiv \sum_{n=0}^{4} I_n = N(N-1)/2 \ , \ I_1 + 2I_2 + 3I_3 \ 4I_4 = 2N - E_{\partial\xi} \ , \quad (14)$$

where $E_{\partial\xi}$ is the number of external $\partial_\mu\xi^a$-lines. The degree of divergence d is given in this case by

$$d = 4(I-N+1) - (4I_0+3I_1+2I_2+I_3) + 2\sum_{ij} \mathrm{Re}z'_{ij}$$

$$= 4 - 2N - E_{\partial\xi} + 2\sum_{ij} \mathrm{Re}z'_{ij} \qquad (15)$$

from (14). Since $\mathrm{Re}z'_{ij}$ can be taken to be as close to zero as possible, the condition for convergence of the supergraph ($d < 0$), $4 - 2N - E_{\partial\xi} < 0$, is surely satisfied for $N \geq 3$. For $N = 2$, we have a single SP which is of course well defined. Therefore, we can conclude that all S-matrix elements are finite for L^{ξ}_{int}.

Finally we discuss the case of the total interaction Lagrangian. The presence of L^{A}_{int} calls for four more kinds of SPs other than those for L^{ξ}_{int} alone: $D_{\mu\nu}e^{\kappa D}$, $(\partial_\mu D)D_{\mu\nu}e^{\kappa D}$, $(\partial_\mu\partial_\nu D)D_{\mu\nu}e^{\kappa D}$ and $(\partial_\mu D)(\partial_\nu D)D_{\mu\nu} \times e^{\kappa D}$. Here, we are interested in their asymptotic behaviors. Noting the asymptotic form of $D_{\mu\nu}$ being as $\sim(k^2)^{-1}$, we can approximate $D_{\mu\nu}$ in them by D and obtain their asymptotic forms. Therefore, our power-counting scheme tells us that the degree of divergence of an N-th order supergraph with respect to the total L_{int} is given by

$$d = 4 - E_A - E_{\partial\xi} - 2N - 2(I^A_0 + I^A_1 + I^A_2) + 2\sum_{ij}\mathrm{Re}z'_{ij} \; . \qquad (16)$$

Here, E_A denotes the number of external A^a_μ-lines and I^A_n are those of SPs containing $D_{\mu\nu}$ and n derivatives of ξ. Recognizing that the term $(4-E_A)$ comes from the renormalizable interactions among A^a_μ, $\bar{C}^a$ and C^a, we find that the divergence of any total supergraph becomes the less, as the more super-vertices are inserted.

Thus, we conclude that the massive YM theory can be dealt with as being renormalizable in view of nonpolynomial Lagrangian theories.

REFERENCES

1) T. Fukuda, M. Takeda, M. Monda, K. Yokoyama, Prog. Theor. Phys. 66 (1981), 1827; 67 (1982), 1206.
2) D.G. Boulware, Ann. of Phys. 56 (1970), 140. K. Shizuya, Nucl. Phys. B121 (1977), 125.
C.N. Ktorides, Phys. Rev. D13 (1976), 2811. Yu.N. Kafiev, Nucl. Phys. B201 (1982), 341.
3) C. Becchi, A. Rouet and R. Stora, Ann. of Phys. 98 (1976), 287.
4) T. Kugo and I. Ojima, Phys. Letters 73B (1978), 459. Prog. Theor. Phys. Suppl. No. 66 (1979), 1.
5) S. Okubo, Prog. Theor. Phys. 11 (1954), 80. M.K. Volkov, Comm. Math. Phys. 7 (1968), 289; 15 (1969), 69.
6) C.J. Isham, Abdus Salam and J. Strathdee, Phys. Rev. D5 (1972), 2548. Abdus Salam, 1971 Coral Gable Conference on Fundamental Interactions at High Energy, Vol. 1, p.3. J.G. Taylor, ibid. p.42.
7) S. Hori, Prog. Theor. Phys. 7 (1952), 578.
8) I.M. Gel'fand and G.E. Shilov, Generalized Functions (Academic Press, New York, 1964), Vol. 1.

ABSENCE OF PARTICLE CREATION AS AN EQUILIBRIUM CONDITION*

Leonard Parker
Department of Physics, University of Wisconsin-Milwaukee
Milwaukee, Wisconsin 53201

1. INTRODUCTION

From the observed range of validity of the Newtonian approximation, the cosmological constant Λ is known to be very small. In units with $\hbar = c = 1$, one has $\Lambda < 10^{-56}\ \mathrm{cm}^{-2}$. In view of the fact that vacuum energy can contribute to the value of Λ,[1] especially in conjunction with symmetry breaking,[2] it is puzzling that the value of Λ is so small. Here I wish to recall and extend an old argument[3-5] which implies that the cosmological constant is zero or very small. The same hypothesis evidently implies also that the expansion of the universe is isotropic and that the early universe is dominated by relativistic particles and radiation.

The basic hypothesis is that the absence of particle creation, in particular, gravitons and minimally coupled massless scalar particles, is a kind of equilibrium condition toward which the evolution of the universe tends, and that the classical Einstein equations must be consistent with that condition. Here we are referring to particle creation as detected by a measuring instrument on one of the preferred geodesics of the expanding universe. In Ref. 5 (Section F), we argued that there must be a deep connection between the Einstein equations and the conditions for zero particle creation. We viewed the Einstein equations as macroscopic equations governing the large scale evolution of the universe. We asserted that the underlying (unknown) microscopic theory, from which the Einstein equations follow, must be such that "in an expansion of the universe in which a particular type of particle (i.e., relativistic or non-relativistic) is predominant, the expansion achieved after a long time will be such as to minimize the average creation rate of that particle" and "the reaction of the particle creation back on the gravitational field will modify the expansion in such a way as to reduce the creation rate." Investigations of particle creation in anisotropic[6-8] as well as isotropic[9] expansions support that hypothesis, but do not address the question of deriving the Einstein equations from an underlying microscopic theory. Candidates for such an underlying unified theory include induced gravity. Adler[10] has noted that the above hypothesis can serve within the context of that theory as a reason for the observed smallness of the cosmological constant.

Our hypothesis may well be a dynamical or statistical consequence of an underlying unified theory, perhaps resulting from a feedback mechanism involving particle creation, or reflecting an underlying equilibrium or consistency condition. Perhaps it is a stability condition on the true vacuum. In any event, we refer to the hypothesis, loosely speaking, as an equilibrium condition. Without being specific about an underlying theory, one can not address the question of the time in the early universe at which the suppression of particle creation is imposed. Possibilities are the Planck time or the times at which the grand unified theory phase transitions occur. It is not even necessary that the usual concepts of space-time or metric should be meaningful "before" the above equilibrium sets in. The underlying microscopic theory is assumed to be such that, at the time and scale when the classical Einstein equations are reasonably well defined, they are consistent with an expansion law in which the creation rate of gravitons and minimally coupled scalar particles is suppressed. In order to find the consequences of that hypothesis, we can regard the background metric and background matter as governed by the classical Einstein equations. Gravitons are taken to be the quantized gravitational wave perturbations of the background metric.

2. GRAVITATIONAL WAVE PERTURBATIONS

It is well known that in an anisotropically expanding universe, particle creation is in general more intense than in an isotropically expanding universe, and that the reaction back of the created particles is such as to bring about isotropic expansion.[6-8] There is no known case in which particle creation vanishes in an anisotropically expanding universe with nonvanishing Riemann tensor. The existence of such a case is all the less likely because we are demanding that the expansion be such that the graviton creation rate is zero at all times. (That excludes global interference effects resulting in no net graviton creation.) Therefore, we turn our attention to isotropic expansions.

The Robertson-Walker universes have line elements

$$ds^2 = -dt^2 + a^2(t)d\sigma^2, \tag{1}$$

where $d\sigma^2 = \tilde{g}_{ij}dx^i dx^j$ is the line element for a space of constant curvature. Define a quantity ε such that $\varepsilon = 1$, -1, or 0 if the spatial curvature is positive, negative, or zero, respectively.

The Einstein equations with cosmological constant Λ are

$$R_{\mu\nu} - \Lambda g_{\mu\nu} = 8\pi G(T_{\mu\nu} - \frac{1}{2} g_{\mu\nu}T) , \tag{2}$$

where $T = T_\lambda{}^\lambda$. The above equation refers to the background metric and background matter, both regarded as classical. The background energy-momentum tensor is that of a perfect fluid

$$T_{\mu\nu} = (\rho+p)u_\mu u_\nu + pg_{\mu\nu} , \tag{3}$$

with four velocity u^μ. The equations governing gravitational wave perturbations on a Robertson-Walker background have been worked out by Lifshitz.[11] Let $\delta\rho = \delta p = \delta u^\mu = 0$ and let the symmetric metric perturbation $\delta g_{\mu\nu} = h_{\mu\nu}$ satisfy the conditions

$$u^\mu h_{\mu\nu} = 0 , \quad h^{\mu\nu}{}_{;\nu} = 0 , \tag{4}$$

which are consistent in a Robertson-Walker space-time and also imply that

$$h_{\mu}^{\ \mu} = 0 \ . \tag{5}$$

(Indices are raised and lowered with the unperturbed metric and covariant differentiation is with respect to that metric.) Two components of $h_{\mu\nu}$ remain independent and correspond to the two polarizations of a gravitational wave.

In the co-moving coordinate system of Eq. (1), only the spatial components $h_i^{\ j}$ do not vanish. Perturbation of the above Einstein equations yields

$$a^{-3}\partial_t(a^3\partial_t h_i^{\ j}) - a^{-2}\tilde{g}^{\ell m}\tilde{\nabla}_\ell\tilde{\nabla}_m h_i^{\ j} + 2\varepsilon a^{-2}h_i^{\ j} = 0 \ , \tag{6}$$

where ∂_t denotes $\partial/\partial t$, $\tilde{\nabla}_\ell$ denotes covariant differentiation with respect to the spatial metric $\tilde{g}_{ij}$, and ε takes values 1, -1, or 0 corresponding to the sign of the spatial curvature. The perturbation equation is independent of the cosmological constant. Let $h_i^{\ j} = G_i^{\ j}(\vec{x})\psi(t)$ where the $G_i^{\ j}$ are tensor spherical harmonics (see Ref. 11) for $\varepsilon = \pm 1$ or plane waves for $\varepsilon = 0$. They satisfy

$$\tilde{g}^{\ell m}\tilde{\nabla}_\ell\tilde{\nabla}_m G_i^{\ j} = -k^2 G_i^{\ j} \ , \quad \tilde{\nabla}_j G_i^{\ j} = 0 \ , \quad \text{and} \quad G_i^{\ i} = 0 \ , \tag{7}$$

with eigenvalues k^2 given by

$$k^2 = \begin{cases} |\vec{k}|^2, & \varepsilon = 0, \quad 0 < |\vec{k}| < \infty \\ n^2-3, & \varepsilon = 1, \quad n = 3, 4, \ldots \\ q^2+3, & \varepsilon = -1, \quad 0 < q < \infty \ . \end{cases} \tag{8}$$

The function $\psi(t)$ satisfies the equation

$$a^{-1}d(a^3 d\psi/dt)/dt + (k^2+2\varepsilon)\psi = 0 \ . \tag{9}$$

This is essentially the same equation as is obeyed by the time-dependent part of a minimally coupled scalar field, so that the methods developed in Refs. 3-5 to study the production of scalar particles are applicable to graviton production. The equation is not conformally invariant. The production of gravitons in Robertson-Walker universes was studied for the case $\varepsilon = 0$ by Grishchuk[12] and for all three cases by Ford and Parker.[13]

3. CONDITIONS FOR ZERO PARTICLE PRODUCTION

In this section, we search for criteria that can be used to infer that the creation rate is zero for gravitons and minimally coupled massless scalar particles. As noted earlier, there evidently are no anisotropically expanding universes in which the particle creation rate vanishes. Even conformally invariant wave equations give rise to particle creation when the expansion is anisotropic. Therefore, the first condition that must hold is that the expansion be isotropic. In seeking other criteria, we can limit our considerations to isotropic expansions.

As a preliminary step, it is helpful to recall the conformally coupled scalar field. It is well known that in a Robertson-Walker universe an unaccelerated observer having no event horizon will find that such particles are not created.[3-5,14] That permits one to unambiguously identify positive frequency solutions of the wave equation. For the conformal scalar field one has

$$-\nabla^{\mu}\nabla_{\mu}\phi + (R/6)\phi = 0 \,, \tag{10}$$

where R is the scalar curvature. In the Robertson-Walker metric of Eq. (1), one can write $\phi = G(\vec{x})\psi(t)$, where

$$\tilde{g}^{\ell m}\tilde{\nabla}_{\ell}\tilde{\nabla}_{m}G = -K^2 G \tag{11}$$

and

$$K^2 = \begin{cases} |\vec{k}|^2, & \varepsilon = 0 \,, \quad 0 < |\vec{k}| < \infty \\ n^2-1, & \varepsilon = 1 \,, \quad n = 1,2, \ldots \\ q^2+1, & \varepsilon = -1, \quad 0 < q < \infty \,. \end{cases} \tag{12}$$

There the time-dependent part of the field satisfies

$$a^{-3}d(a^3 d\psi/dt)/dt + [a^{-2}K^2 + (R/6)]\psi = 0 \,. \tag{13}$$

Using the expression for the scalar curvature,

$$R = 6(a^{-2}\dot{a}^2 + a^{-1}\ddot{a} + \varepsilon a^{-2}) \tag{14}$$

one finds that a solution of Eq. (13) is

$$\psi_K(t) = a^{-1}(t)\exp[-i\int^t dt' a^{-1}(t')(K^2+\varepsilon)^{1/2}] \,. \tag{15}$$

As no particle creation is occurring and this solution is clearly positive frequency for sufficiently large K, we will identify it as a purely positive frequency solution. An instrument measuring the particle creation rate couples to the field ϕ and is not directly influenced by the coefficient of R appearing in the wave equation (10). Therefore, for a massless scalar field with a different coupling to the scalar curvature, such as the minimally coupled field, we assume that the creation rate is zero for particles of that field if and only if a purely positive frequency solution of the form of Eq. (15) exists. The same condition was used by us (for $\varepsilon = 0$) in Ref. 3-5.

We proceed in the same way for gravitons. If one adds a term $(R/6)h_i{}^j$ to Eq. (6), the equation satisfied by the time-dependent part of $h_i{}^j$ is

$$a^{-3}d(a^3 d\psi/dt)/dt + [a^{-2}K^2 + 2a^{-2}\varepsilon + (R/6)]\psi = 0 \,. \tag{16}$$

This has the purely positive frequency solution

$$\psi_k(t) = a^{-1}(t)\exp[-i\int^t dt'a^{-1}(t')(k^2+3\varepsilon)^{1/2}] \ . \qquad (17)$$

An instrument measuring the graviton creation rate will not be directly sensitive to the coefficient of R in Eq. (16). Therefore, for the actual graviton equation, (9), we assume that the creation rate is zero if and only if a purely positive frequency solution of the form of Eq. (17) exists.

The special cases in which the metric of Eq. (1) describes a flat space-time are given by $\varepsilon = 0$, $a(t) = \text{constant}$ and $\varepsilon = -1$, $a(t) = t$. In those cases, the above criteria give zero particle creation. Another much studied example is the linearly expanding spatially flat universe ($\varepsilon = 0$, $a(t) = t$).[15,16,17] The above criteria imply that there is particle creation in that case, in agreement with the other methods. Finally, in the spatially curved static universes ($\varepsilon = \pm 1$, $a(t) = \text{constant}$), our criteria yield different results depending on the coupling to the curvature. There would be no creation of conformally coupled massless scalar particles, but there would be creation of minimally coupled scalar particles and gravitons. It is certainly conceivable that spatial curvature can create particles in a globally static space-time. On the other hand, it is possible that the above criteria for zero particle creation rate may have to be generalized to include those static universes. However, until it becomes clear that such a generalization is needed and exactly how it should be done, we will assume that Eqs. (15) and (17) are the necessary and sufficient criteria for zero particle creation rate of massless scalar particles and gravitons in Robertson-Walker universes.

4. IMPLICATIONS OF ZERO PARTICLE CREATION

Suppose now that the classical background metric and background matter obey the Einstein equations given in Eq. (2). The requirement that gravitons are not created places constraints on the form of the Einstein equations and on the equation of state of the matter. (We will deal with gravitons, but it should be understood that minimally coupled massless scalar particles would give the same results.)

The absence of graviton production requires that Eq. (9) for $\psi(t)$ has a solution of the form in Eq. (17). Substitution of (17) into (9) yields the condition that

$$a^{-2}\dot{a}^2 + a^{-1}\ddot{a} + a^{-2}\varepsilon = 0 \ , \qquad (18)$$

or

$$R = 0 \ . \qquad (19)$$

With the background energy-momentum tensor given by Eq. (3), contraction of the Einstein equations yields

$$R - 4\Lambda + 8\pi G(3p-\rho) = 0 \ , \qquad (20)$$

or with Eq. (19),

the behavior of a(t) is essentially arbitrary.

The energy-momentum tensor

$$T^{(\phi)}_{\mu\nu} = \partial_\mu\phi\partial_\nu\phi - \frac{1}{2} g_{\mu\nu}\partial_\alpha\phi\partial^\alpha\phi$$
$$+ \xi[g_{\mu\nu}\nabla_\alpha\nabla^\alpha(\phi^2) - \nabla_\mu\nabla_\nu(\phi^2) + G_{\mu\nu}\phi^2] \tag{28}$$

satisfies the local conservation law $\nabla_\mu T^{(\phi)\mu\nu} = 0$. Furthermore, in the "in" and "out" regions, the factor multiplying ξ is a four-divergence. For $\xi = 1/6$, $T^{(\phi)}_{\mu\nu}$ is the "improved" energy-momentum tensor of Callan, Coleman, and Jackiw.[18] Let us take $\xi = 1/6$.

In a Robertson-Walker metric of Eq. (1), it can be shown that[14]

$$\rho^{(\phi)}(t_2)a^4(t_2) - \rho^{(\phi)}(t_1)a^4(t_1) = g(t_2) - g(t_1) , \tag{29}$$

where

$$g(t) = 6(4\pi^2)^{-2}[B(\dot{a}^4 + 2\varepsilon\dot{a}^2)$$
$$+ C(-a^2\dot{a}\dddot{a} - a\,\dot{a}^2\ddot{a} + \frac{1}{2} a^2\ddot{a}^2 + \frac{3}{2}\dot{a}^4 + \varepsilon\dot{a}^2)] \tag{30}$$

and $\rho^{(\phi)} = \langle T^{(\phi)00}\rangle$, the expectation value being taken in a state having the symmetries of the space-time. In Eq. (30), $B = -1/360$ and $C = 1/180$. Let t_1 be in the "in" region and t_2 be in the "out" region. Then clearly $g(t_1) = g(t_2)$, and

$$\rho^{(\phi)}(t_2)a^4(t_2) = \rho^{(\phi)}(t_1)a^4(t_1) . \tag{31}$$

The energy density of the particles present initially is merely red-shifted, with no real particles being created by the expansion. As the behavior of a(t) in the interpolating regions is arbitrary, one concludes that the creation rate of minimally coupled massless particles must be zero during the radiation dominated stage of the expansion. Although interpolating regions and "in" and "out" regions were used to derive that result, it must be valid when there are no such regions, in agreement with our earlier method.

6. CONCLUSIONS

We have shown that if one requires as an equilibrium condition that the creation rate for gravitons or minimally coupled massless scalar particles vanishes in the early universe, then the expansion of the universe must be isotropic, the cosmological constant must be zero, and the early universe must be dominated by relativistic particles and radiation. Such an equilibrium does not necessarily imply that gravitons will be absent, since they may have been created prior to equilibrium.

The linearized equation for gravitational wave perturbations was used. In higher order, the presence of self interactions among gravitons may give a small graviton creation rate when $R = 0$, but one would nevertheless expect the graviton creation rate to be near its minimum in the radiation dominated universe. In the case of a self interacting scalar field which has been studied,[19,20] it is interesting that the explicit term shown in Ref. 19 to cause particle creation vanishes when $R = 0$.

REFERENCES

1. Ya. B. Zel'dovich, Usp. Fiz. Nauk. 95, 209 (1968) [Sov. Phys.-Uspekhi 11, 381 (1968)].
2. A. H. Guth, Phys. Rev. D 23, 347 (1981).
3. L. Parker, Ph.D. Thesis, Harvard University, 1966 (available from Xerox University Microfilms, Ann Arbor, Michigan, Na DCJ 73-31244).
4. L. Parker, Phys. Rev. Lett. 21, 562 (1968).
5. L. Parker, Phys. Rev. 183, 1057 (1969).
6. Ya. B. Zel'dovich, Pisma v. Zh. ETF 12, 443 (1970) [JETP Lett. 12, 307 (1970)]; Ya. B. Zel'dovich and A. A. Starobinsky, Zh. ETF 61, 2161 (1971) [Sov. Phys.-JETP 34, 1159 (1972); V. N. Lukash and A. A. Starobinsky, Zh. ETF 66, 1515 (1974) [Sov. Phys.-JETP 32, 742 (1974)].
7. S. A. Fulling, L. Parker, and B. L. Hu, Phys. Rev. D 10, 3905 (1974); B. L. Hu and L. Parker, Phys. Rev. D 17, 933 (1978).
8. J. B. Hartle and B. L. Hu, Phys. Rev. D 21, 2756 (1980).
9. J. B. Hartle, Phys. Rev. D 23, 2121 (1981).
10. S. L. Adler, Revs. Mod. Phys. 54, 729 (1982).
11. E. M. Lifshiftz, Zh. ETF 16, 587 (1946) [also in J. Phys. USSR 10, 116 (1946)].
12. L. P. Grishchuk, Zh. ETF 67, 824 (1974) [Sov. Phys.-JETP 40, 409 (1975)].
13. L. H. Ford and L. Parker, Phys. Rev. D 6, 1601 (1977).
14. L. Parker in *Recent Developments in Gravitation, Cargese 1978*, edited by M. Levy and S. Deser (Plenum Press, N.Y., 1978).
15. B. M. Chitre and J. B. Hartle, Phys. Rev. D 15, 251 (1977).
16. H. Nariai and T. Azuma, Progr. Theor. Phys. 59, 1532 (1978).
17. Ch. Charach and L. Parker, Phys. Rev. D 24, 3023 (1981).
18. C. G. Callan, Jr., S. Coleman, and R. Jackiw, Ann. Phys. (N.Y.) 59, 42 (1970); S. Coleman and R. Jackiw, *ibid.* 67, 552 (1971).
19. N. D. Birrell and P.C.W. Davies, Phys. Rev. D 22, 322 (1980).
20. N. D. Birrell and J. G. Taylor, J. Math. Phys. 21, 1740 (1980).

* Work supported by the National Science Foundation.

RENORMALIZATION AND SCALING OF NON-ABELIAN

GAUGE FIELDS IN CURVED SPACE-TIME

Leonard Parker
Department of Physics, University of Wisconsin-Milwaukee
Milwaukee, Wisconsin 53201

It is well known that in flat space-time non-abelian gauge field theories are renormalizable and asymptotically free. In view of the absence of a theorem stating that flat-space renormalizability implies renormalizability in more general space-times, one must examine each theory individually. This is a brief summary of work by my student Todd K. Leen showing that non-abelian gauge fields in curved space-time are renormalizable at the one-loop level; and that the property of asymptotic freedom is preserved.[1] Renormalizability of these theories has been proved independently by David J. Toms.[2]

In the present work, it is shown that gauge invariance, as expressed through the Taylor-Slavnov identities, insures that no curvature-dependent divergences occur in the vector two-point function. The divergences in the ghost two-point function and the ghost-vector-ghost vertex are extracted using local momentum space expansions for the propagators. As with the vector two-point function, no curvature-dependent divergences are found. Thus, these three Green functions are rendered finite by the usual Minkowski space counterterms. It then follows from gauge invariance, that the three- and four-vector vertex functions are finite. Finally, renormalization group arguments are used to show that the theory remains asymptotically free in curved space-time.

We first establish notation. The gauge group is denoted Ω and its associated Lie algebra T_Ω. The gauge covariant derivative is written

$$D_\mu \equiv \nabla_\mu - gA_\mu \tag{1}$$

where ∇_μ is the covariant derivative on the space-time and g the coupling constant of the theory. The field strength (or curvature) tensor is given by

$$F_{\mu\nu} \equiv \nabla_\nu A_\mu - \nabla_\mu A_\nu + g[A_\mu, A_\nu] \ . \tag{2}$$

The potentials are decomposed along a basis $\{T^a\}$ on T_Ω as

$$A_\mu = A^a_\mu T_a \ . \tag{3}$$

The gauge invariant classical action is then written as

$$S = \int d\tau(x) \ (-\frac{1}{4} F^{\mu\nu}_a F^a_{\mu\nu}) \tag{4}$$

where $d\tau(x)$ is the covariant volume element on the space-time. The theory is quantized via the path integral formalism. The procedure of Faddeev and Popov[3] allows the elimination of the redundancy due to integration over gauge related field configurations. The resulting generating functional with ghost fields $\bar{C}$ and C is written

$$W[j,\bar{\xi},\xi] = \int D[A]D[\bar{C}]D[C] \exp i\int d\tau(x)\Big\{-\frac{1}{4}F^{\mu\nu}_a F^a_{\mu\nu} - \frac{1}{2\alpha}(\nabla_\mu A^\mu)^2$$

$$-(\nabla_\mu \bar{C}^a)(D^\mu C)_a + j^\mu_a A^a_\mu + \bar{\xi}^a C_a + \bar{C}^a \xi_a \quad . \tag{5}$$

The perturbative expansion of this generating functional and the extraction of Green functions leads to a diagrammatic expansion as in flat space-time. The bare propagators for the ghost and vector fields satisfy, respectively

$$\Box D_{ab}(x,x') = -\delta_{ab}\,\delta(x,x') \tag{6}$$

and

$$\Box D^{ab}_{\nu\sigma}(x,x') - (1-\frac{1}{\alpha})\nabla_\nu\nabla^\mu D^{ab}_{\mu\sigma}(x,x') - R_\nu{}^\mu D^{ab}_{\mu\sigma}(x,x') = -g_{\mu\sigma}\delta^{ab}\delta(x,x') \tag{7}$$

where $\delta(x,x')$ is the covariant delta function and the derivatives and Ricci tensor are at the point x.

The gauge invariance of the classical action Eq. (4) is reflected in relations between the Green functions of the theory revealed in the Taylor-Slavnov identities. For the corrected vector propagator one recovers the relation ($\alpha = 1$ hereafter)

$$\nabla'_\nu\nabla_\mu \tilde{D}^{\mu\nu}_{ab}(x,x') = \delta_{ab}\delta(x,x') \;. \tag{8}$$

In lowest order we write the corrected vector propagator as

$$\tilde{D}^{\mu\nu}(x,x') = D^{\mu\nu}(x,x') + \int d\tau(y)d\tau(y')D^{\mu\sigma}(x,y)\pi_{\sigma\rho}(y,y')D^{\rho\nu}(y',x') \tag{9}$$

where the vacuum polarization tensor $\pi_{\sigma\rho}(y,y')$ contains one-loop contributions only. Making use of Eqs. (8) and (7) we recover

$$\nabla_\mu\nabla'_\nu\pi^{\mu\nu}(x,x') = 0 \;. \tag{10}$$

Thus the vacuum polarization (to lowest order) remains transverse as in Minkowski space. The pole part of $\pi^{\mu\nu}$ may be written as

$$\pi^{\mu\nu}(x,x')\Big|_{\text{divergent}} = Ag^{\mu\nu}\delta(x,x') + B\nabla^\mu\nabla^{\nu'}\delta(x,x') + E\,\Box g^{\mu\nu}\delta(x,x')$$

$$+ (aR^{\mu\nu}+bg^{\mu\nu}R)\delta(x,x') \tag{11}$$

where A represents the quadratic divergence while B, E, a and b carry logarithmic divergences. This may be understood as follows. The divergences in $\pi^{\mu\nu}(x,x')$ arise in the coincidence limit of the arguments, hence the delta functions. The $\nabla^\mu\nabla^{\nu'}$, $\Box$, $R^{\mu\nu}$ and $g^{\mu\nu}R$ factors carry dimension $(\text{length})^{-2}$ so that their corresponding coefficients must carry two additional powers of momentum in the denominator of

Feynman integrals. Since A carries the leading quadratic divergence, the remaining terms are, at most, logarithmically divergent. The transversality of the polarization Eq. (10) determines relations between the coefficients in Eq. (11). One finds

$$\pi^{\mu\nu}(x,x')\Big|_{\text{div.}} = B(g^{\mu\nu}\Box\delta(x,x') + \nabla^{\mu}\nabla^{\nu'}\delta(x,x') - R^{\mu\nu}\delta(x,x')) \ . \tag{12}$$

Substituting this form for the polarization tensor into Eq. (9) for the corrected vector propagator leaves

$$\tilde{D}^{\mu\nu}(x,x') = (1-B)D^{\mu\nu}(x,x') - B\int d\tau(y)(\nabla'_{\rho_y}\nabla_{\sigma_y}D^{\mu\sigma}(x,y))D^{\rho\nu}(y,x') \tag{13}$$

Since B is a curvature-independent space-time constant, we see that curvature-dependent divergences in the vector propagator are absent. The divergence in the first term of Eq. (13) is removed by wavefunction renormalization while that in the second term is removed by renormalizing the gauge fixing parameter. We define renormalized quantities via

$$A^{\mu} = Z_3^{1/2}\, A_R^{\mu} \tag{14}$$

$$\alpha = Z_3\ \alpha_R \ . \tag{15}$$

The renormalization constant is identical to that in Minkowski space

$$Z_3 = 1 + \frac{g^2 C_2}{16\pi^2}\left(\frac{5}{3}\right)\frac{2}{\varepsilon} \tag{16}$$

where $\varepsilon = \frac{-N}{2} + 2$ is the dimensional parameter appearing in the regularization and C_2 is the value of the quadratic Casimir operator for Ω.

We have seen that the gauge invariance is sufficient to insure that no divergences not present in Minkowski space arise in a general curved space-time. The appearance of explicitly curvature-dependent divergent corrections to the vector two-point function would have necessitated the introduction of renormalized couplings between the gauge field and the background. Such couplings would spoil the gauge properties of the theory.

The divergences in the ghost propagator are handled using the local momentum space expressions of Refs. (4,5,6) for the bare propagators. Here again we find no divergences not present in Minkowski space. The latter are removed by defining renormalized ghost fields

$$C = \tilde{Z}_3^{1/2}\ C_R \tag{17}$$

with the usual renormalization constant

$$\tilde{Z}_3 = \left(1 + \frac{g^2 C_2}{16\pi^2}\frac{1}{\varepsilon}\right) \ . \tag{18}$$

Finally, the vector-ghost-ghost vertex is computed using the momentum space expansions. Again we find no curvature-dependent divergences. This vertex is thus renormalized as in flat space-time leading to the definition of the renormalized coupling constant

$$g = \frac{\tilde{Z}_1}{\tilde{Z}_3 Z_3^{1/2}} g_R \, \mu^{\varepsilon/2} \tag{19}$$

where μ is the mass parameter required to maintain proper dimensions during the regularization. The constant $\tilde{Z}_1$ is as in Minkowski space

$$\tilde{Z}_1 = 1 - \frac{g^2 C_2}{16\pi^2} \frac{1}{\varepsilon} \, . \tag{20}$$

Gauge invariance insures that the above renormalizations are sufficient to render finite the three- and four-vector vertices.

We have seen that the divergences in the theory are removed by the same counter-terms that render the flat space theory finite. This suggests that the remarkable feature of asymptotic freedom remains in the presence of space-time curvature. To verify this we derive a renormalization group equation and exhibit the behavior of the effective coupling constant. The renormalized one particle irreducible Green functions are given by

$$\Gamma^{(n)}_{\mu,\ldots}(x_1,\ldots x_n;\, g_R,\alpha_R,\mu,g_{\alpha\beta}) = Z_3^{(n/2)} \Gamma^{UN}_{\mu,\ldots}(x_1,\ldots x_n;\, g,\alpha,g_{\alpha\beta}) \tag{21}$$

where Γ^{UN} are the unrenormalized n-point functions. Differentiating the above with respect to μ and multiplying by μ leaves

$$(\mu\, \partial/\partial_\mu + \beta\, \partial/\partial g_R - \gamma(\alpha_R\, \partial/\partial\alpha_R + \tfrac{n}{2}))\Gamma^{(n)}_{\mu,\ldots}(x_1\ldots x_n;\, g_R,\alpha_R,\mu,g_{\alpha\beta}) = 0 \tag{22}$$

where

$$\beta \equiv \mu \frac{\partial g_R}{\partial\mu} \tag{23}$$

$$\gamma \equiv \mu \frac{\partial \ln Z_3}{\partial\mu}$$

The usual procedure is to eliminate $\mu\, \partial/\partial\mu$ by scaling the coordinates x_i or the momenta p_i of the external legs. In curved space-times the natural approach is to scale the metric tensor.[7] We consider a one-parameter family of metrics $g_{\alpha\beta}/K^2$ and scale the parameter K. The resulting renormalization group equation reads.

$$[(D - \tfrac{n}{2}\gamma) - K\, \partial/\partial K + \beta\, \partial/\partial g_R - \gamma\alpha_R\, \partial/\partial\alpha_R]\Gamma^{(n)}_{\mu,\ldots}(x_1\ldots;\, g_R,\alpha_R,\mu,g_{\alpha\beta}/K^2) = 0. \tag{24}$$

where D is the mass dimension of Γ.

Since, in lowest order, the β function is independent of α,[8,9] we are free to discuss the solution to this equation for $\alpha = 0$; and calculate the β function using our previous results (calculated using $\alpha = 1$) for the renormalization constants. The solution to Eq. (24) is

$$\Gamma^{(n)}_{\mu,\ldots}(x_1,\ldots x_n; g_R, \alpha_R=0, \mu, g_{\alpha\beta}/K^2)$$

$$= K^D \Gamma^{(n)}_{\mu,\ldots}(x_1\ldots,x_n;\ g(K),\mu,g_{\alpha\beta}) \exp - \frac{n}{2}\int_1^K \frac{dK'}{K'}\gamma(g(K')) \tag{25}$$

with $$K\frac{\partial g}{\partial K} = \beta(g(K)) \qquad g(K=1) = g_R \tag{26}$$

Using Eqs. (16), (18), (19) and (20) we find

$$\beta = -\frac{11}{3}\frac{C_2}{16\pi^2} g^3(K) \tag{27}$$

Because β is negative this indicates that the theory remains asymptotically free in curved space-times.

REFERENCES

1. T. K. Leen (to be published).
2. D. J. Toms, preprint (Imperial College, London, 1982).
3. L. D. Faddeev and V. N. Popov, Phys. Lett. 25B, 29 (1967).
4. T. S. Bunch and L. Parker, Phys. Rev. D 20, 2499 (1979).
5. P. Panangaden, Phys. Rev. D 23, 1735 (1981).
6. T. S. Bunch, Liverpool University report (unpublished).
7. B. L. Nelson and P. Panangaden, to appear in Phys. Rev. D (1982).
8. D. Gross and F. Wilczek, Phys. Rev. D 8, 3633 (1973); Phys. Rev. Lett. 26, 1343 (1973).
9. H. D. Politzer, Phys. Rev. Lett. 26, 1346 (1973).

TWO-POINT FUNCTIONS AND RENORMALIZED OBSERVABLES

S. A. Fulling
Mathematics Department
Texas A&M University
College Station, Texas 77843 USA

This is a victory declaration in the theory of a quantized field propagating in a given curved background space-time. We now have an unambiguous, internally consistent quantum theory of such a system, in which any physical quantity can in principle be calculated. Whether this kind of model is relevant to the real world is a separate question. I shall present the theory rather dogmatically.

Doctrine 1: The physical interpretation of a quantum field theory in curved space must be sought in the stress-energy-momentum tensor and other local field observables. (Particle observables are not meaningful, in general.)

As a concrete example, let's keep in mind the minimally coupled neutral scalar field, whose stress tensor is

$$T_{\mu\nu}(x) = \nabla_\mu \phi(x)\nabla_\nu \phi(x) - \frac{1}{2} g_{\mu\nu}\nabla_\alpha\phi\nabla^\alpha\phi + \frac{1}{2} m^2 g_{\mu\nu}(x)\phi(x)^2. \qquad (1)$$

For a field satisfying a linear wave equation, there is little difficulty in defining Heisenberg-picture field operators rigorously. Making sense of the stress tensor, as a quantum operator, is much more of a problem. In my opinion, the most profound clarification of this problem came in a series of papers by Wald (1977, 1978a,b). The ingredients of the solution go back to the work of Utiyama and DeWitt (1962). The bibliography lists many (but not all) other papers which have contributed to our present understanding.

Doctrine 2: The stress tensor is conserved $[\nabla^\mu T_{\mu\nu} = 0]$, it depends causally on the metric, and the difference of its expectation values with respect to any two quantum states can be correctly calculated from the classical formula (1). ("The divergent part of $T_{\mu\nu}$ is a c-number.")

(These are three of Wald's five axioms. The other two have been supplanted by an improved understanding of the renormalization problem. See Wald's papers for a definition of "causally." I leave as an exercise the equivalence of Wald's axiom about orthogonal matrix elements to mine about differences of expectation values.)

Theorem 1 (Wald): The stress tensor operator of a given field theory is uniquely determined by these requirements, up to c-number terms proportional to conserved, covariant, local, polynomial functionals of the metric and curvature tensors, such as

$$c_1 g_{\mu\nu} + c_2(R_{\mu\nu} - \tfrac{1}{2}Rg_{\mu\nu}) + c_3 g^{-1/2}\frac{\delta}{\delta g^{\mu\nu}}\int R^2 g^{1/2}dx + c_4 g^{-1/2}\frac{\delta}{\delta g^{\mu\nu}}\int C^2 g^{1/2}dx. \tag{2}$$

(C^2 is the square of the Weyl tensor).

Wald observed that the procedure of "point-splitting" seemed to provide a $T_{\mu\nu}$ satisfying his axioms. The evidence (e.g., from calculations in simple models) suggested: (1) The two-point function

$$G(x,y) = \langle\psi|\phi(x)\phi(y)|\psi\rangle$$

is well-defined as a distribution for a large class of states ψ. (2) As $y \to x$, G has an asymptotic expansion consisting of terms which are singular where the geodesic separation between x and y is null (lightlike). These singular terms are (a) c-numbers (have the same value in all quantum states) and (b) purely local and polynomial in their dependence on the geometry. For example, a typical singular term in $G(x,y)$ is

$$\frac{R_{\mu\nu}(x)\sigma^\mu\sigma^\nu}{g_{\alpha\beta}(x)\sigma^\alpha\sigma^\beta} ,$$

where $-\sigma^\mu(x,y)$ is the vector at x tangent to the geodesic from x to y, with length equal to the geodesic separation (in other words, the Riemann normal coordinates of y relative to x). [A similar description applies to an expectation value of the point-split stress tensor, $T_{\mu\nu}(x,y)$, obtained by applying a suitable differential operator to $G(x,y)$, so that $T_{\mu\nu}(x)$ [Eq. (1)] is formally recovered when $y = x$:

$$T_{\mu\nu}(x,y) = e_\nu{}^{\nu'}(x,y)\nabla_\mu\phi(x)\nabla_{\nu'}\phi(y) + \dots ,$$
$$\langle\psi|T_{\mu\nu}(x,y)|\psi\rangle = e_\nu{}^{\nu'}\nabla_\mu\nabla_{\nu'}G(x,y) + \dots ,$$

where $e_\nu{}^{\nu'}$ is a parallel-transport matrix. See Christensen (1976) for details.] (3) The symmetric part (under the interchange $x \leftrightarrow y$) of this singular series coincides with what is called the Hadamard solution of the hyperbolic field equation ($\nabla^\mu\nabla_\mu + m^2\phi = 0$ in our example). For the definition of this object, see DeWitt and Brehme (1960), Hadamard (1952), Garabedian (1964), Friedlander (1975),

Adler et al. (1977). (The antisymmetric part of G is the familiar commutator distribution, which is entirely c-number and local and hence can be disregarded in the rest of our discussion.

I emphasize that the remainder in the symmetrized G (which can be made arbitrarily smooth by subtracting off enough terms of the series) is not a local functional of the metric and is not a c-number -- it depends on the quantum state ψ. Precisely for this reason it contains the most interesting physics in any concrete problem.

For consistency of the theory it was necessary to prove that the picture I've just described holds in general. In 1978, Sweeny, Wald, and I proved:

Theorem 2: If a two-point function, $\langle\psi|\phi(x)\phi(y)|\psi\rangle$, is a distribution of Hadamard form at one instant of time, then it remains so for all time.

In 1981, Narcowich, Wald, and I closed the remaining hole by proving:

Theorem 3: In a static background geometry, if ψ is the "natural" vacuum state and the mass is positive, then $G(x,y)$ is a distribution of the Hadamard form.

Thus it now makes sense to state:

Doctrine 3: A "physically reasonable" state ψ of a quantum field system in curved space is one for which the two-point function G is a distribution with singularity of the Hadamard form.

For, as a corollary of Theorems 2 and 3, we have:

Theorem 4: In an arbitrary globally hyperbolic space-time (i.e., one where the Cauchy problem is well-posed) there exist many "physically reasonable" states. (They form a dense subspace of a Hilbert space.)

If $m > 0$, the vacuum in a static background $g_{\mu\nu}$ is in this class of states. If $m = 0$, strangely enough, the traditional vacuum sometimes does not qualify as physically reasonable because its two-point function does not exist, as a distribution. In that case the "good" states contain, in some sense, lots of particles in the infrared modes.

The reason why the Hadamard states are regarded as "physically reasonable" is that they yield finite expectation values for physical observables after renormalization:

Doctrine 4: The renormalized stress tensor, $T^{ren}_{\mu\nu}(x)$, is obtained by subtracting from $T_{\mu\nu}(x,y)$ a c-number equal to sufficiently many terms of its Hadamard series, and taking the limit $y \to x$. The ambiguity in this prescription must be reduced by requiring that (1) $T^{ren}_{\mu\nu}$ is conserved, and (2) terms involving derivatives of $g_{\mu\nu}$ of degree higher than fourth are not to be subtracted.

This procedure is manifestly covariant, since the Hadamard expansion is. The result is ambiguous since the terms to be subtracted could be changed by any covariant, local, polynomial functional of the curvature tensor at x. Even after requirements (1) and (2) are imposed, some ambiguity remains (cf. Theorem 1).

Doctrine 5: The ambiguous terms in $T^{ren}_{\mu\nu}$ can be absorbed into the coupling constants in the gravitational field's equation of motion. In other words, they represent the nonexistence of any clear division of the physical energy density into matter energy and gravitational energy.

Indeed, the ambiguous terms are precisely those listed in Eq. (2); c_1 renormalizes the cosmological constant, c_2 renormalizes the gravitational constant, and c_3 and c_4 renormalize the coefficients of terms involving fourth derivatives of the metric tensor. These coupling constants must be determined by experiment. The appearance of fourth-order terms is one aspect of the nonrenormalizability of the gravitational interaction.

Theorem 5 (Wald): In a scale-invariant theory, it is meaningless to require *a priori* that c_3 and c_4 be 0.

The point is that c_3 and c_4 appear in contexts of the general nature of

$$c_3 + \ell n(L^2R).$$

The value of c_3 depends on the arbitrary length L. We could *define* L so that $c_3 = 0$, but then L would be a new fundamental constant with units of length. It could be determined (in principle) by experiment, but it cannot be predicted by a scale-invariant theory. Similar phenomena are known to particle physicists under the headings of "renormalization group" and "dimensional transmutation."

The point I have tried to make by this show of orthodoxy is this: Within its own terms, the theory is completely well defined, and it is uniquely determined, *I* think, by accepted physical principles (except for the numerical values of coupling constants). The calculation of the stress tensor, or some other observable, in any given quantum state is today a problem of ordinary applied mathematics, not a matter of the individual investigator's choice of ad hoc mumbo-jumbo as was the case seven years ago.

However, I do not regard it as a substitute for a full quantum theory of gravity. The physical circumstances under which such an external-field model is a valid approximation to reality have not yet been established; they are the subject of active investigation and debate [Horowitz (1981), Duff (1981), Kay (1981), Ford (1982), Fulling (1983)].

BIBLIOGRAPHY

Adler, S. L., Lieberman, J., and Ng, Y. J., Ann. Phys. (N.Y.) 106, 279, (1977).

Bunch, T. S., Phys. Rev. D 18, 1844 (1978).

Bunch, T. S., Christensen, S. M., and Fulling, S. A., Phys. Rev. D 18, 4435 (1978).

Christensen, S. M., Phys. Rev. D 14, 2490 (1976).

Christensen, S. M., and Fulling, S. A., Phys. Rev. D 15, 2088 (1977).

Davies, P. C. W., Proc. Roy. Soc. A 354, 529 (1977a).

Davies, P. C. W., Phys. Lett. B 68, 402 (1977b).

Davies, P. C. W., and Fulling, S. A., Proc. Roy. Soc. A 354, 59 (1977).

Davies, P. C. W., Fulling, S. A., Christensen, S. M., and Bunch, T. S., Ann. Phys. (N.Y.) 109, 108 (1977).

Davies, P. C. W., Fulling, S. A., and Unruh, W. G., Phys. Rev. D 13, 2720 (1976).

Davies, P. C. W., and Unruh, W. G., Phys. Rev. D 20, 388 (1979).

DeWitt, B. S., Dynamical Theory of Groups and Fields, Gordon and Breach, New York (1965).

DeWitt, B. S., Phys. Reports 19, 295 (1975).

DeWitt, B. S., and Brehme, R. W., Ann. Phys. (N.Y.) 9, 220 (1960).

Dimock, J., Commun. Math. Phys. 77, 219 (1980).

Dowker, J. S., and Critchley, R., Phys. Rev. D 15, 1484 (1977).

Duff, M. J., in Quantum Gravity 2: A Second Oxford Symposium, ed. by C. J. Isham, R. Penrose, and D. W. Sciama, Oxford University Press, Oxford (1981), pp. 81-105.

Ford, L. H., Phys. Rev. D 11, 3370 (1975).

Ford, L. H., "Gravitational Radiation by Quantum Systems," to appear (1982).

Friedlander, F. G., The Wave Equation on a Curved Space-Time, Cambridge University Press, Cambridge (1975).

Fulling, S. A., in Quantum Theory and Gravitation, ed. by A. R. Marlow, Academic Press, New York (1980), pp. 187-197.

Fulling, S. A., in a volume in honor of B. S. DeWitt, ed. by S. M. Christensen, Institute of Physics, to appear (1983).

Fulling, S. A., Narcowich, F. J., and Wald, R. W., Ann. Phys. (N.Y.) 136, 243 (1981).

Fulling, S. A., and Parker, L., Ann. Phys. (N.Y.) 87, 176 (1974).

Fulling, S. A., Sweeny, M., and Wald, R. W., Commun. Math. Phys. 63, 257 (1978).

Garabedian, P. R., Partial Differential Equations, Wiley, New York (1964).

Hadamard, J., Lectures on Cauchy's Problem in Linear Differential Equations, Dover, New York (1952).

Horowitz, G. T., in Quantum Gravity 2: A Second Oxford Symposium, ed. by C. J. Isham, R. Penrose, and D. W. Sciama, Oxford University Press, Oxford (1981), pp. 106-130.

Horowitz, G. T., and Wald, R. M., Phys. Rev. D 17, 414 (1978).

Horowitz, G. T., and Wald, R. M., Phys. Rev. D 25, 3408 (1982).

Isham, C. J., in Differential Geometrical Methods in Mathematical Physics II, ed. by K. Bleuler, H. R. Petry, and A. Reetz, Springer, Berlin (1978), pp. 459-512.

Kay, B. S., Commun. Math. Phys. 62, 55 (1978).

Kay, B. S., Phys. Lett. B 101, 241 (1981).

Nariai, H., Prog. Theor. Phys. 46, 433 (1971).

Parker, L., Phys. Rev. 183, 1057 (1969).

Parker, L., and Fulling, S. A., Phys. Rev. D 9, 341 (1974).

Sakharov, A. D., Dokl. Akad. Nauk SSSR 177, 70 (1967) [Sov. Phys. --Dokl. 12, 1040].

Sakharov, A. D., Teor. Mat. Fiz. 23, 178 (1975) [Theor. Math. Phys. 23, 435].

Schwinger, J., Phys. Rev. 82, 664 (1951).

Utiyama, R., and DeWitt, B. S., J. Math. Phys. 3, 608 (1962).

Vilenkin, A., Nuovo Cim. A 44, 441 (1978).

Wald, R. M., Commun. Math. Phys. 54, 1 (1977).

Wald, R. M., Ann. Phys. (N.Y.) 110, 472 (1978a).

Wald, R. M., Phys. Rev. D 17, 1477 (1978b).

VACUUM ENERGY IN THE BAG MODEL

P. Candelas
Center for Theoretical Physics
The University of Texas at Austin
Austin, Texas 78712 U.S.A.

Abstract

The vacuum energy of the Yang-Mills field is examined for the conditions of the bag model. The dominance of high frequency effects results in a vacuum energy that decomposes naturally into a volume energy, a surface energy and higher shape energies. These quantities are identified with the parameters of the bag model. The imposition of confining boundary conditions for all frequencies is shown to be inconsistent since this would result in the bag constant and certain of the shape tensions being infinite. The manner in which the boundary conditions should be relaxed at high frequency is discussed. The most naive procedure for relaxing the boundary conditions, which is to apply confining conditions only on modes of frequency less than some cutoff frequency, results in a negative bag constant and surface tension and would render the vacuum unstable against the spontaneous breaking of Poincaré invariance. Consideration of the manner by which the interacting electromagnetic field avoids a similar instability suggests that a more realistic way to relax the boundary conditions on the bag surface is to endow the vacuum exterior to the bag with a frequency dependent dielectric constant and magnetic permeability.

Introduction

The aim of this report is to examine the energy of the Yang-Mills vacuum under the conditions of the bag model and to show that a consideration of the shift in the vacuum energy of the field due to its confinement may yield important insight into the nature of the vacuum state for non-Abelian gauge theories.

The bag model [1] has achieved reasonable phenomenological success in the explanation of hadron spectroscopy. This model pictures the quantum chromodynamic vacuum as coexisting in two phases. One of these, the 'ordinary' vacuum, is impenetrable to colour while the other, corresponding to the interior of the hadron, is such that the gluon fields that are the carriers of colour are able to propagate freely. These two phases are taken to be separated by a sharp boundary on the interior of which the gluon fields satisfy confining boundary conditions

$$\underline{n} \cdot \underline{E}^a = 0 , \qquad \underline{n} \times \underline{B}^a = 0 . \tag{1}$$

The model was first formulated by the MIT school who proposed that the two phases of the vacuum be taken to differ by an amount of energy $\mathcal{B}$ per unit volume. Good phenomenology results if the energy of the bag is taken to be given by

$$\mathcal{E} = \mathcal{B}V - \frac{Z}{R} \tag{2}$$

with V the volume of the bag and R its radius. A variant of the model, due to the Budapest school, adds to the right hand side of (2) the effect of surface tension so that

$$\mathcal{E} = \mathcal{B}V + \mathcal{E}^S S - \frac{Z}{R} \tag{3}$$

with S the surface area of the bag and $\mathcal{E}^S$ a surface tension. Fitting the data with the MIT expression (2) yields the values $Z \simeq 1.8$ and $\mathcal{B}^{\frac{1}{4}} \simeq 0.145$ GeV. The Budapest energy (3) achieves a similarly good fit with the same value of Z and a variety of pairs of values for $\mathcal{B}$ and $\mathcal{E}^S$.

The origin of the Z/R term is not well understood though a contribution to Z of approximately 0.75 is explained as a center of mass effect [2]. It has been suggested [3] that the remainder of this term represents the change in the vacuum energy due to the confinement of the field. The inspiration for this suggestion would seem to be the result obtained first by Boyer [4] and subsequently by several authors for the change in the energy of the electromagnetic vacuum due to the introduction of a perfectly conducting spherical shell. A recent calculation by Milton, deRaad and Schwinger [5] yields the value

$$\mathcal{E}_{MDS} = \frac{\hbar c}{2R}(0.09235) \quad . \tag{4}$$

The primary aim of this report is to assert that:
(i) The change in the vacuum energy occasioned by the confining surface is radically different in its effect from the simple 1/R dependence that has been suggested on the basis of (4) and
(ii) that the vacuum energy resides near the boundary. This has the effect that, for a sharp boundary, the vacuum energy possesses a geometrical expansion and may be expressed in the form

$$\mathcal{E}_{vac} = \mathcal{E}^V V + \mathcal{E}^S S + \mathcal{E}^C \int dS(\kappa_1 + \kappa_2) + \mathcal{E}^C_I \int dS(\kappa_1 - \kappa_2)^2 + \\ + \mathcal{E}^C_{II} \int dS\kappa_1\kappa_2 + \cdots \tag{5}$$

where κ_1 and κ_2 denote the principal curvatures of the surface and the coefficients $\mathcal{E}^S, \mathcal{E}^C, \cdots$ are shape tensions (the first of these is the surface tension) which are independent of the geometrical configuration of the surface. The fact that the vacuum energy decomposes naturally in this way suggests strongly that the coefficients in the geometrical expansion should be identified with the parameters of the bag model. On the basis of this identification we find

(iii) that it is inconsistent to apply confining boundary conditions for all frequencies since, as a result of high frequency effects, the bag constant and the surface tension turn out to be not only infinite but of the wrong sign and hence would be such as to render the vacuum unstable against the spontaneous breaking of Poincaré invariance.

Some comments are perhaps in order regarding the totally different appearances of expressions (4) and (5). In order to relate them we remark that for the case of pure electromagnetism, *i.e.* in the absence of interaction with say the Dirac field, the surface tension $\mathcal{E}^S$ vanishes while, for the case of thin shells, the first integral on the right hand side of (5) whose coefficient is the curvature tension $\mathcal{E}^C$ vanishes owing to a cancellation between the two sides of the shell. Furthermore the first of the integrals quadratic in the curvatures vanishes since for a sphere $\kappa_1 = \kappa_2$. This leaves us with the term whose coefficient is $\mathcal{E}^C_{II}$. For a sphere this term takes the value

$$\mathcal{E}^C_{II} \int dS\kappa_1\kappa_2 = 8\pi \; \mathcal{E}^C_{II} \tag{6}$$

independent of the radius of the sphere (this integral is in fact a topological invariant and takes the same value for any surface topologically equivalent to a sphere). The remaining terms in (5) are the terms cubic in the curvatures which, again for the case of thin shells, cancel between the inside and the outside and a term which is cutoff independent, in the limit of large cutoff, and which corresponds to the energy (4) computed by M.D.S. The energy (6) however, although independent of the radius of the sphere, is not zero and in fact depends linearly on a cutoff. Thus the vacuum energy of a perfectly conducting spherical shell differs from the value (4), which seems to have been generally accepted, by a term which is independent of the radius of the shell. This possibility was known to Boyer who was scrupulous to point out that his calculation determines the derivative of the vacuum energy only up to an additive constant. Subsequent calculations have overlooked this term for a variety of technical reasons [6].

For the conditions appropriate to the bag model, however, it transpires that all the coefficients $\mathcal{E}^S$, $\mathcal{E}^C$, $\mathcal{E}^C_I$ and $\mathcal{E}^C_{II}$ are present. The coefficients $\mathcal{E}^C_I$ and $\mathcal{E}^C_{II}$ are present just as they are in the electromagnetic case and to lowest order in the Yang-Mills coupling g their values can be inferred by multiplying the corresponding electromagnetic quantities by eight. The curvature tension is present since the situation envisioned in the bag model is that of a cavity in an infinite medium rather than the vacuum in the presence of a thin shell hence there is no cancellation between the inside and outside. Perhaps the most striking difference between the Yang-Mills field and the electromagnetic field is the presence of a non-zero surface tension $\mathcal{E}^S$. The surface tension is brought about by the self interaction of the Yang-Mills field and since it depends cubically on a cutoff it has important effects. As for the volume energy, we expect on dimensional grounds that it should vary quartically with a cutoff. It is these last two terms those associated with $\mathcal{E}^S$ and $\mathcal{E}^V$ that we shall principally be concerned with here.

The following estimate [7] may be obtained for the surface tension of the Yang-Mills field due to its self interaction

$$\mathcal{E}^S_{YM} \simeq - \frac{1}{36\pi^4}\left(11 - \frac{2}{3}N_F\right) g^2\Lambda^3 \quad \text{(confining boundary conditions)} \tag{7}$$

where N_F denotes the number of fermion species and Λ a high frequency cutoff and in deriving this estimate we have worked to leading order in the coupling g and we have assumed that the fermion masses may be neglected in comparison with Λ.

A similar estimate, which proves useful for the purpose of comparison, may be derived for the electromagnetic field for the case of perfect conductor boundary conditions we have

$$\mathcal{E}^S_{EM} \simeq - \frac{e^2\Lambda^3}{216\pi^4} \quad \text{(perfect conductor boundary conditions)} \tag{8}$$

The negative surface tension (7) would seem to indicate an instability that would lead to bag fragmentation. Equally serious is the result of computing the bag constant $\mathcal{B}$ (if the confining conditions (1) are taken literally then this is just the volume energy $\mathcal{E}^V$) from the non-linear boundary condition

$$\mathcal{B} = -\frac{1}{4} F_{\mu\nu}{}^a F^{\mu\nu a} + i\frac{\partial}{\partial x}\bar{\psi}\psi \tag{9}$$

which dictates the response of the bag wall to the gluon pressure. We may estimate $\mathcal{B}$ by taking a vacuum expectation value of this equation for the case of a plane boundary. The contribution of the second term on the right hand side of this equation is small in comparison with that of the first term if the fermion masses are small compared with the cutoff, neglecting this term we find

$$\mathcal{B}_{YM} = -\frac{\Lambda^4}{\pi^2} \quad . \tag{10}$$

The fact that $\mathcal{B}$ and $\mathcal{E}^S$ turn out to be negative indicates that the vacuum is unstable against dissolving into foam. Clearly something is seriously amiss.

Let us pursue our *reductio ad absurdum* a little further since the manner in which the interacting electromagnetic field avoids a similar instability indicates, I believe, the resolution to the difficulty. Consider for definiteness the result of taking a large box on the surface of which we impose the confining conditions (1) and subdividing the box into smaller ones by creating new surfaces on which the field is also subject to the same boundary conditions. Since the surface tension (7) is negative this process is energetically favourable. Along with the new surfaces we will create edges and corners where these surfaces intersect. This turns out to be energetically favourable also.

It is significant that the interacting electromagnetic field subject to perfect conductor boundary conditions would be subject to a similar instability. Of course charges and currents are required to enforce perfect conductor boundary conditions but these are available to the interacting field which will create electron-positron pairs from the vacuum if it is energetically favourable to do so. The point is that if the field were to create particle-antiparticle pairs in an attempt to create a perfectly conducting surface which, if possible, would be energetically advantageous then the created particles would form not a perfect conductor but rather a medium akin to an electron gas the electromagnetic properties of which is described by a *dielectric constant*. This is an important point since, as a consequence of the analyticity properties enjoyed by the dielectric constant, the *surface tension due to a dielectric boundary is always positive* thereby restoring the stability of the vacuum against partitioning. The positivity of the surface tension for a dielectric boundary derives ultimately from the fact that when account is taken of the energy of the sources required to enforce the boundary conditions the sum of the field energy proper and the energy of the sources is always positive.

Dielectric boundary conditions also resolve the difficulty associated with the sign of the bag constant at least to zeroth order in the coupling. This can be seen either by appealing to the appropriate generalization of (9) *viz*

$$\frac{1}{2}\,[\langle \underline{E}\cdot\underline{D} - \underline{B}\cdot\underline{H}\rangle] + i\,\frac{\partial}{\partial x}\,\bar{\psi}\psi = \mathcal{B} - \mathcal{E}^S(\kappa_1 + \kappa_2) + \cdots\,, \quad (11)$$

where the square bracket denotes the discontinuity of the enclosed quantity across the interface, or by the following elementary argument.

The dispersion relation for modes outside the bag is $k^2 = \mu\varepsilon\omega^2$, but since we require $\mu\varepsilon = 1$ to preserve Lorentz invariance this relation is just $k^2 = \omega^2$ which is the same as the dispersion relation inside. It follows that the volume energies inside and outside are equal and hence that

$$\mathcal{B} = \mathcal{E}^V_{in} - \mathcal{E}^V_{out} = 0$$

at least to zeroth order in the coupling. Clearly the calculation of $\mathcal{B}$ should be pursued to higher order. It is an important point, however, that the coefficients $\mathcal{B}$ and $\mathcal{E}^S$ can be calculated either from the boundary condition (10) or directly from the energy density and that the results agree. This is not the case if confining boundary conditions are employed.

In conclusion: we have shown that confining boundary conditions cannot be applied for all frequencies since otherwise the bag constant and the surface tension would be infinite and we have suggested a way in which these boundary conditions might be relaxed at high frequencies by supposing that the external vacuum can be viewed as a dielectric medium. These boundary conditions suffer from the serious deficiency that in all probability they fail to confine. It is an interesting question whether it is possible to find boundary conditions that both confine and yield physically acceptable values for the bag constant and the surface tension.

References

[1] J. Kuti and P. Hasenfratz, Phys. Rep. 40, 45 (1978).
[2] J.F. Donoghue and K. Johnson, Phys. Rev. D21, 1975 (1980).
[3] K. Johnson, in Particles and Fields - 1979 (APS/DPF Montreal). B. Margolis and D.G. Stairs, eds., AIP, New York, 1980.
[4] T.H. Boyer, Phys. Rev. 174, 1764 (1968).
[5] K.A. Milton, L.L. DeRaad, Jr., and J. Schwinger, Ann. Phys. (N.Y.) 115, 388 (1978).
[6] P. Candelas, "Vacuum energy in the presence of dielectric and conducting surfaces." To appear in Ann. Phys. (N.Y.).
[7] P. Candelas, "Vacuum energy in the bag model," Center for Theoretical Physics preprint.

Stochastic Quantization and Gribov Problem

Minoru Horibe[1], Akio Hosoya[2] and Jiro Sakamoto[3]

Faculty of Education, Fukui University[1],

Department of Physics, Osaka University[2],

Department of Physics, Shimane University[3].

§1. Introduction

In my talk I would like to report a modest progress in the Gribov problem(1) in the framework of stochastic quantization method.

According to Paris: and Wu(2), the Langevin equation for a gauge field A_μ has a form:

$$\frac{\partial A_\mu}{\partial t} + \frac{\delta S_{cl}}{\delta A_\mu} = \eta_\mu \quad ,$$

$$S_{cl} = 1/4 \int d^4x \; tr(F_{\mu\nu}^2) \, , \qquad (1)$$

where η_μ is the white noise. The advantage of the stochastic quantization will be the unnecessity of the gauge fixing. So our natural question will be: Is the stochastic quantization method with no gauge fixing equivalent to the well-established quantized gauge theory with gauge fixing ? In order to answer such a problem it is natural to consider an intermediate step: stochastic quantization with gauge fixing. The Langevin equation (1) can be deformed by t -dependent gauge transformation $U(t)$ as

$$\frac{\partial A_\mu}{\partial t} + \frac{\delta S_{cl}}{\delta A_\mu} - D_\mu V = U\eta_\mu U^{-1},$$

$$V = -i \, \partial U/\partial t \cdot U^{-1} \, . \qquad (2)$$

As far as gauge invariant quantities are concerned, Eq. (2) gives the same results as Eq. (2) does.

Baulieu and Zwanziger(3) claimed to find V such that the probability density $P(t)$ approaches the Faddeev-Popov measure:

$$P = \int dc\, d\bar{c} \; e^{-S_{tot}} \, ,$$

where $S_{tot} = S_{cl}$ + gauge fixing and Faddeev-Popov ghost terms and

c and $\bar{c}$ are the ghost fields. That is, the Faddeev-Popov measure P is a static solution of the Fokker-Planck equation associated with the Langevin equation (2), if we choose the functional V as

$$V^a(x) = -P^{-1}\int dy \int dB\, dc\, d\bar{c}\; c^a(x) \frac{\delta}{\delta A^b_\nu(y)} \left[\frac{\delta K}{\delta A^b_\nu(y)} e^{-S_{tot}} \right], \tag{3}$$

where K is such that

BRS transform of K = gauge fixing and Faddeev-Popov terms,

A popular choice of K will be

$$K = \mathrm{tr} \int dx \left\{ -\frac{\alpha}{2} B\bar{c} + F(A)\bar{c} \right\} \tag{4}$$

with F being a gauge fixing function (e.g. $F = \partial^\mu A_\mu$)[4].

§2. Singular Langevin Equation and Gribov Problem

The attention should be paid to the factor P^{-1} in front of the expression (3). As Gribov pointed out Faddeev-Popov measure has zeros. Therefore the Langevin equation has singularities in the drift force. The Brownian motion of gauge field { $A(t)$ } may or may not cross the Gribov boundary $\partial\Omega$ where $P = 0$, depending on the sign and strength of the drift force. Let us consider the time development $P[A]$ itself regarding it as a functional of the stochastic variable $A(t)$.

Namely,

$$\frac{dP[A]}{dt} = \int dx \frac{\delta P[A]}{\delta A^a_\mu} \dot{A}^a_\mu$$

$$\doteq P^{-1}\int dx \frac{\delta P[A]}{\delta A^a_\mu(x)} g^a_\mu(x) + \int dx \frac{\delta P[A]}{\delta A^a_\mu(x)} \eta^a_\mu(x) \qquad \text{near } \partial\Omega,$$

where g_μ is a regular functional: $P D_\mu V$. Defining

$$\tilde{P} = P / \sqrt{\int dx\, (\delta P/\delta A)^2} \quad \text{and} \quad \tilde{\eta} = \eta / \sqrt{\int dx\, (\delta P/\delta A)^2},$$

we obtain

$$\frac{d\tilde{P}}{dt} = \frac{c}{\tilde{P}} + \tilde{\eta}(t), \tag{5}$$

where

$$C = \int dx \frac{\delta P}{\delta A^a_\mu} P D^{ab}_\mu V^b \Big/ \int dx \left(\frac{\delta P}{\delta A^a_\mu} \right)^2 . \qquad (6)$$

Equation (5) actually gives a projection of the Brownian motion to the normal direction to the Gribov boundary. According to the theorem by Feller[5], the Brownian motion will never reach the boundary $\partial\Omega$ if the repulsive drift force is strong enough: $C \gtrsim 1$. Otherwise it will go through.

In the case of "entrance" $C \gtrsim 1$, the Langevin equation by Baulieu and Zwanziger gives the equilibrium distribution Gribov suggested; the path-integral should be limitted within the Gribov region,

$$\int_\Omega [dA]\, P[A] . \qquad (7)$$

In the other cases ($|C| < 1$, $C < -1$), we must impose the boundary condition at $\partial\Omega$ to solve the Fokker-Planck equation, which implies the non-equivalence of (1) and (2). Unfortunately we have not yet succeeded in the evaluation of C for the covariant gauge. (We obtained a trivial result $C=0$ for the axial gauge.)

References

(1) V.N. Gribov, Nucl. Phys. B139, 1, (1981).

(2) G. Parisi and Wu, Y.-S., Scientia Sinica 24, 483, (1981).

(3) L. Baulieu and D. Zwanziger, Nucl. Phys. B192, 259, (1981).

(4) T. Kugo and S. Uehara, Prog. Theor. Phys. 64, 1395, (1980).

(5) Any mathematical textbook on stochastic process.

The geometry of the configuration space of non abelian gauge theories

C.M. Viallet

Laboratoire de Physique Théorique et Hautes Energies
Université Pierre et Marie Curie
Tour 16 - 1er étage
4 place Jussieu
75230 PARIS CEDEX 05

We present here some results on the geometry of the configuration space of non abelian gauge theories [1, 2, 3]. It is on this space that a Schrödinger equation (or equivalently a path integral) is to be defined. The study of the geometry of the configuration space is a necessary step for a proper non perturbative quantization of the theory. Two basic ingredients enter our study : (1) a volume cut-off (space is compact without boundary e.g. a sphere or a torus) and (2) the theory is non abelian.

I. SOME NOTATIONS [4, 5]:

The basic objects of the theory are gauge potentials (connections) on which acts a group of gauge transformations. Let $\mathcal{E}$ be the space of connections on a principal fibre bundle $P(M,G)$ where $\mathcal{M}$ = compact metric space without boundary and G = compact semisimple group.

On $\mathcal{E}$ acts the group of gauge transformations $\mathcal{G}$. Locally

$$A_\mu \rightarrow {}^g A_\mu = g^{-1}\partial_\mu g + g^{-1} A_\mu g$$

To any ω in $\mathcal{E}$ is associated a covariant derivative ∇_ω acting on covariant objects. In a small gauge transformation $g = \exp \xi \approx 1 + \xi$, we have $\delta\omega = \nabla_\omega \xi$.

$\mathcal{E}$ is an affine space.

$\mathcal{E}$ is equipped with a gauge invariant scalar product (,) :

$$(\tau, \eta) = \int_{\mathcal{M}} dv \quad tr(\tau_\mu \eta^\mu).$$

II. LOCAL STRUCTURE OF $\mathcal{E}$:

Through a point ω we may draw the orbit formed with all gauge related points in $\mathcal{E}$. Tangent vectors to this orbit at ω will be called <u>vertical</u> at ω (all vertical vectors are of the form $\nabla_\omega \xi$). By definition, a vector at ω is said to be <u>horizontal</u> if it is perpendicular to all vertical vectors at ω. If ∇_ω^* is the adjoint of ∇_ω with respect to (,), horizontal vectors verify :

$$\nabla_\omega^* \tau = 0.$$

Generically the covariant laplacian $\Box_\omega = \nabla_\omega^* \nabla_\omega$ has trivial kernel. We denote by G_ω its inverse.

The local structure of $\mathcal{E}$ around a generic point may then be described by the splitting of the tangent space at ω into vertical and horizontal space.

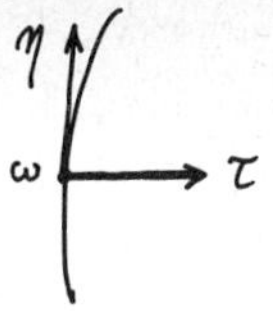

$T_\omega(\mathcal{E}) = H_\omega \oplus V_\omega$.

V_ω = space of vertical vectors at ω = tangent through the orbit through ω

H_ω = space of horizontal vectors.

For generic points, there is a projection operator $\Pi_\omega = \mathbb{1} - \nabla_\omega G_\omega \nabla_\omega^*$ (orthogonal projection on H_ω along V_ω).

III. GAUGE CONDITION

To fix the gauge is to cut all orbits once. Around a point ω_0 define the set $\mathcal{S}_0 = \{\omega \in \mathcal{E} \mid \tau = \omega - \omega_0$ is horizontal at $\omega_0\}$. For generic connections this is locally a good gauge condition [6]. We then have a local coordinate system around ω_0 for the quotient space $\mathcal{E}/\mathcal{G}$, given by the covariant background gauge condition at ω_0.

IV. ORBIT SPACE. METRIC ON THE ORBIT SPACE

Modulo certain restrictions on the connections and the gauge transformations (taking some away and imposing some regularity conditions) the quotient space $\mathcal{E}/\mathcal{G}$ is a C^∞ manifold of infinite dimension and the projection $p : \mathcal{E} \longrightarrow \mathcal{M} = \mathcal{E}/\mathcal{G}$ is a principal fibration [1, 2, 7].

Notice that the horizontality condition introduced above in $\mathcal{E}$ yields a connection in $\mathcal{E}$ with connection form $\chi_\omega = G_\omega \nabla_\omega^*$, which is not flat : we cannot construct a horizontal section of $\mathcal{E}$.

Since $\mathcal{E}$ is equipped with a gauge invariant scalar product, there is an induced scalar product g on $\mathcal{M}$ computed as follows :

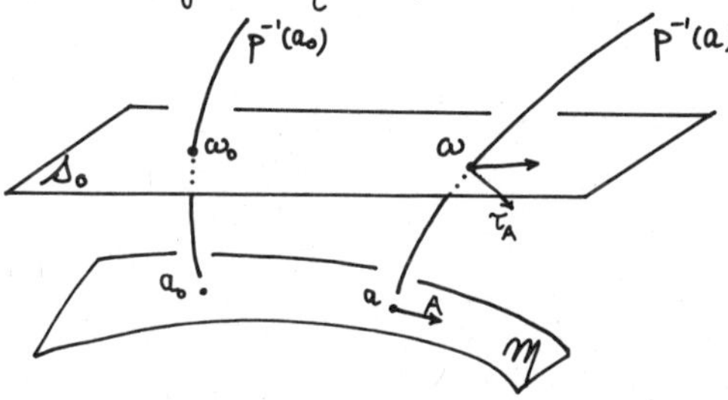

Suppose A and B are vectors tangent to $\mathcal{M}$ at a. Let ω be a point of $p^{-1}(a)$. The vectors A and B have horizontal lifts τ_A and τ_B at ω. By definition $g(A, B) = (\tau_A, \tau_B)$. In the coordinate system defined by $\omega_0 \in p^{-1}(a_0)$, (supposing a is not too far from a_0) the metric is given by $g = \Pi_0 \Pi_\omega \Pi_0$.

V. NON SINGULAR LAGRANGIAN OF GAUGE THEORIES [8, 9].

M = 3-dimensional space, e.g. S^3, T^3 ... On $P(M, G)$ are defined time dependant potentials $A(t)$. The Lagrangian of the theory is $L = \frac{1}{2}(\dot{A} - \nabla A_0, \dot{A} - \nabla A_0) - V(A)$, where $\dot{A} = \frac{\partial A_i}{\partial t} dx^i$ and $V = \frac{1}{4}\int_M dv \, \mathrm{tr}\, F_{ij}^2$ $(i,j = 1,2,3)$, $F_{ij} = \partial_i A_j - \partial_j A_i + [A_i, A_j]$.

This lagrangian is singular. The use of Dirac's analysis for singular lagrangians yields the definition of the proper configuration space, which is nothing but the orbit-space $\mathcal{E}/\mathcal{G} = \mathcal{M}$. The non singular lagrangian we get is just $\mathcal{L} = \frac{1}{2}(\pi_A \dot{A}, \pi_A \dot{A}) - \mathcal{V}$. In other words if a is a point on $\mathcal{M}$ and $\dot{a} = \frac{da}{dt}$, we have $\mathcal{L} = \frac{1}{2} g(\dot{a}, \dot{a}) - \mathcal{V}(a)$, where g is just the natural metric introduced earlier [10].

VI. METRIC AND FADDEEV-POPOV DETERMINANT

In a given system of coordinates (say the covariant background gauge condition at ω_0) we may compare det g and det γ where γ = Faddeev Popov operator = $\nabla_0^* \nabla_\omega$, and $g = \Pi_0 \Pi_\omega \Pi_0$. Formally we have [10] .

$$\sqrt{\det g} = \frac{\det \gamma}{\sqrt{\det \Box_0}\,\sqrt{\det \Box_\omega}} .$$

VII. RIEMANNIAN CALCULUS ON $\mathcal{M}$.

We may write down covariant derivative, curvature, equation for geodesics. If we use the background gauge condition, everything can be expressed in terms of simple operators.

Set $P = 1\!\!1 - \nabla_0 \gamma^{-1} \nabla_\omega^*$ (resp. $P^* = 1\!\!1 - \nabla_\omega \gamma^{-1} \nabla_0^*$)

$K_\tau : \xi \to K_\tau(\xi) = [\tau, \xi]$, and K_τ^* its adjoint

$\chi_\omega = G_\omega \nabla_\omega^*$ and χ_ω^* its adjoint.

Suppose X , Y, Z denote vector fields.

The covariant derivative is

$$D_X Z = \frac{1}{2} P^* P \left(-\chi_\omega^* K_X^* \pi_\omega Z - \Pi_\omega K_X \chi_\omega Z - \chi_\omega^* K_Z^* \pi_\omega X - \Pi_\omega K_Z \chi_\omega X + [\chi_\omega X, \pi_\omega Z] + [\chi_\omega Z, \pi_\omega X] \right).$$

The Riemannian curvature tensor $R(X,Y) = [D_X, D_Y] - D_{[X,Y]}$ is

$$R(X,Y)Z = \Pi_0 \left(-2 K_Z G_\omega K_X^*(Y) - K_Y G_\omega K_X^*(Z) + K_X G_\omega K_Y^*(Z) \right)$$

The sectional curvature in the 2-plane generated by two orthogonal vectors X, Y is $\mathcal{H}(X,Y) = g(R(X,Y)Y, X) = 3\,(K_X^*(Y), G_\omega K_X^*(Y))$.

<u>The sectional curvature is everywhere non negative</u>

Notice that any straight line in $\mathcal{E}$ which cuts one orbit perpendicularly cut all orbits it meets perpendicularly. In other words there are horizontal straight lines. They project on geodesics i.e. <u>the background gauge condition at ω_0 gives normal coordinates at ω_0</u> .

VIII. CONJUGATE POINTS

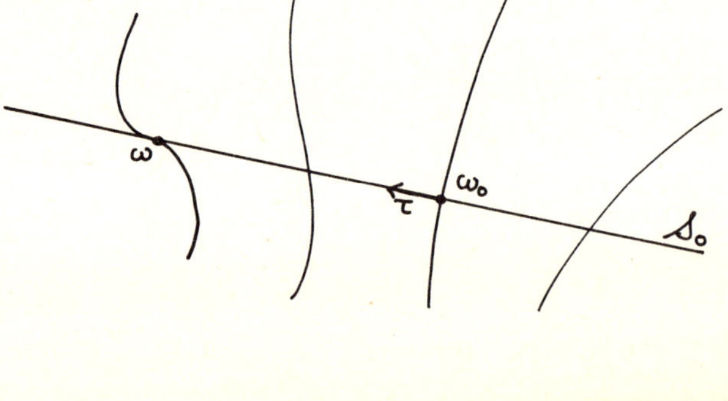

Starting from ω_0 along the horizontal line $\omega_0 + \lambda\tau$, we reach a conjugate point ω of ω_0 when some vector which is vertical at ω, verifies the gauge condition. This is equivalent to saying that there exists ξ such that $\nabla_0^* \nabla_\omega \xi = \gamma \xi = 0$ i.e. the Faddeev-Popov operator has non trivial kernel. For any ω_0 and τ, there is a finite λ for which this happens [1]. There are conjugate points at finite distance in all directions. Moreover if we consider the region Ω around ω_0 where γ is a positive operator, this region is convex and has the "Gribov horizon" as a boundary [11].

We thus get to the following conclusion : the configuration space of gauge theories is an infinite dimensional space, but the volume cut-off and the non abelian character of the theory makes it look like a "sphere" (positive curvature, possibly finite diameter). On that space is defined the potential term coming from the magnetic part of the Lagrangian. The next step toward quantization will be to write down a Schrödinger equation, which includes both, and control the removing of necessary cut-offs. Some hope is reasonable for 2 + 1 dimensions especially about about the existence of a mass gap. The 3 + 1 dimensional case is still out of reach, but in any case, the non trivial geometry of the configuration space is a saliant feature of non abelian pure gauge theory, and it will matter for any non perturbative result.

References

[1] I.M. Singer, Com. Math. Phys. 60, 7 (1978) and Physica Scripta 24, 817 (1981).
[2] M.S. Narasimhan, T.R. Ramadas, Com. Math. Phys. 67, 21 (1979).
[3] O. Babelon, C.M. Viallet, Com. Math. Phys. 81, 515 (1981).
[4] M. Daniel, C.M. Viallet, Rev. Mod. Phys. 52, 175 (1980).
[5] T. Eguchi, P.B. Gilkey, A. Hanson, Phys. Rep. 66, 6 (1980).
[6] M. Daniel, C.M. Viallet, Phys. Lett. 76B, 458 (1978).
[7] P.K. Mitter, C.M. Viallet, Com. Math. Phys. 79, 457 (1981).
[8] P.A.M. Dirac in "Lectures on Quantum Mechanics" Belfer Series (1964).
[9] L.D. Faddeev, Theor. Math. Phys. 1, 3 (1969).
[10] O. Babelon, C.M. Viallet, Phys. Lett. 85B, 246 (1979).
[11] D. Zwanziger, Preprint. NY U/TR 5/82.

LATTICE GAUGE THEORY - A PROGRESS REPORT*

Junko Shigemitsu
Department of Physics
Brown University
Providence, RI 02912, USA

and

Department of Physics
The Ohio State University
174 West 18th Avenue
Columbus, Ohio 43210, USA†

Abstract

Some recent calculations in lattice gauge theory are reviewed. These include estimates of the heavy quark potential, the hadron spectrum, and scales of chiral symmetry breaking.

1. Introduction

Lattice gauge theory[1] provides an important nonperturbative approach to quantum gauge field theories and it promises to continue serving this purpose very productively in the years to come. In particular with the advent of Monte Carlo calculations with fermions, a much wider range of problems can now be investigated.[2-9] The next couple of years should witness a wealth of new quantitative and qualitative results which will hopefully help unravel the intricate dynamics of non-abelian gauge theories

In today's talk I would like to give a progress report on some of the calculations that are currently being carried out on the lattice. Since time is limited, I will unfortunately not be able to touch upon a large part of the research in pure gauge theories. I apologize to the many physicists who have made important contributions to the field for not mentioning their work. To list just some of the topics that I will have to omit, they include:

Estimates of glueball masses

Studies of the restoration of rotational invariance

Experiments with different lattice actions

Investigations of the role of monopoles, vortices, etc.

The above work has contributed to our understanding of lattice gauge theory and its continuum limit and the topics that I will be discussing are built on the knowledge

---

*Talk presented at the International Symposium on Gauge Theory and Gravitation "g and G", Nara, Japan, August 20-24, 1982.

†Permanent address after September 1, 1982.

accumulated through such efforts.

First, I would like to describe a recent calculation of the heavy quark potential in Ref. 10 by John Stack. This will give an example of how continuum physics is extracted from the lattice theory by working in the scaling region. I will then go on to discuss Monte Carlo calculations with fermions. A major objective there is to obtain a QCD prediction for the hadron spectrum. Many groups have embarked on this ambitious program and initial results are encouraging.[2-8] However, these calculations should still be regarded as being at a preliminary stage. I do not believe that there is sufficient understanding of the approximations involved in order to be able to estimate errors reliably. Undoubtedly, future work will be devoted to gaining control over and improving on the approximations.

Fermion Monte Carlo techniques are also being used to investigate more general properties of gauge fields coupled to fermions. The spectrum calculations mentioned above indicate that chiral symmetry is broken spontaneously in non-abelian gauge theories with fermions in the fundamental representation. If one wants to go beyond the strong interactions to gauge theories of the electro-weak plus strong interactions, one must understand chiral symmetry breaking (or the lack thereof) in a more general setting. What is the scale of chiral symmetry breaking relative to the confinement scale? Is chiral symmetry breaking sensitive to the center of the gauge group? Does one see evidence for "tumbling" ideas? At the end of this talk I would like to describe some attempts to answer these questions using lattice techniques.[9]

Before turning to the specific topics, let me remind you of the basic features of lattice gauge theory. In the lattice approach one replaces continuum space-time by a hypercubic discrete lattice. The gauge degrees of freedom are unitary matrices $U(x,x+\mu)$ that reside on the links between neighboring sites x and $x+\mu$. The U-matrices can be viewed as the lattice analogues of the path ordered non-abelian phase.

$$P\cdot\exp\left[ig\int_x^{x+\mu} A_\mu d\ell\right] \to e^{iagA_\mu} \sim U(x,x+\mu) \ . \tag{1.1}$$

The matter degrees of freedom live on lattice sites. One defines, for instance, four component spinors $\psi(x)$ and $\overline{\psi}(x)$ at each site x. The lattice action splits up into two parts

$$S = S_G + S_F \ . \tag{1.2}$$

The pure gauge part S_G is the Wilson action

$$S_G = \frac{1}{g^2}\sum_p \mathrm{tr}\left\{2 - \left[U_p + U_p^\dagger\right]\right\} \tag{1.3}$$

where $\sum_p$ is the sum over unoriented plaquettes and U_p is the product of four U's on links bordering the plaquette. For the fermionic part S_F, one takes

$$S_F = m \sum_x \bar{\psi}(x)\psi(x) + \frac{1}{2} \sum_x \sum_\mu \bar{\psi}(x)\gamma_\mu \Big[U(x,x+\mu)\psi(x+\mu) - U^\dagger(x-\mu,x)\psi(x-\mu)\Big] + \frac{r}{2} \sum_x \sum_\mu \bar{\psi}(x)\Big[2\psi(x) - U(x,x+\mu)\psi(x+\mu) - U^\dagger(x-\mu,x)\psi(x-\mu)\Big] \quad (1.4)$$

Eq.(1.4) is often rewritten as:

$$S_F = \sum_x \bar{\psi}'(x)\psi'(x) + K \sum_x \sum_\mu \bar{\psi}'(x) \Big\{(\gamma_\mu - r)U(x,x+\mu)\psi'(x+\mu) - (\gamma_\mu + r)U^\dagger(x-\mu,x)\psi'(x-\mu)\Big\} \quad (1.5)$$

In going from (1.4) to (1.5) we have introduced rescaled spinors,

$$\psi'(x) \equiv \sqrt{\frac{1}{2K}}\ \psi(x) \quad (1.6)$$

and a new parameter,

$$K \equiv \frac{1}{2m+8r} \ . \quad (1.7)$$

The connection with continuum physics is established by going to the scaling region, i.e., towards a continuous phase transition point of the lattice theory. In order to make contact with an asymptotically free continuum field theory, one is interested in the critical point at $g_{crit} = 0$. As one approaches $g \to 0$ the renormalization group tells us that physical dimensional quantities such as masses M_i obey the following relationship:

$$M_i = \frac{C_i}{a}\left(b_o g^2\right)^{-b_1/2b_o^2} \exp\left[\frac{-1}{2b_o g^2}\right] \equiv C_i \Lambda_L \ . \quad (1.8)$$

In Eq.(1.8), b_o, b_1 are the coefficients of the two loop perturbative β-function and C_i is a pure number. The identity symbol defines the quantity Λ_L. Ratios between C_i's for different masses are calculable predictions of the theory. Physical distance-dependent quantities such as the interquark potential V(R) obey

$$\xi \cdot V(R) = f(R/\xi) = \text{function only of } R/\xi \quad (1.9)$$

where,

$$\xi = \text{correlation length} = (\text{typical mass})^{-1} \tag{1.10}$$

Work on the pure gauge theory has verified that Eq.(1.8) is obeyed by quantities such as the string tension σ, the glueball mass M_G, and the deconfinement temperature T_{dec}. One now believes that the scaling region extends to relatively large values of g^2, namely

$$\begin{aligned} \beta &\equiv 4/g^2 \gtrsim 2.2 \qquad SU(2) \\ \beta &\equiv 6/g^2 \gtrsim 5.5 \qquad SU(3) \end{aligned} \quad . \tag{1.11}$$

2. The Heavy Quark Potential

An important quantity that can be calculated in the quarkless theory is the static potential between two external color sources. I would like to show you a recent result obtained in Ref. 10 for the gauge group SU(2) (the discrete 120 element icosahedral subgroup was actually used, but this should provide an excellent approximation to the full SU(2) group for the values of the coupling constant that were involved).[12,13] Using previous calculations of the string tension σ to fix the scale,[11-13] the continuum heavy quark potential was obtained up to an overall additive constant (other calculations of $V(R)$ have been reported in Ref. 14).

The static potential can be extracted from an evaluation of the Wilson loop around a rectangular contour Γ of extension $R \times t$.

$$W(\Gamma) = \langle \mathrm{tr}\, P \exp\left[ig \oint_\Gamma A\, d\ell\right]\rangle \to \langle \mathrm{tr}(UU...U)_\Gamma\rangle \quad . \tag{2.1}$$

For $t \gg R$ one has

$$W(\Gamma) \sim \exp\left[-t(V_o + V(R))\right] \quad . \tag{2.2}$$

V_o represents the self-energy of the color sources. In Ref. 10, Wilson loops were evaluated for $R = a, 2a, 3a$ and $4a$ at several β-values, $2.2 \le \beta \le 3.1$. A correlation length ξ was then introduced.

$$\xi \equiv .012\, \Lambda_L^{-1} = .012\, a \left[b_o g^2\right]^{b_1/2b_o^2} \exp\left[\frac{1}{2b_o g^2}\right] \quad . \tag{2.3}$$

The coefficient .012 was chosen so that ξ is given by $\xi = 1/\sqrt{\sigma}$, if one uses the string tension σ measured in Refs. 11, 12, and 13. Although R/a takes on only four values by combining data at different β, one can work at many more values of

$X \equiv R/\xi$. As mentioned before, in the scaling region all the data points for $\xi \cdot V(R)$ versus R/ξ should fall on a single curve. The result is shown in Fig. 1 and one sees that scaling holds quite well.

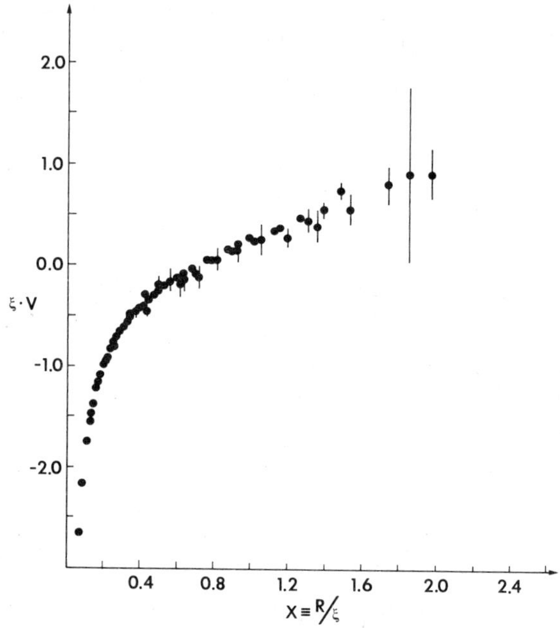

Fig. 1. $\xi \cdot V(R)$ versus $X \equiv R/\xi$. ξ is the correlation length defined in Eq.(2.3).

In order to obtain Fig. 1, one had to subtract $V_o(\beta)$ for each β separately. I refer to the original paper for details on how this was accomplished. Fig. 1 still contains a single overall additive constant that remains undetermined.

Although Fig. 1 represents the heavy quark potential for SU(2) and not for SU(3), it is still tempting to try to compare with phenomenological charmonium potentials. I have made a rough comparison in Fig. 2. ξ^{-1} (or $\sqrt{\sigma}$) has been set to 400 MeV. This corresponds to renormalizing the theory such that the $R \gg 1$ behavior of $V(R)$ reproduces the experimentally measured Regge slope. Once the scale has been set one can express R and $V(R)$ in physical units, fm and GeV respectively. The full curve in Fig. 2 is a phenomenological potential taken from Ref. 15 multiplied by 9/16, the ratio of the fundamental representation quadratic Casimirs of SU(2) and SU(3). This factor 9/16 will hopefully take into account the bulk of the changes necessary in going from SU(3) to SU(2). Keeping in mind that the Monte Carlo results can be shifted by an overall additive constant, one sees that the agreement between phenomenology and quarkless QCD is fairly good.

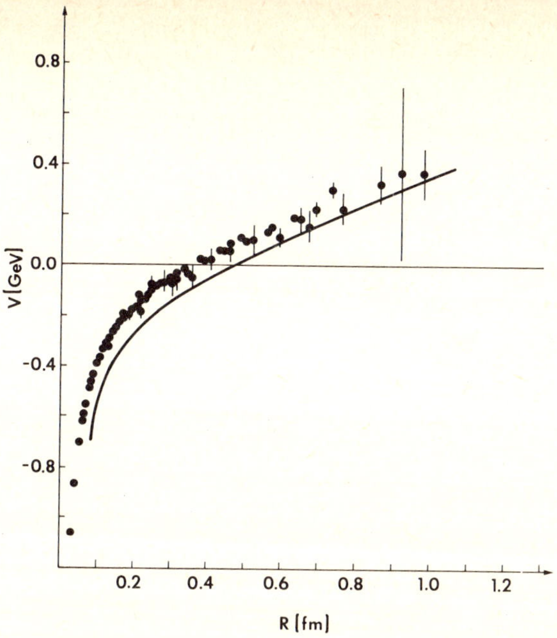

Fig. 2. Same as Fig. 1 with ξ^{-1} set to 400 MeV. The full curve is a phenomenological potential from Ref. 15 converted approximately from SU(3) to SU(2).

3. Lattice Fermions and Spectrum Calculations

Recent advances in lattice gauge theory have led to the first round of Monte Carlo spectrum calculations for the full non-abelian gauge theory with quarks.[2-8]

One must now work with the full action $S_G + S_F$. In order to be able to perform Monte Carlo calculations, one formally integrates out the fermionic variables and ends up with an effective action,

$$S_{eff} = S_G - \mathrm{tr}\, \ell n(\Delta[U]) \quad . \tag{3.1}$$

$\Delta[U]$ is the lattice Euclidean Dirac operator defined in Eq.(1.4) or (1.5). In most of the 4D fermion Monte Carlo calculations to date (except for Refs. 6 and 7) one has ignored the second term in (3.1), namely one sets $\det(\Delta[U]) \rightarrow 1$. This amounts to neglecting virtual quark loops. In the loopless approximation (also called the quenched approximation) there is no feedback from the fermions on the gauge degrees of freedom. One can first evaluate the quark propagator $G(x,U)$ in a fixed background gauge field configuration. One then builds physical, gauge invariant quantities such as meson or baryon propagators and averages over U with weight $dU\, e^{-S_G}$.

These calculations require in addition to a pure gauge Monte Carlo also efficient numerical matrix inversion methods to invert $\Delta[U]$ and obtain $G(x,U)$. Most calculations in 4D gauge theories have utilized the Gauss-Seidel method although other methods have also been proposed.[16,17] Hadron masses are extracted by writing the meson or baryon propagators $D(x)$ as

$$D(x)\Big|_{\substack{\text{averaged over}\\ \text{spatial positions}}} = \sum_n A_n e^{-m_n t} . \tag{3.2}$$

One then hopes that it is possible to go to sufficiently large t so that only the lightest state contributes to (3.2). Periodic boundary conditions actually restrict the allowed values of t to be $\leqslant N_t/2$, where N_t is the number of lattice sites in the temporal direction.

As mentioned in the Introduction, the first estimates of the hadron spectrum are in reasonable agreement with experiment. Considering that these are first principles calculations of the spectrum of a very complicated theory, this is very encouraging news. On the other hand people are still struggling to understand and bring down the error bars. I will not attempt to show you table after table of hadron masses as they are quoted in the many references. Different groups have worked with different lattice sizes, at different values of β and used different fermion methods. They have also averaged over different numbers of U configurations and have different criteria for extrapolating to the physical values of m or K (it turns out one can never work exactly at K_{phys} or m_{phys} since the matrix inversion methods break down there). In order to be able to compare various references one must first understand the dependence of the final result on the many variables in the calculation. The next round of fermion Monte Carlo spectrum calculations must answer the following questions:

1. How do mass estimates change when one varies N_s and/or N_t?
2. Has one averaged over a sufficient number of independent gauge field configurations?
2. Do hadron masses scale (i.e., obey Eq.(1.8))?
3. Do we understand how to extrapolate to physical values of m or K?
4. Are there matrix inversion methods that enable us to work closer to m_{phys} or K_{phys}?
5. How do virtual fermion loops affect the final results?

Most of the points raised above are technical ones. There are also more basic questions concerning lattice fermions that still need to be checked. Just as one must verify scaling and try to test the restoration of Lorentz (rotational) invariance in as many quantities as possible, a sensible continuum limit of theories with fermions must exhibit all continuous chiral symmetries for $m = 0$.

It is well known that there are many difficulties with lattice fermions and chiral symmetry. Let me discuss separately the cases $r = 0$ and $r \neq 0$.

For $r = 0$, S_F of Eq.(1.4) becomes the "naive" lattice fermion action. It describes 16 species (flavors) of Dirac fermions. For finite lattice spacing, the $m = 0$ theory does not have the full $U(16) \times U(16)$ flavor symmetry of the continuum model. It has only a subset of the continuous axial and vector symmetries. The remaining continuous symmetries are supposed to be restored in the $a \to 0$ limit. However even at finite lattice spacing, there are enough discrete symmetries to prevent quark bilinear counter terms from developing. This is the main advantage of the $r = 0$ formulation.

One can actually reduce the number of flavors from 16 to 4 by spin diagonalizing the fermionic action. A convenient way to achieve this is to perform the following canonical transformation.[18]

$$\begin{aligned} \psi(x) &\to T(x)\,\psi(x) \equiv \chi(x) \\ \bar{\psi}(x) &\to \bar{\psi}(x)\,T^{\dagger}(x) \equiv \bar{\chi}(x) \end{aligned} \tag{3.3}$$

where

$$T(x) \equiv \gamma_0^{x_0}\gamma_1^{x_1}\gamma_2^{x_2}\gamma_3^{x_3} . \tag{3.4}$$

Then

$$S_F = m\sum_x \bar{\chi}(x)\,\chi(x) + \frac{1}{2}\sum_x\sum_\mu \bar{\chi}(x)\,\eta_\mu(x)(U(x,x+\mu)\,\chi(x+\mu) - U^{\dagger}(x-\mu)\,\chi(x-\mu)) \tag{3.5}$$

where

$$\eta_0(x) = 1,\quad \eta_1(x) = (-1)^{x_0},\quad \eta_2(x) = (-1)^{x_0+x_1},\quad \eta_3(x) = (-1)^{x_0+x_1+x_2} . \tag{3.6}$$

Although the $\chi(x)$'s start out as four component spinors, Eq.(3.5) shows that the different components decouple and one can work with single component fermionic variables. This leads to the "staggered" fermion method.

The Euclidean staggered fermion action describes 4 flavors. At finite lattice spacing the $m = 0$ theory has only one continuous axial symmetry (nonsinglet) in addition to the $U(1)$ vector symmetry. The full $U(4) \times U(4)$ symmetry is restored only in the $a \to 0$ limit (up to anomalies). Again discrete symmetries prevent mass counter terms from developing.

The staggered fermion method has been used to evaluate $\langle\bar{\psi}\psi\rangle_{m=0}$ in the quenched approximation.[2,3,9] There were clear indications that $\langle\bar{\psi}\psi\rangle \neq 0$ both at strong coupling and into the scaling region. Since staggered fermions have one continuous symmetry that is spontaneously broken by $\langle\bar{\psi}\psi\rangle \neq 0$, one should find a Goldstone boson even at strong coupling. This particle will be one out of the 15 pions expected in the 4 flavor continuum theory. This massless state has been observed in Refs. 2, 3 and 7. The real challenge now is to show that the remaining 14 pions also become massless as one approaches the continuum limit. This will verify that the full $SU(4) \times SU(4) \times U(1)$ flavor symmetry is being restored, realized however in the Nambu-Goldstone mode. Finally at the same time that the 14 pions are becoming massless, one would like to observe that the flavor singlet meson, the η', remains massive. In order to achieve this however one must put back quark annihilation graphs. Although both the naive and the staggered Euclidean fermions are believed to have the correct $U(1)$ anomaly in weak coupling perturbation theory,[19] it is not known how easily the anomaly will show up in a Monte Carlo calculation.

For $r \neq 0$, one obtains the Wilson fermion formulation which has been used in most spectrum calculations.[2,4,5,6] The moment r is nonzero, 15 of the 16 species encountered in the naive fermion method acquire large masses as $a \to 0$. The Wilson formulation avoids the "doubling" problem. However the term in S_F proportional to r breaks chiral symmetry completely. There is no symmetry to prevent mass terms from developing. With this method one must fix the parameter K of Eq.(1.7) by some prescription before being able to extract hadron masses. One has traditionally fixed K so that the pion mass comes out right. In particular if one adjusts K to K_c such that the pion becomes massless, one argues that the theory has been finetuned to its chirally symmetric point. I believe it is important to devise as many independent tests as possible that one is indeed dealing with a chirally symmetric theory at $K = K_c$.

The Wilson method has also been shown to have the correct $U(1)$ anomaly.[20] There are in principle no obstacles in obtaining a correct pion-eta splitting once annihilation graphs have been taken into account. Some work in this direction has already been carried out in Ref. 2.

4. Scales of Chiral Symmetry Breaking

As the last topic, I would like to describe some further studies of chiral symmetry breaking using lattice techniques. My collaborators, J. Kogut, Steve Shenker, D. Sinclair, M. Stone, W. Wyld and I are investigating chiral properties of theories with quarks in different representations of the gauge group. We hope that such studies will shed new light on the mechanism responsible for chiral symmetry breaking.

According to the picture that we have of chiral symmetry breaking, the vacuum of QCD should be unstable with respect to the formation of quark-antiquark pair condensates. Once a condensate has formed chirality of the vacuum becomes indefinite and it is possible to have nonzero vacuum expectation values of operators such as $\bar{\psi}\psi$. This picture tells us that in order for chiral symmetry breaking to occur, at the very least pairs must bind. Attractive, maybe relatively strong, binding forces must exist. It is, however, not at all clear whether long range confining forces are necessary. In fact, initial models in 4D of chiral symmetry breaking were theories with very short range interactions.[21] If an effective four fermion interaction theory is ever a good guide to what goes on in QCD, maybe chiral symmetry breaking is insensitive to the long distance behavior of the theory.

The question raised above can be studied by considering theories with quarks in N-ality zero representations (e.g., the adjoint representation). Such quarks do not experience a confining force at large distances since they can be screened by gluons. Consequently if chiral symmetry breaking is sensitive to the force law at large distances one would expect the chiral properties of theories with adjoint fermions to be very different from in models with fundamental representation fermions. It could also be that adjoint quarks prefer to bind with glue degrees of freedom rather than form pairs.

My collaborators and I have performed Monte Carlo evaluations of $\langle\bar{\psi}\psi\rangle_\ell$ for several representations ℓ of SU(2). We have used the staggered fermion formulation (Eq.(3.5)) and work within the quenched approximation described in the previous section. We find that quarks in screenable representations also lead to chiral symmetry breaking. We conclude from this that chiral symmetry breaking is independent of confinement and it occurs in general, (depending on the representation) at shorter distances than confinement.

Having established that chiral symmetry breaking is not associated with confinement, one might ask the following questions: Is it possible to introduce disparate length scales into a theory with a single gauge group? Does one see evidence for "tumbling"?[22]

The authors of Ref. 22 have pointed out that in an asymptotically free theory one can obtain a large hierarchy in length scales by allowing different types of condensates to form sequentially. We have attempted to use fermion Monte Carlo calculations to estimate the relevant scales of chiral symmetry breaking for different representations. We find that indeed a hierarchy of scales emerges. Let me be a little bit more specific about what is meant by the "relevant scale of chiral symmetry breaking".

Consider a system at nonzero real temperature T. As T increases one expects all symmetries that are spontaneously broken at zero temperature to eventually be restored. This should also be true of chiral symmetry breaking (the simplest picture of such a phase transition would have the pairs in the condensate unbind as the system becomes too hot). Thus if one observes the order parameter $\langle\bar{\psi}\psi\rangle$ at different temperatures one should see,

$$\langle\bar{\psi}\psi\rangle \begin{cases} \neq 0 & T < T_c \\ = 0 & T > T_c \end{cases} . \tag{4.1}$$

The critical temperature T_c sets the relevant scale for chiral symmetry breaking. We find that if one compares two representations such that their quadratic Casimirs $C_2(\ell)$ obey

$$C_2(\ell_2) > C_2(\ell_1) \tag{4.2a}$$

then

$$T_c(\ell_2) > T_c(\ell_1) \tag{4.2b}$$

which is consistent with tumbling ideas.

It is also interesting to compare T_c for the fundamental representation with the deconfining temperature T_{dec} of the quarkless theory.[23-26] One finds

$$T_{dec} \leqslant T_c(\text{fundamental}) \ . \tag{4.3}$$

In Fig. 3 we compare our curve for T_c(fundamental), denoted T_F, with a curve for T_{dec} which is taken from Ref. 23. Since temperature is a physical quantity with dimension of mass, it obeys the renormalization group relation Eq.(1.8) once one has entered the scaling region $(4/g^2 \gtrsim 2.2)$.

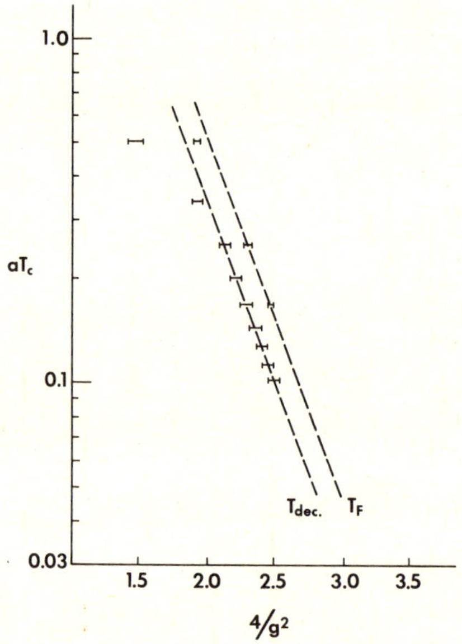

Fig. 3. Comparison of the chiral symmetry restoration temperature T_F for fundamental quarks with the deconfining temperature T_{dec}. T_{dec} is taken from Ref. 23.

From Fig. 3 one reads off the continuum result

$$T_F/T_{dec} \sim 1.6 \pm .2 \ . \tag{4.4}$$

(Other estimates of T_{dec} [24-26] tend to place it slightly higher than in Ref. 23, so T_{dec} and T_F could be closer than indicated by Eq.(4.4).)

We also have data for the representation $\ell = 1$ (adjoint), $\ell = \frac{3}{2}$ and $\ell = 2$ and they are consistent with Eq.(4.2) with large ratios between the T_c's. We are currently working very hard to nail down T_c more precisely for the adjoint representation. We are finding that T_A/T_F is at least 7 or larger. However, the error bars are still too big and uncertain to enable us to quote a reliable number. Needless to say, many of the reservations and difficulties with fermion Monte Carlo calculations that were mentioned in the previous section also apply here. We are trying to understand possible sources of systematic errors in our calculations. For instance previously when we had less data on different sized lattices, our analysis gave $T_A/T_F \sim 56$. We now believe that this number will be reduced when finite size effects are properly taken into account. In a forthcoming paper we will give a detailed discussion of finite size effects and other aspects of our calculations. We will also report on results for SU(3) and U(1) gauge theories.

5. Concluding Remarks

I have tried to give you a flavor of some current work on the lattice. Most of the projects are very time consuming (and often also very CPU time consuming) long term endeavors. We may still have some ways to go before we can be truly satisfied with the accuracy of the results. However it is gratifying that more and more questions are starting to lend themselves to the lattice analysis and that this nonperturbative method is producing a continuous flow of new results and ideas.

Acknowledgements

The author is supported by a grant from the U.S. Department of Energy, DE-AC02-76ER 03130.A009 Task A-Theoretical. She thanks her collaborators, J. Kogut, Steve Shenker, D. Sinclair, M. Stone and W. Wyld, for sharing their insights about lattice gauge theory with her. Many thanks are also due to the organizers of "g and G" for their warm hospitality at Nara.

References

1. K.G. Wilson, Phys. Rev. D 10, 2445 (1975); Erice Lectures (1975).
2. H. Hamber and G. Parisi, Phys. Rev. Lett. 47, 1792 (1981); Brookhaven preprint, May (1982).
3. E. Marinari, G. Parisi and C. Rebbi, Phys. Rev. Lett. 47, 1795 (1981).
4. D. Weingarten and D. Petcher, Phys. Lett. 99B, 333 (1981).
5. D. Weingarten, Phys. Lett. 109B, 57 (1982); Indiana preprint IUHET-82, June (1982).
6. A. Hasenfratz, P. Hasenfratz, Z. Kunszt and C.B. Lang, Phys. Lett. 110B, 289 (1982).
7. A. Duncan, R. Roskies and H. Vaidya, Pittsburgh preprint, May (1982).
8. F. Fucito, G. Martinelli, C. Omero, G. Parisi, R. Petronzio and F. Rapuano, CERN preprint, TH.3288-CERN, April (1982).
9. J. Kogut, M. Stone, H.W. Wyld, J. Shigemitsu, S.H. Shenker and D.K. Sinclair, Phys. Rev. Lett. 48, 1140 (1982).
10. J.D. Stack, U. of Illinois preprint, ILL-(TH)-82-21, June (1982).

11. M. Creutz, Phys. Rev. Lett. 45, 313 (1980); Phys. Rev. D21, 2308 (1980).
12. D. Petcher and D.H. Weingarten, Phys. Rev. D22, 2465 (1980).
13. G. Bhanot and C. Rebbi, Nucl. Phys. B180, 469 (1981).
14. E. Kovacs, Rockefeller U. preprint, RU82/B/18 (1982).
15. K. Gottfried, Cornell U. preprint, CLNS 80/467.
16. J. Kuti, Phys. Rev. Lett. 49, 183 (1982).
17. F. Fucito, E. Marinari, G. Parisi and C. Rebbi, Nucl. Phys. B180 [FS2], 369 (1981).
18. N. Kawamoto and J. Smit, Nucl. Phys. B192, 100 (1981).
19. H.S. Sharatchandra, H.J. Thun and P. Weisz, Nucl. Phys. B192, 205 (1981).
20. L.H. Karsten and J. Smit, Nucl. Phys. B183, 103 (1981).
21. Y. Nambu and G. Jona-Lasinio, Phys. Rev. 122, 345 (1961).
22. S. Raby, S. Dimopoulos and L. Susskind, Nucl. Phys. B169, 373 (1980).
23. J. Kuti, J. Polonyi and K. Szlachany, Phys. Lett. 98B, 199 (1981).
24. L. McLerran and B. Svetitsky, Phys. Lett. 98B, 195 (1981).
25. J. Engels, F. Karsch, H. Satz and I. Montvay, Phys. Lett. 101B, 89 (1981); University of Bielefeld preprint BI-TP 81/29, Dec. (1981).
26. F. Green, Northeastern U. preprint, NUB #2556, May (1982).

Recent Developments in the Theory of Large N Gauge Fields

Tohru Eguchi and Hikaru Kawai

Department of Physics
University of Tokyo
Bunkyo-ku, Tokyo 113

It has been known for some time that quantum field theories with global or local U(N) (SU(N), O(N),,,) symmetry greatly simplify in the limit $N\to\infty$. The well-known example is the so-called O(N) vector model which is a theory of massless scalar fields $\phi_i(i=1,,N)$ taking values on S^{N-1}. In two dimensions lagrangian is given by

$$\mathcal{L} = \int \frac{1}{2} \sum_{i=1}^{N} (\partial_\mu \phi_i)^2 d^2x, \qquad \sum_{i=1}^{N} \phi_i^{\,2} = 1. \tag{1}$$

It is known that the interaction caused by the constraint $\sum\phi_i^{\,2}=1$ creates a finite mass gap in this system ($N \geq 3$). In fact the model becomes exactly soluble in the large N limit and becomes a free theory of massive scalar particles. When N is finite, these massive fields begin to interact weakly with the strength 1/N. Thus the $N = \infty$ limit yields an exact solution which gives a qualitatively correct description of the system also for finite N.

In the case of gauge theories it is also known that a considerable simplification takes place in the limit of large gauge group. Both in the continuum and the lattice formulation of gauge theory the following characteristic properties have been known of the $N = \infty$ gauge fields.

1. Dominance of planar Feynman diagrams[1] (planar surfaces) in the weak coupling (strong coupling) perturbation theory;
2. The factorization property of Wilson loop amplitudes.[2]

These properties are derived from a power-counting analysis of Feynman (strong coupling) diagrams. In the case of weak coupling perturbation theory a simple combinatorial analysis shows that the weight of a Feynman diagram becomes N^χ where χ is the Euler characteristic of the diagram (interaction vertices, propagators and color loops are regarded as vertices, edges and faces of a polyhedron, respectively) when we let g^2N to be independent of N (or let $g^2=O(1/N)$). Thus in the limit of $N = \infty$ with g^2N fixed graphs with the heighest Euler number give the dominant contributions. In the case of a Wilson loop amplitude, for instance,

$$\langle \exp i \int_C A_\mu dx_\mu \rangle \approx O(N) \qquad (2)$$

the leading contributions come from the planar graphs with the topology of a disc while non-planar graphs with h handles are down by N^{-2h}.

Moreover, when we consider the case of more than one quark loop, say C_1 and C_2, the leading contribution to the correlation function

$$\langle \exp i \int_{C_1} A_\mu dx_\mu \; \exp i \int_{C_2} A_\mu dx_\mu \rangle \qquad (3)$$

come from the disconnected piece,

$$\langle \exp i \int_{C_1} A_\mu dx_\mu \rangle \langle \exp i \int_{C_2} A_\mu dx_\mu \rangle \approx O(N^2). \qquad (4)$$

This is because when there exist gluon exchanges between C_1 and C_2 we obtain a topology of an annulus and hence of order only $O(N^0)$.

In this way in the large N limit with g^2N fixed there exist no correlation between quark loops and the amplitudes always factor

$$\langle \prod_i \exp i \int_{C_i} A_\mu dx_\mu \rangle = \prod_i \langle \exp i \int_{C_i} A_\mu dx_\mu \rangle \qquad (5)$$

Since a Wilson loop amplitude may be interpreted as a meson propagator, factorization implies that at $N = \infty$ gauge interactions are exhausted to form bound states and mesons do not scatter from each other. In this respect $N = \infty$ gauge theory is analogous to O(N) vector model where interactions are also exhausted at $N = \infty$ in creating a dynamical mass to scalar fields. The dominance of planar surfaces and factorization are also shown order by order in the strong coupling lattice perturbation theory.

Reduction

Now the recent developments on the theory of large N gauge and spin system[3,4,5,6] have uncovered a further property of N = ∞ gauge fields;

3. Reduction of dynamical degrees of freedom.

By means of reduction one may replace the N = ∞ gauge field theory by a much simpler system, a model with only a finite number (=space-time dimensionality) of U(N) (or N×N hermitian) matrices, without loosing information of the original theory. This is a remarkable result in the sense that a quantum field theory may be reduced to a kind of dynamical system with a finite number of dynamical variables.
In the lattice formulation of gauge fields the argument goes as follows[3]; we start from the standard Wilson theory defined by the partition function

$$Z = \prod_y \prod_{\rho=1}^{d} dU_{y,\rho} \exp\{\beta \sum_y \sum_{\rho\neq\sigma=1}^{d} \mathrm{tr}\, U_{y,\rho} U_{y+\rho,\sigma} U^{+}_{y+\sigma,\rho} U^{+}_{y,\sigma}\} \tag{6}$$

where $U_{y,\rho}$ is an U(N) matrix lying on a link connecting the lattice sites y and y+ρ (ρ is the unit vector in the ρ-direction) and d is the dimensionality of space-time.
The Wilson loop amplitude is defined by

$$W(C) = \langle \mathrm{tr}\, U_{x,\mu} U_{x+\mu,\nu} U_{x+\mu+\nu,\lambda} \cdots U_{x-\sigma,\sigma} \rangle \tag{7}$$

for a contour C which connects lattice sites x,x+μ,x+μ+ν,x+μ+ν+λ,,,x-σ, x successively. We then reduce the model by identifying all link variables in the same direction

$$U_{y,\rho} \Longrightarrow U_\rho \tag{8}$$

and thereby shrink the entire space-time lattice to a single hypercube. The reduced model is then defined in terms of matrices $U_1, U_2,,, \cdot\cdot, U_d$ and its action is given by

$$S_r = \sum_{\rho\neq\sigma=1}^{d} \mathrm{tr}(U_\rho U_\sigma U^{+}_\rho U^{+}_\sigma). \tag{9}$$

The analogue of the Wilson loop amplitude is defined by

$$W_r(C) = \langle \mathrm{tr}(U_\mu U_\nu U_\lambda \cdots U_\sigma) \rangle_r \tag{10}$$

where the averaging is taken with respect to the weight Eq.(9) and we have identified the contour C with the sequence of directions ($\mu,\nu,\lambda,,,\sigma$)

$$C \; ; \; (x, x+\mu, x+\mu+\nu, x+\mu+\nu+\lambda, \cdots, x-\sigma, x) \approx (\mu,\nu,\lambda,\cdots,\sigma). \qquad (11)$$

The above correspondence is one to one when we ignore over-all translations.

We remark that the reduced model Eq.(9) is ivariant under the phase transformation

$$U_\rho \longrightarrow e^{i\theta}\, U_\rho \qquad (12)$$

and this symmetry implies

$$W_r(C) = \langle \mathrm{tr}\, U_\mu U_\nu U_\lambda \cdots U_\sigma \rangle = 0 \qquad (13)$$

for every open contour C. This is because in the case of an open contour there exists at least one direction ρ for which U_ρ and U_ρ^+ appear different number of times and $W_r(C)$ has to vanish. It was noticed[3] that if the symmetry Eq.(12) is left intact, i.e. not spontaneously broken, the equation of motion for the Wilson loop amplitudes in the original and reduced models become identical in the limit $N = \infty$ with g^2N fixed. Consequently the Wilson loop amplitudes agree

$$W(C) = W_r(C). \qquad (14)$$

Also the free-energy per unit volume of the original theory agrees with the free-energy of the reduced model. Thus the infinite-volume Wilson theory Eq.(6) and the one-site reduced model Eq.(9) become equivalent to each other in the large N limit.

In order to elucidate the physical meaning of the reduction let us consider a "partial reduction" of $N = \infty$ gauge field where we shrink the size of one particular direction, say, the time-direction, of the d-dimensional lattice to a unit distance while keeping the size of the other directions unchanged. We compare the free-energy of the original and partially reduced systems,

$$\begin{aligned} Z &= \mathrm{Tr}\, T^L = e^{-LF} \\ Z_r &= \mathrm{Tr}\, T = e^{-F_r} \end{aligned} \qquad (15)$$

where we have introduced the transfer matrix and L is the original size of the time-direction. Making use of eigenstates of the transfer matrix with eigenvalues E_i $(i=0,1,2,..)$ we obtain

$$F = E_0 ,$$
$$F_r = E_0 - \ln\left(1 + \sum_i e^{-(E_i - E_0)}\right). \qquad (16)$$

If the system confines, there exist only color-singlet excitations and the sum over i will not generate N dependent factors. Thus

$$F_r = E_0 + O(N^0). \qquad (17)$$

On the other hand we know that the vacuum energy E_0 is $O(N^2)$ and consequently

$$F_r = F, \qquad N = \infty. \qquad (18)$$

Therefore the free-energy of the $N = \infty$ gauge field is independent of the lattice size and we may reduce the theory so far as the system confines.

U(1) symmetry

In the partial reduction of the lattice we have squeezed the time-direction to a unit distance. This creates a finite-temperature situation since the periodic boundary condition is imposed in the reduced model. Thus the reduction effectively heats up the gauge system. It is known that gauge fields, when heated, will undergo a deconfining transition into a plasma phase at a certain critical temperature T_c via the spontaneous violation of the invariance under the center of the gauge group.[7,8] Therefore if this transition temperature stays finite at $N = \infty$ gauge fields, we would expect that there exists a critical coupling λ_c in the reduced model such that for λ $(= N/\beta) > \lambda_c$ U(1) (the center of U(N)) symmetry Eq.(12) is unbroken, however, it will break spontaneously below λ_c (T_c and λ_c are related as $1/T_c = a(\lambda_c)$ where a is the lattice constant dependent upon the coupling strength λ). Monte Carlo simulation of the reduced model[4,9] in fact shows the spontaneous violation of U(1) symmetry below $\lambda_c \simeq 2$. This is signaled by the non-zero expectation value for open loops. Below the transition point the eigenvalues of the reduced link variables U_ρ's are no longer

uniformly distributed but are concentrated around an arbitrary point on the unit circle. The transition is smooth and appears to be 2nd order. This is consistent with our interpretation of its being the finite-temperature deconfining transition.
Thus the original reduced model is equivalent to the standard Wilson theory only in the strong coupling regime $\lambda > \lambda_c$ and the equivalence will be lost below λ_c.

Quenching
In this situation it is possible to restore the broken symmetry by integrating over the location of the concentration of the eigenvalues. In the quenching procedure of Bhanot-Heller-Neuberger link variables are diagonalized as

$$U_\rho = V_\rho D_\rho V_\rho^+ \qquad (D_\rho)_{ij} = e^{i\theta_\rho^j} \delta_{ij}. \tag{19}$$

The angular variables θ_ρ^j's are held fixed when we first average over marices V_ρ's

$$W(C;\theta) = \frac{\int \prod_\rho dV_\rho \; \mathrm{tr}(V_\mu D_\mu V_\mu^+ V_\nu D_\nu V_\nu^+ \cdots V_\sigma D_\sigma V_\sigma^+) \; e^{+\beta S(V,\theta)}}{\int \prod_\rho dV_\rho \; e^{+\beta S(V,\theta)}} . \tag{20}$$

We then take an averaging over θ's

$$W_q(C) = \int d\mu(\theta) \quad W(C;\theta) \tag{21}$$

with a suitable measure μ. Using the method of Parisi[5] it was shown[6] that when one makes a change of variable

$$V_\mu D_\mu V_\mu^+ = D_\mu W_\mu . \tag{22}$$

and expands

$$W_\mu = \exp iag A_\mu \tag{23}$$

in terms of g, the quenched model reproduces the planar perturbation theory of the continuum gauge fields. Here the eigenvalue $\theta_\mu^i - \theta_\mu^j$ has the meaning of the momentum carried by a gluon line with a pair of

color indices (i,j).
Thus the quenched model reproduces the weak coupling expansion of the continuum gauge theory for small λ while it agrees with the original reduced model for large λ. Hence it is now believed to be equivalent to the standard theory throughout the range of coupling constants. Here the possible trouble is that the quenching procedure is justified only with recourse to the weak coupling perturbation theory and it is not completely clear if the non-perturbative information of the theory is coded correctly into the quenched model.

Twist

Another promising method of avoiding the problem of the degeneracy of the eigenvalues of U_ρ is to introduce a twist to the system.[10] For instance, we introduce a phase factor $e^{i2\pi/N}$ into the action. Then the minimum energy configuration for a plaquette

$$e^{i2\pi/N}\,\mathrm{tr}(U_\rho U_\sigma U_\rho^+ U_\sigma^+) + \mathrm{h.c.} \tag{24}$$

is no longer given by $U_\rho = U_\sigma = 1$ but by

$$\begin{aligned} &U_\rho = P, \quad U_\sigma = Q \\ &PQ = QP\, e^{i2\pi/N} \end{aligned} \tag{25}$$

where P, Q are matrices of 't Hooft[11]

$$P = \begin{pmatrix} 0 & 1 & & \\ & \ddots & \ddots & \\ & & \ddots & 1 \\ 1 & & & 0 \end{pmatrix}, \qquad Q = \begin{pmatrix} 1 & & & 0 \\ & e^{2\pi i/N} & & \\ & & \ddots & \\ 0 & & & e^{2\pi i(N-1)/N} \end{pmatrix} \tag{26}$$

Eigenvalues of P and Q are uniformly distributed on the circle and thus U(1) symmetric. Therefore the introduction of a twist lifts the degeneracy of eigenvalues and would restore U(1) symmetry. In fact there exists some numerical indication[10] that the original reduced model has no U(1) symmetry breaking in the region of negative coupling constant λ and agrees with the Wilson theory.

References

1. G. 't Hooft, Nucl. Phys. B75 461 (1974).
2. E. Witten, Cargese Lectures (1979).
Yu. M. Makeenko and A. A. Migdal, Phys. Lett. 88B 135 (1979).
3. T. Eguchi and H. Kawai, Phys. Rev. Lett. 48 1063 (1982).
4. B. Bhanot, U. Heller and H. Neuberger, Phys. Lett. 113B 47 (1982).
5. G. Parisi, Phys. Lett. 112B 463 (1982).
6. D. Gross and Y. Kitazawa, Princeton preprint May (1982).
7. A. Polyahov, Phys. Lett. 72B 477 (1978),
L. Susskind, Phys. Rev. D20 2610 (1978).
8. L. McLerran and B. Svetitsky, Phys. Lett. 98B 195 (1981),
J. Kuti, J. Polonyi and K. Szlachanyi, Phys. Lett. 98B 199 (1981),
E. D'Hoker, Nucl. Phys. B200 [FS4] 517 (1982).
9. M. Ohkawa, Phys. Rev. Lett. 49 353 (1982).
10. M. Ohkawa, Private communication.
11. G 't Hooft, Caltech preprint, April 1981.

Topological Excitations on a Lattice

Y. Iwasaki and T. Yoshiè
Institute of Physics, University of Tsukuba, Ibaraki 305, JAPAN

We study by numerical methods the existence (or the non-existence) of instantons and the role played by instantons on a lattice, by taking the CP^1 model in two dimensions as an example.

The CP^1 (non-linear O(3) σ) model is defined by

$$L = \frac{\beta}{2} \sum_{\mu=1}^{2} \sum_{i=1}^{3} (\partial_\mu \sigma_i)(\partial_\mu \sigma_i) \quad : \qquad \sum_i \sigma_i^2 = 1 \tag{1}$$

An O(3) invariant regularization of this model gives the classical O(3) Heisenberg model defined by

$$H = \sum_n \sum_{\hat{\mu}} \vec{S}(n)\vec{S}(n+\hat{\mu}) \ , \tag{2}$$

where $\vec{S}$ is a 3-component unit vector, n is a lattice site and $\hat{\mu}$ is a unit vector on the lattice. This model will be referred to as the standard model.

However there are infinitely many other choices for Hamiltonians which give the same naive (classical) continuum limit. For example, taking three couplings (nearest, next-nearest, and third-nearest neighbor), we write[1)] in the form

$$H = \sum \{\alpha_1(\nabla_\mu \vec{S}(n)\nabla_\mu \vec{S}(n)) + \alpha_2(\nabla_\mu \nabla_\nu \vec{S}(n))(\nabla_\mu \nabla_\nu \vec{S}(n))$$
$$+ \alpha_3[(\nabla_x^2 \vec{S}(n))(\nabla_x^2 \vec{S}(n)) + (\nabla_y^2 \vec{S}(n))(\nabla_y^2 \vec{S}(n))]\} \tag{3}$$

where

$$\nabla_\mu f(n) = f(n+\hat{\mu}) - f(n) \tag{4}$$

The correlation functions are given by

$$\langle F(S) \rangle = \frac{1}{Z} \int \prod_n d\vec{S}(n)\,\delta(S^2(n) - 1)F(S)\exp(-\beta H) \ . \tag{5}$$

If we take $\alpha_1 = 1/2$, any Hamiltonian given by eq.(3) reduces to the non-linear O(3) σ model defined by eq.(1) in the classical continuum limit.

Let us investigate topological properties of the models. Berg and Lüscher[2)] have provided a suitable definition of topological charge on the lattice in the form

$$Q = \sum_{n^*} q(n^*) \tag{6}$$

where n* is a dual lattice site. They have found that the topological susceptibility

$$\chi_t = \sum_{n^*} \langle q(0)q(n^*)\rangle \tag{7}$$

does not scale as a renormalization invariant $(\text{mass})^2$ in the case of the standard model. Further Lüscher[3)] has explained this phenomenon by pointing out that there exist short range fluctuations of the topological charge with such a small action that they overwhelm the contribution of the slowly varying fields, which otherwise dominate in the continuum limit. The point is that the minimum energy configurations with $Q = 1$ are not the solutions of the lattice field equation. The spin configurations are planar. They are at the boundary of configurations with $|Q| = 1$ and those with $Q = 0$; they are "exceptional" configurations due to the terminology in ref.2). We call these exceptional configurations "dislocations", following Berg.[4),5)] These dislocations dominate the topological susceptibility χ_t at low temperatures.

Up to this point they are all well known. Now, let us take the Hamiltonians (3) instead of the standard one. With positive α_2 and α_3, the energy of a short range fluctuation becomes large. However, the energy of a slowing varying field does not change much. Therefore by increasing the parameter α_2 and/or α_3, we can make the energy of a dislocation much larger than 4π, the energy of an instanton of the continuum theory.

If the energy of a dislocation is much larger than 4π, we naturally expect that some configurations with $Q = 1$ are solutions of the lattice field equation. We indeed find such solutions by increasing α_2 and/or α_3. We find them by two ways: One way is to start from random spin configurations and to lower systematically the energy of spin configurations until a solution of the classical lattice field equation is obtained. The other way is to start from a discretized-instanton

$$\omega(n) = \frac{S_1(n) + iS_2(n)}{1 + S_3(n)} = c\,\frac{z-a}{z-b} \tag{8}$$

and to lower systematically the energy. Here $z = n_1 + in_2$ and a, b, c are complex numbers.

For a 25×25 lattice, with $c = 1$, $a = 7.5 + 11.5i$, $b = 16.5 + 11.5i$, the energy of the discretized-instanton is systematically lowered by replacing a spin in such a way to minimize its local energy. Even for $\alpha_2 = 0.1$ and $\alpha_3 = 0.0$ where the energy of the dislocation is smaller than 4π, we obtain a stable instanton. For $\alpha_2 = 0.25$ and $\alpha_3 = 0.25$ (these numbers have no special meanings. We choose them arbitrary) where the energy of the dislocation is about twice that of the continuum instanton, we certainly obtain a stable instanton.

Thus we conclude that the existence (or the non-existence) of instantons depends on the form of lattice action. Among theories which are identical in the naive continuum limit, some theories where short range fluctuations are suppressed prossess stable instanton solutions, while others do not.

Because the Hamiltonians (3) reduce to the same non-linear O(3) σ model in the naive (classical) continuum limit irrespective of α_2 and α_3, all of them are equivalent in perturbative theory, only by redefining the coupling constant β. However, they may give inequivalent non-perturbative effects. We will discuss it below.

First let us give the relation between various coupling constants in the framework of perturbation theory. We follow the method which was first used by Parisi[6] when deriving the relation between the coupling constant of the standard model and that of the continuum theory.

The relation may be given in terms of the scale parameter

$$\Lambda = \frac{\beta}{a} \exp(-2\pi\beta) \tag{9}$$

After some calculation we obtain

$$\begin{aligned} \Lambda(\alpha_2=0.1,\ \alpha_3=0) &= 8.9\ \Lambda(\alpha_2=0,\ \alpha_3=0) \\ \Lambda(\alpha_2=0.25,\ \alpha_3=0.25) &= 124.3\ \Lambda(\alpha_2=0,\ \alpha_3=0) \end{aligned} \tag{10}$$

If we neglect β in the numerator in eq.(9) we have rough relations from eq.(10)

$$\begin{aligned} \beta(\alpha_2=0.1,\ \alpha_3=0) &\cong \beta(\alpha_2=0,\ \alpha_3=0) - 0.35 \\ \beta(\alpha_2=0.25,\ \alpha_3=0.25) &\cong \beta(\alpha_2=0,\ \alpha_3=0) - 0.77 \end{aligned} \tag{11}$$

At any rate eqs.(10) hold for $\beta >> 1$ and therefore eqs.(11) are enough for our later use. Thus the inverse-temperature $\beta = 1.3$ in the standard model corresponds to $\beta \cong 0.85$ when $\alpha_2 = 0.1$, $\alpha_3 = 0$ and $\beta \cong 0.53$ when $\alpha_2 = 0.25$, $\alpha_3 = 0.25$.

We will mainly discuss two theories with $\alpha_2 = 0$, $\alpha_3 = 0$ and with $\alpha_2 = 0.25$, $\alpha_3 = 0.25$. The case with $\alpha_2 = 0$, $\alpha_3 = 0$ corresponds to the case where an instanton is unstable, while the cases with $\alpha_2 = 0.25$, $\alpha_3 = 0.25$ to the case where an instanton is stable.

Next we measure χ_t defined by eq.(7) by Monte Carlo simulations on 50×50 spins. In the case of $\alpha_2 = \alpha_3 = 0$, Berg and Lüscher[2] have already measured χ_t for a 100×100 lattice to find that χ_t does not scale as expected. We find that for $\alpha_2 = \alpha_3 = 0.25$, χ_t is consistent with the scaling for $0.4 \leq \beta \leq 0.6$ [where the correlation length ξ varies from 2 to 6]. This is expected because the energy of an instanton is much lower than that of the dislocation. The reason why we have considered χ_t for the range of β, $0.4 \leq \beta \leq 0.6$, is the following; for $\beta < 0.4$ the correlation length is too small to expect the scaling behavior of χ_t and for $\beta > 0.6$ finite size effect becomes significant.

Qualitative differences between the two theories in physical quantities which are connected with the topological charge are naturally expected. However, the differences between the two theories may be more deep. If instantons are stable at $\beta = \infty$, their effects will remain in the limit $\beta \to \infty$, while if instantons are unstable at $\beta = \infty$, their effects will become weak in the limit $\beta \to \infty$.

As already noticed by Berg and Lüscher[2)], and Martinelli, Parisi and Petronzio[7)], the magnetic susceptibility χ_m itself does not scale as expected, although the deviation from the scaling is not so large as in χ_t. Our results are consistent with the previous results for $\alpha_2 = \alpha_3 = 0$ and the data for $\alpha_2 = \alpha_3 = 0.25$ are consistent with RG.

The deviation from RG for the standard model might be due to the higher order power corrections. However, we rather interpret the fact that χ_m does not scale as consistent with RG for $\alpha_2 = \alpha_3 = 0$, while it does scale for $\alpha_2 = \alpha_3 = 0.25$, as results from the fact that short range fluctuations dominate for $\alpha_2 = \alpha_3 = 0$ at low temperature, while slowly varing fields (instantons) dominate for $\alpha_2 = \alpha_3 = 0.25$.

If our interpretation is correct, it means that in the continuum limit for $\alpha_2 = \alpha_3 = 0.25$ χ_t and χ_m have their limits, for $\alpha_2 = \alpha_3 = 0$ they do not have non-trivial limits. This further implies that theories which are equivalent in perturbation theory do not necessarily give equivalent non-perturbative effects.

Now let us try to find indications of the topological symmetry breaking proposed previously by the present authors: In previous papers[8)] we have conjectured that the vacuum of the non-linear O(3) σ model (CP^1 model) is two-fold degenerate and that a spontaneous symmetry breaking occurs from the requirement of the cluster property of the vacuum. We add an external source term proportional to Q

$$- 4\pi\theta Q \tag{12}$$

to the Hamiltonian (3). We use the Metropolis method to measure hysteresis effects by changing θ. We measure hysteresis curves for $\alpha_2 = \alpha_3 = 0$ and for $\alpha_2 = \alpha_3 = 0.25$. Two cases show completely different patterns. In the case of $\alpha_2 = \alpha_3 = 0$, at $\theta = 0$ the residual topological charge is zero, while in the case of $\alpha_2 = \alpha_3 = 0.25$, at $\theta = 0$ the residual topological charge is about ten. We make the same measurement by changing random numbers as well as the number of steps. In every cases the patterns of hysteresis curves are the same, only the residual topological charge at $\theta = 0$ for $\alpha_2 = \alpha_3 = 0.25$ changes slightly. This behavior is consistent with our assertion that the topological symmetry breaking occurs for $\alpha_2 = \alpha_3 = 0.25$.

We also investigate the size dependence of χ_t for both cases with $\alpha_2 = \alpha_3 = 0$

and $\alpha_2 = \alpha_3 = 0.25$. If a system has the spontaneous topological symmetry breaking, the expectation value of the topological charge density is not zero;

$$<q(i)> = q \tag{13}$$

Therefore the topological susceptivility χ_t has a size dependence $\chi_t \sim N^2 q^2$ where the N^2 is the size of the lattice. This is analogue of the spontaneous magnetization. On the other hand, if a system has no spontaneous topological symmetry breaking, χ_t is expected to be size independent.

It should be noted that χ_t is dominated by dislocations in the standard model. Dislocations are short range fluctuations and therefore they are not influenced each other; a dislocation with positive topological charge and a dislocation with negative topological charge can coexist. The situation is similar to that of the dilute instanton gas picture. Contrary to the standard model, in the model with $\alpha_2 = \alpha_3 = 0.25$, instantons dominate χ_t; instantons are slowly varing fields and they influence each other. In the continuum limit we cannot apply the dilute instanton gas picture to this system due to the analysis of the continuum theory.[9),10)]

AT $\beta = 1.3$ with $\alpha_2 = \alpha_3 = 0$, the topological susceptibility χ_t does not change even if the size of the lattice is changed from 25×25 to 100×100 (via 50×50). On the other hand, at $\beta = 0.6$ with $\alpha_2 = \alpha_3 = 0.25$, χ_t increases if the size of the lattice changed from 25×25 to 50×50. For a 100×100 lattice we have also a preliminary result for χ_t; it shows also a tendency to increase. These results strongly support our assertion.

More details are discussed elsewhere.[11)]

References

1) S. H. Shenker and J. Tobochnik, Phys. Rev. B22 (1980) 4462.
2) B. Berg and M. Lüscher, Nucl. Phys. B190 (1981) 412.
3) M. Lüscher, Nucl. Phys. B200 (1982) 61.
4) B. Berg, CERN preprint TH.3147 (1981).
5) See also G. Martinelli, R. Petronzio and M. Virasoro, CERN preprint TH.3074 (1981).
6) G. Parisi, Phys. Lett. B92 (1980) 133.
7) G. Martinelli, G. Parisi and R. Petronzio, CERN priprint TH.3000 (1980).
8) Y. Iwasaki, The Structure of the Vacuum I, Prog. Theor. Phys. (in press); For a brief report, see Y. Iwasaki, Phys. Lett. 104B (1981) 458.
9) V. A. Fateev, I. V. Frolov and A. S. Schwarz, Nucl. Phys. B154 (1979) 1.
10) B. Berg and M. Lüscher, Comm. Math. Phys. 69 (1979) 57.
11) Y. Iwasaki and T. Yoshiè, UTHEP-94.

OBSERVATION OF AHARONOV-BOHM EFFECT BY ELECTRON HOLOGRAPHY

A.Tonomura, T.Matsuda, R.Suzuki, A.Fukuhara, N.Osakabe, H.Umezaki,
J.Endo, K.Shinagawa, Y.Sugita and H.Fujiwara
Central Research Laboratory, Hitachi Ltd., Kokubunji, Tokyo 185, Japan

We feel really honoured to give a talk before active researchers in this frontier field of physics, gauge theory and gravity. Although the member of our group are not familiar with the details of concepts and theoretical approachs in this field, we understand the importance of the Aharonov-Bohm effect in the electromagnetism, i. e. the first example of gauge fields.

Since the theoretical work by Aharonov and Bohm[1] in 1959, several experiments[2)-4)] have been performed to prove this effect and these experiments have been fairly famous also among electron-microscopist.

We thought the effect has the sound basis beyond doubt but we noticed also that a few people[5)] still insisted on its non-existance or doubted the validity of the experiments and that the controversy still continued.[6)] Therefore it seemed worth while to try an experiment in a newly designed form to confirm the effect again. This was our motivation.

Before going into our experiment, let us explain briefly about those in the past.

The schematic diagram in Fig.1 shows the idea of the elaborate experiment by Möllenstedt group.[2)] The lens and bi-prism are, of cource, electro-magnetic ones in fact. They fabricated a fine solenoid coil whose diameter was unbelievably small, 4.7 μm.

Two electron waves from the same source travel around the solenoid and are overlapped coherently to cause interference fringes on the film below. Even if the waves never touch the magnetic flux inside the solenoid, the fringe must be shifted with the change in the phase difference between the waves owing to the Aharonov-Bohm effect when the coil current i changes. In order to confirm the fringe shift, they set a slit over the recoreding film and moved the film with changing the coil current i. The result is reproduced in Fig.2. The fringe shift is clearly recorded.

Other experiments[3)-4)] are similar to this one in principle except that ferromagnetic needles were used instead of solenoids.

All these experiments were very elaborate ones for the technology of those days but we must admit that they have one defect in common. That is, the lack of experimental verifications that there is no magnetic flux leakage into the electron paths.

To improve this points, Kuper[7)] proposed in 1980 the idea of perfect confinement of magnetic fluxon by a hollow torus of super-conductive material, as shown in Fig.3.

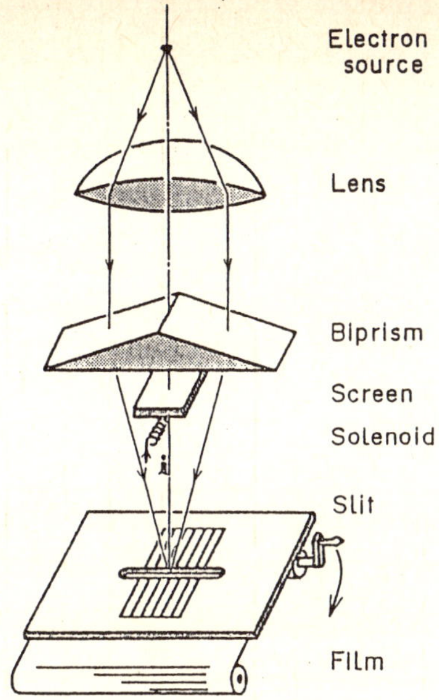

Fig.1 Interference experiment by Mollenstedt 1962.

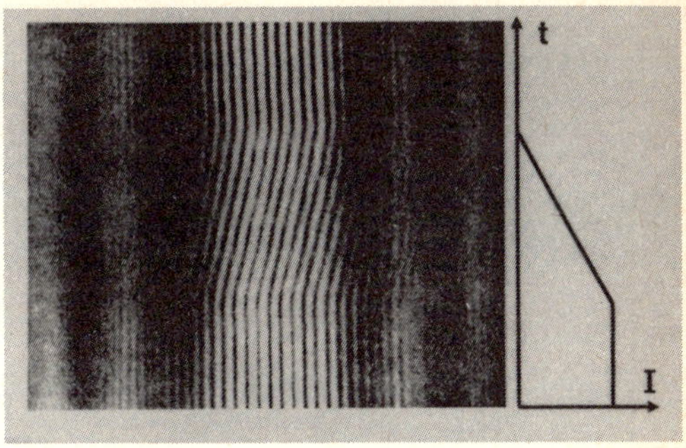

Fig.2 Observed phase change due to the coil current.

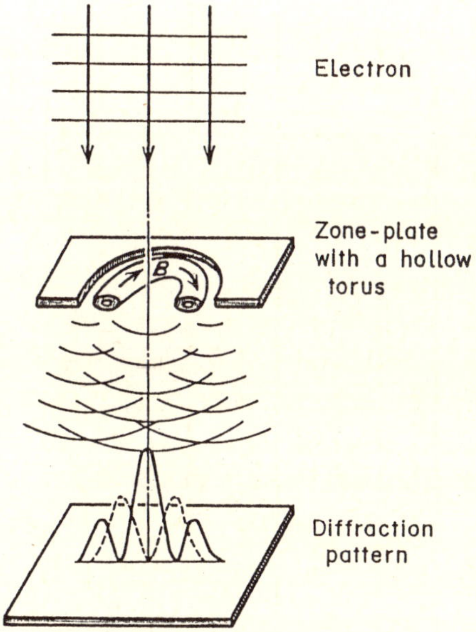

Fig.3 An experiment proposed by Kuper 1980.

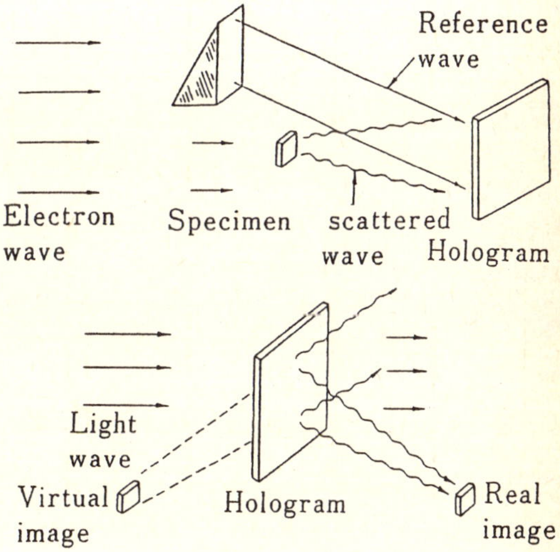

Fig.4 Electron hologram recording and optical reconstruction.

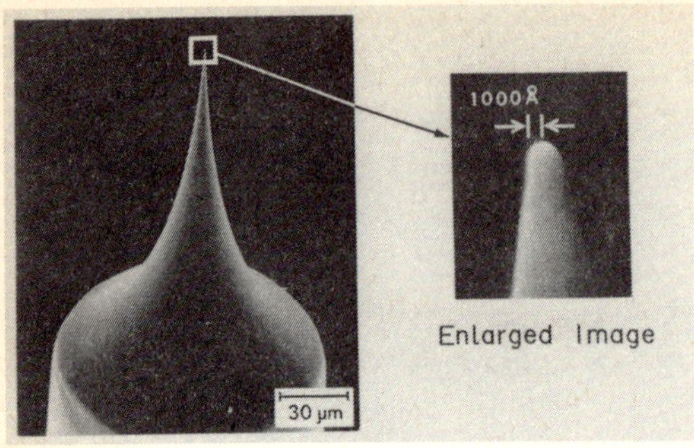

Fig.5 Tungsten tip as electron emitting source.

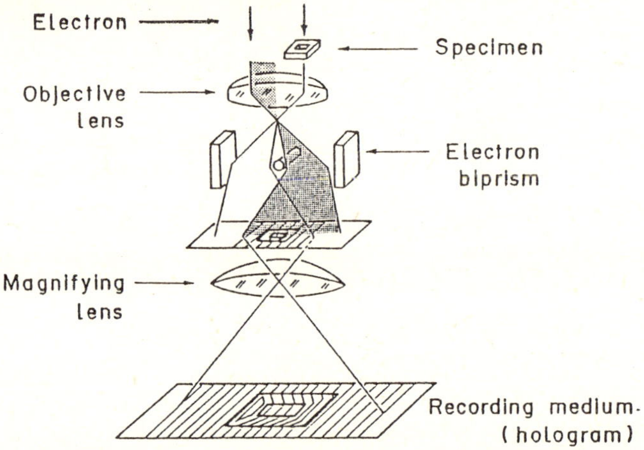

Fig.6 Hologram recording of a toroidal magnet.

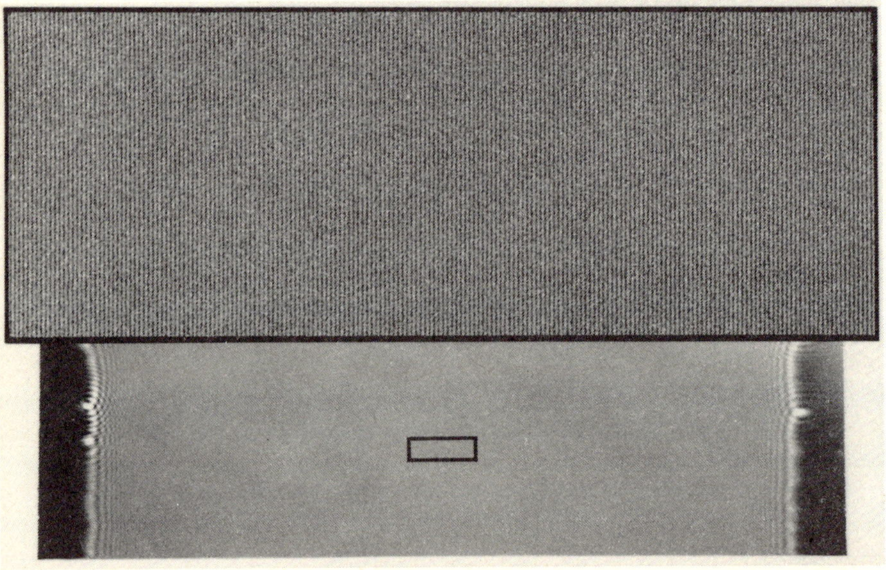

Fig.7 Interference fringe pattern of electron waves.

From the profile of the diffraction pattern below, one can tell the phase difference between waves having passed through the inner and outer field-free spaces around the torus. However, a tiny specimen of such a structure seems desperately difficult to be realized for an experiment.

We would like to describe our experiment[8] hereafter. What was new of our experiment is two-hold.

One point is making use of a very small toroidal magnet which has a closed circuit structure of magnetism so as to leak no magnetic flux outside of itself. The usage of a toroidal ferromagnet was independently recommended by Greenberger in 1981. Our toroids are made of permalloy and its fabrication was made possible by the optical lithography of microelectronics together with magnetic material technology, which are both available now in our research laboratories.

The second point of our experiment is an application of holographic technique to measure the phase difference of electron waves having passed around the toroid.

Let us explain the principle of holography very briefly with Fig.4. Suppose a monochromatic electron wave irradiates a specimen and eventually reaches a screen, where only its intensity distribution of this object wave is recorded at first. Then a part of the wave, called a reference wave, from the same source but free from the effect of the specimen is overlapped on the screen and we have a different pattern, which includes not only the intensity of the object wave but also its phase information in a form of interference fringes. The pattern recorded on the medium in the latter way is called a hologram.

The hologram is then irradiated by a chohèrent optical wave usually from a laser after the hologram is enlarged roughly by the wave length ratio. The interference fringes on the hologram act as a mixture of optical gratings and diffract the incident light so as to reproduce[9] the whole field of the electron object wave with a visible light. This is the principle of the holography, for which D. Gabor[10] got the Novel Prize in 1971.

The most serious technical problem in electron holography is to assure good coherence of wide electron waves. We have been successful in this point by using field-emission type electron sources[11]; electrons are emitted from a tungusten tip like the one shown in Fig.5 and the sharpness of the tip ensures good spatical coherence. Fig.6 illustrates schematically the horogram formation in our apparatus. The hatched part shows the reference wave. When the specimen is empty, we have only a fringe pattern of equal spacings as shown in the enlarged part of Fig.7. The number of fringes is a measure of good coherence and is some 10 times as many as that by conventional electron sources.

Fig.8 shows the view of an arrangement for the optical reconstruction. The optical system may look complicated, but its function is only to adjust the size and position of the reconstructed real image.

Fig.9 shows pictures taken by this apparatus from the same horogram of a single crystal of cobalt.

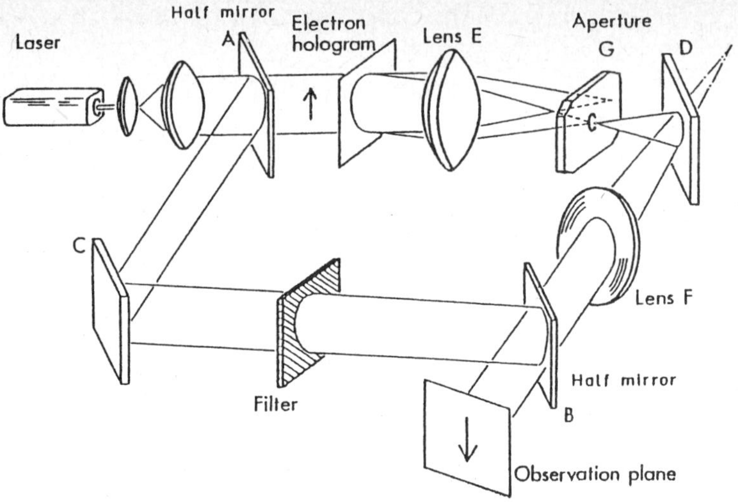

Fig.8 Optical arrangement for image reconstruction and interference microscopy.

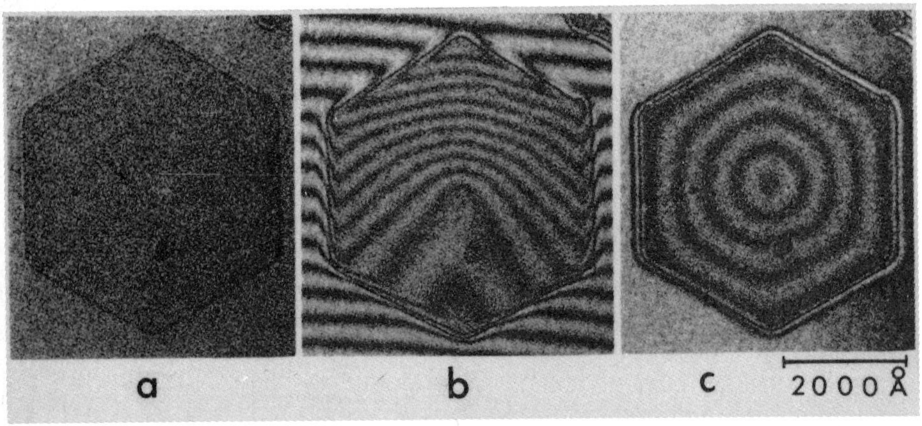

Fig.9 Image and interferograms optically formed from a hologram of a cobalt single crystal.
(a)ordinaly electron-microscopic image, (b)interferogram, (c)contour map of phase change.

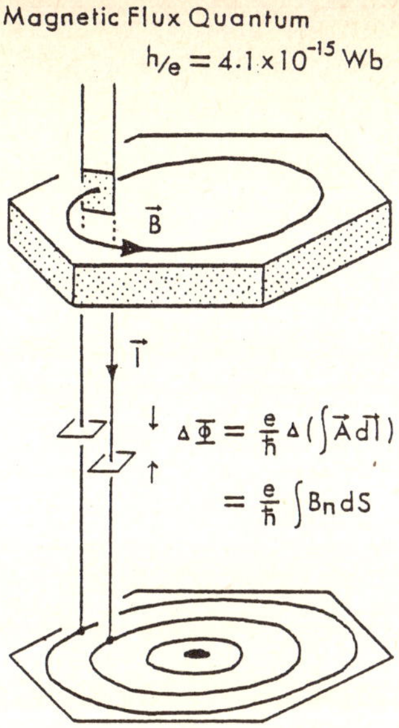

Magnetic Flux Quantum

$$h/e = 4.1 \times 10^{-15} \, Wb$$

$$\Delta \overline{\Phi} = \frac{e}{\hbar} \Delta \left(\int \vec{A} \, \vec{dl} \right)$$

$$= \frac{e}{\hbar} \int B_n \, dS$$

Fig.10 Magnetic flux enclosed by electron trajetories.

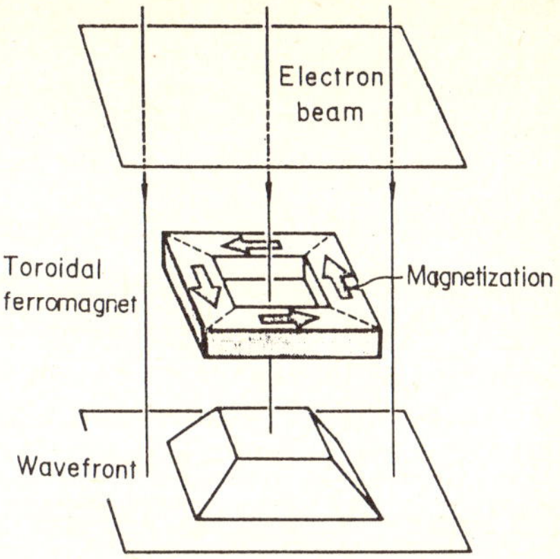

Fig.11 Phase shift due to a toroidal magnet.

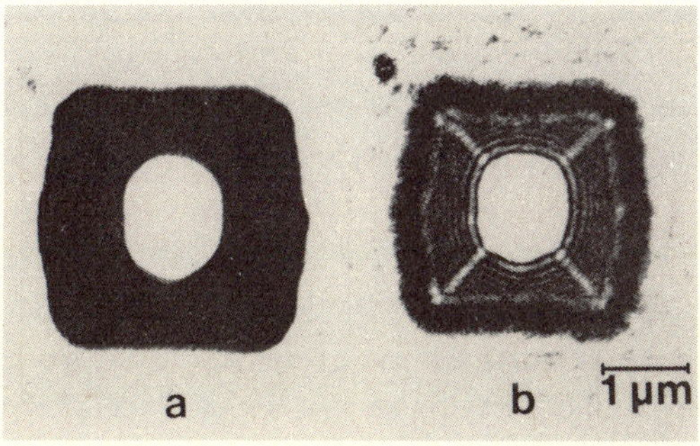

Fig.12 Ordinary(a) and Lorentz(b) electron-micrographs of a toroidal magnet.

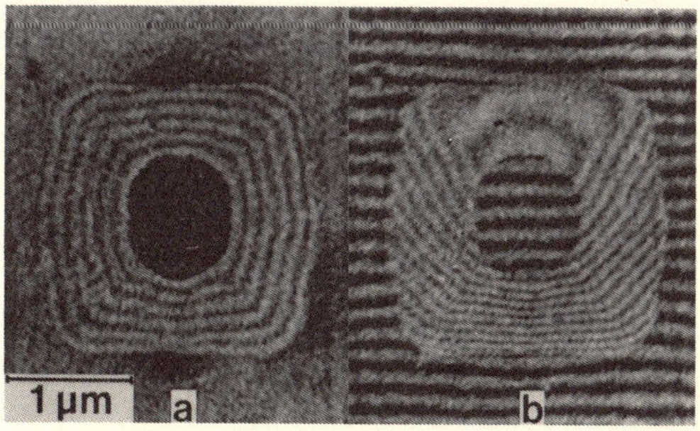

Fig.13 Interferogram of the toroidal magnet in Fig.12. (a) for parallel comparison wave, (b) for tilted comparison wave.

First, the usual electron microscopic image on the left(a) is reproduced according to the holography principles. In this case the comparison light beam, reflected by mirror C in Fig.8, is omitted. The image shows only the outline of the hexagonal shape and the uniformity of its thickness, but no phase information.

When the comparison light beam is superposed, we have an interferogram shown in the middle(b). Furthermore, we can adjust the direction of the comparison wave so that the fringes on the background vanish as shown on the right(c). This corresponds to the coincidence of axes of two light waves, i.e. the comparison wave is parallel. In this final case(c) the concentric pattern is, in fact, a contour map of the phase of the object wave caused by the magnetization within the crystal.

Suppose two electron trajectories within the object wave as in Fig.10. Then the phase difference between these two is described by the magnetic flux enclosed between them. If the two trajectories are those reaching adjacent dark lines on the contour map, the phase difference must be just 2π and the enclosed flux equals to the flux quantum h/e, irrespective of the electron energy.

Thus we can say that the contour map for the parallel comparison wave is equivalent to a projection of the magnetic flux field normalized by a unit of h/e.

Now we will show the results about toroidal magnets(Fig.11). The photographs in Fig.12 were taken with an ordinary electron microscope. Under usual experimental conditions the electron deflection due to the Lorenze force is too small to affect the image at just focusing(a). However, in the image far out of focus(b) the effect of the deflection can be recognized as white contrast at magnetic domain boundaries. Thus the closure structure of magnetization or equivalently of magnetic flux is just as expected.

Fig.13 shows the interference fringes obtained by our holographic technique just like those for the cobalt crystal. The pattern(a) for the parallel comparison wave, i.e. the projection of the magnetic flux, also shows the closed circuit structure more accurately; the phase difference between the inner and outer space of the toroid is equivalent to some 6 wavelengths. From this we can estimate the total magnetic flux inside the toroid at same 6h/e, which was consistent with the magnetization of permalloy 9500G,the thickness 400 Å and the width 6400 Å.

An interferogram(b) obtained from the same hologram and tilted comparison wave can decide the direction of magnetization as well as the phase difference. Here the phase shift is measured as the discrepancy of fringes between the outer and the inner spaces around the toroid.

We found also the sample with the opposite direction of magnetization(Fig.14). In this case the sample width is larger so that the phase difference is larger, some 8 wavelengths.

We intentionally chose the thickness such that the sample is half transparent for the electron beam in order to count the number of fringes or to trace them. Thus a part of the electron wave has passed through the toroid and has felt the magnetic flux inside. It should be remembered that our hologram is taken just at the image plane of the electron optical system. The electron wave having passed through the toroid does not contribute the image of the inner or outer space around it within the accuracy of electron optical aberrations. In the experiment in the past the interference to be observed was that between electron waves both having travelled around a specimen; however, in our experiment, these waves are made interfere with the third one, called the reference wave, so that the relative phase difference between the former two can be determined where they do not overlap each other.

As for the flux leakage out of troids, these contour maps themselves reveal that the leakage was less than one flux quantum so that it does not affect the conclusion. When the leakage is serious, we can recognize it on the contour map as an example of unsuccessful preparation(Fig.15).

In conclusion, we have succeeded in measuring the phase difference of the electron wave around magnetic toroids, which are consistent with the prediction by Aharonov and Bohm. We would like to believe that we have added solid experimental confirmation to this famous effect.

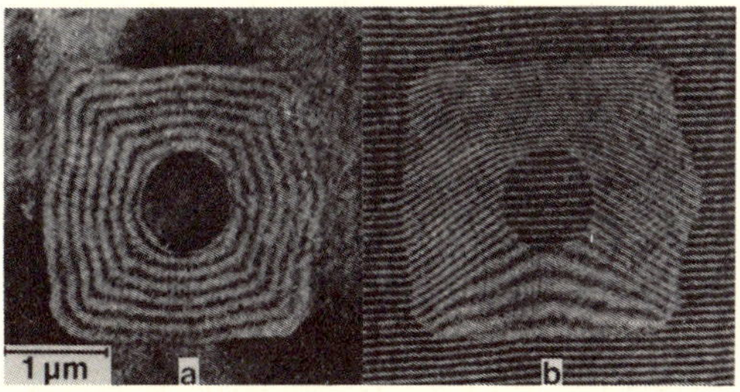

Fig.14 Interferograms of a toroidal magnet with magnetization direction opposite to that in Fig.13.

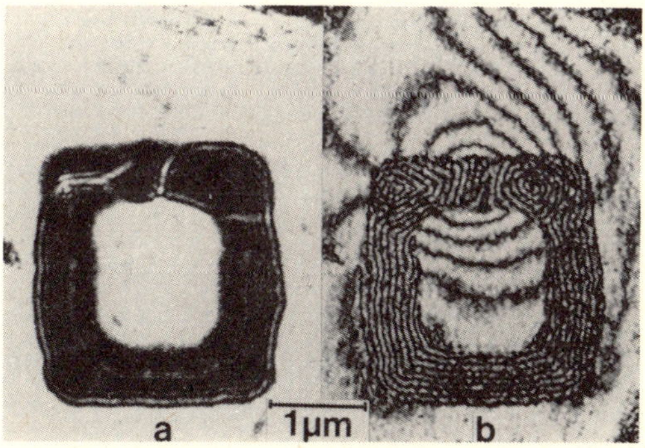

Fig.15 Toroidal magnet with flux leakage. (a)Lorentz micrograph, (b)phase contour map.

References

1) Y.Aharonov and D.Bohm : Phys. Rev. 115 (1959) 485.
2) G.Mollenstedt and W.Bayh : Phys. Bl. 18 (1962) 299.
3) R.G.Chambers : Phys. Rev. Let. 5 (1960) 3.
4) H.A.Fowler, L.Marton, J.A.Simpson and J.A.Sudeth : J.A.P. 32 (1961) 1153
5) D.S.DeWitt : Phys. Rev. 125 (1962) 2189.
S.M.Roy: Phys. Rev. Let. 44 (1980) 111.
Bocchieri et al : Nuovo Cimento 47 (1978) 475 ; 51 (1979) 1 ; 56 (1980) 55 ; 59 (1980) 121.
6) For example, D.Bohm and B.J.Hiley : Nuovo Cimento 52A (1979) 295 ; H.J.Lipkin: Phys. Rev. D23 (1981) 1466 ; U.Klein : Phys. Rev. D23 (1981) 1463.
7) C.G.Kuper : Phys. Lett. 794 (1980) 413.
8) A.Tonomura et al : Phys. Rev. Lett. 48 (1982) 1443.
9) D.M.Greenberger : Phys. Rev. D23 (1981) 1460.
10) D.Gabor : Proc. Roy. Soc. London A197 (1949) 454 ; Proc. Phys. Soc. B64 (1951) 449
11) A.Tonomura et al : J.Electron Microsc. 28 (1979) 1.;

BPS Transformation and Color Confinement*)

K.Nishijima
Department of Physics,University of Tokyo
Tokyo, Japan 113

Interpretation of hadrons in terms of quarks and antiquarks has been so successful that one can no longer think of its substitute. The hadron spectrum and high energy hadron reactions are believed to be described by means of quantum chromodynamics (QCD). Thus, we feel the existence of quarks so real on one hand, but we have never detected isolated quarks on the other hand. In this way the explanation of the confinement of quarks and also of gluons became one of the central problems in QCD.

In the lattice gauge theory the condition for quark confinement is given by the area law for the Wilson loop [1]. In the present paper we shall look for the corresponding condition within the framework of the conventional continuum field theory. As we shall see later this condition is given by the existence of certain bound states between a pair of Faddeev-Popov ghosts.

$$\mathcal{L} = -\frac{1}{4}F_{\mu\nu}\cdot F_{\mu\nu} + A_\mu\cdot\partial_\mu B + \frac{\alpha}{2}B\cdot B$$

$$+ i\partial_\mu\bar{c}\cdot D_\mu c - \bar{\psi}(\gamma_\mu D_\mu + m)\psi, \qquad (1)$$

where covariant derivatives D_μ are defined by

$$D_\mu c = \partial_\mu c + gA_\mu \times c,$$

$$D_\mu\psi = (\partial_\mu - igT\cdot A_\mu)\psi, \qquad (2)$$

$$F_{\mu\nu} = \partial_\mu A_\nu - \partial_\nu A_\mu + gA_\mu \times A_\nu .$$

We have made use of the abbreviations, $S\cdot T = S^a T^a$ and $(S\times T)^a = f_{abc}S^b T^c$.

Next we introduce the Becchi-Rouet-Stora (BRS) transformtion of Heisenberg-fields [2].

$$\delta A_\mu = D_\mu c,$$

$$\delta B = 0,$$

$$\delta c = -\frac{1}{2}g\, c \times c, \qquad (3)$$

$$\delta\bar{c} = iB,$$
$$\delta\psi = ig(c\cdot T)\psi.$$

This supersymmetric transformation can be expressed in terms of its generator Q_B as

$$\delta O = i[Q_B, O]_{\mp}, \tag{4}$$

where we choose the -(+) sign when O involves an even (odd) number of the hermitian anticommuting ghost fields c and $\bar{c}$. Kugo and Ojima [3] have introduced another charge Q_c satisfying

$$i[Q_c, c(x)] = c(x), \; i[Q_c, \bar{c}(x)] = -\bar{c}(x). \tag{5}$$

It commutes with all other fields, and it defines the ghost number, namely +1 for c and -1 for $\bar{c}$. These two charges satisfy the relations

$$i[Q_c, Q_B] = Q_B, \; Q_B^2 = 0. \tag{6}$$

The second relation implies that the BRS transformation is nilpotent, namely, $\delta^2 = 0$.

Then we shall introduce asymptotic fields and their BRS transformation. Because of infrared singularities in QCD the existence of asymptotic fields might be doubtful, nevertheless we shall simply assume it in the present paper. Then the BRS transformation for the asymptotic fields is linear. When $\delta a(x) = b(x) \neq 0$ so that $\delta^2 a(x) = 0$, $\{a(x), b(x)\}$ is called a BRS doublet. When $\delta a(x) = 0$ but its parent $f(x)$, satisfying $\delta f(x) = a(x)$, does not exist, $a(x)$ is called a BRS singlet. Doublets and singlets are the only irreducible representations of the BRS transformation.

By extending the assumed existence of asymptotic fields we shall further postulate the asymptotic completeness. The state vector space spanned by asymptotic fields in QCD will be denoted by $\mathcal{V}$. Kugo and Ojima [3] introduced a physical subspace $\mathcal{V}_{phys}$ by

$$\mathcal{V}_{phys} = \{|\;\rangle \,\big|\, |\;\rangle \in \mathcal{V}, \; Q_B|\;\rangle = 0\} \tag{7}$$

Then, by applying only the singlet asymptotic fields to the vacuum state a subspace of $\mathcal{V}$, denoted by $\mathcal{V}_S$, is generated. Obviously we have

$$\mathcal{V} \supset \mathcal{V}_{\rm phys} \supset \mathcal{V}_S. \tag{8}$$

The S matrix exists as a consequence of the asymptotic completeness and commutes with Q_B. When $|\alpha\rangle$ and $|\beta\rangle$ belong to $\mathcal{V}_S$, the unitarity condition of the S matrix can be expressed as

$$\langle\beta|\alpha\rangle = \langle\beta|S^+S|\alpha\rangle = \langle\beta|S^+P(\mathcal{V}_S)S|\alpha\rangle, \tag{9}$$

and similarly for $SS^\dagger$. This relation is a consequence of the Kugo-Ojima theorem [3]. $P(\mathcal{V}_S)$ stands for the projection operator to the subspace $\mathcal{V}_S$, so that no doublets show up in the intermediate states of the unitarity condition. In this sense, doublets in QCD are analogous to longitudinal and scalar photons in QED and are confined in the unphysical state vector space. Interpreting that singlets represent hadrons, the problem of color confinement reduces to that of demonstrating that both quarks and gluons are BRS doublets.

We have already assumed that the vacuum state $|0\rangle$ belongs to $\mathcal{V}_S$ and hence to $\mathcal{V}_{\rm phys}$. Thus we have $Q_B|0\rangle = 0$ and consequently the BRS identity

$$\langle 0|\delta T(\cdots)|0\rangle = 0. \tag{10}$$

In what follows we shall abbreviate $\langle 0|T(\cdots)|0\rangle$ as $\langle\cdots\rangle$. Then, by making use of the BRS identities we find the following Ward-Takahashi (W-T) identities [4]:

$$\partial_\lambda\langle(D_\lambda\bar{c})^a(x),\ \delta\psi(y),\ \bar{\psi}(z)\rangle + \partial_\lambda\langle(D_\lambda\bar{c})^a(x),\ \psi(y),\ \delta\bar{\psi}(z)\rangle$$

$$= ig\ T^a(\delta^4(x-y) - \delta^4(x-z))S_F(y-z), \tag{11}$$

$$\partial_\lambda\langle(D_\lambda\bar{c})^a(x),\ \delta A^b_\mu(y),\ A^c_\nu(z)\rangle + \partial_\lambda\langle(D_\lambda\bar{c})^a(x),\ A^b_\mu(y),\ \delta A^c_\nu(z)\rangle$$

$$= igM^a_{bc}(\delta^4(x-y) - \delta^4(x-z))D_{F\mu\nu}(y-z), \tag{12}$$

where $M^a{}_{bc} = if_{bac}$, and S_F and $D_{F\mu\nu}$ denote propagators of the quark and gluon fields, respectively. We shall write

$$\langle(D_\lambda\bar{c})^a(x),\ \delta\psi(y),\ \bar{\psi}(z)\rangle = g\int d^4z' G^a_\lambda(yz':x)S_F(z'-z), \tag{13}$$

$$\langle(D_\lambda\bar{c})^a(x),\ \psi(y),\ \delta\bar{\psi}(z)\rangle = g\int d^4y' S_F(y-y')\bar{G}^a_\lambda(y'z:x). \tag{14}$$

The Fourier-transform of Eq. (11) may be expressed as

$$(p-q)_\lambda \cdot G^a_\lambda(p, q) S_F(q) + S_F(p)(p-q)_\lambda \cdot \bar{G}^a_\lambda(p, q)$$
$$= iT^a(S_F(p) - S_F(q)). \tag{15}$$

In this equation we may replace G^a_λ and $\bar{G}^a_\lambda$ by their spin zero projection defined by

$$G^a_\lambda(p, q)^{(0)} = \frac{(p-q)_\lambda (p-q)_\mu}{(p-q)^2} G^a_\mu(p, q). \tag{16}$$

In order to simplify our argument we shall choose the Landau gauge ($\alpha = 0$) in what follows. In this gauge we have $\partial_\lambda D_\lambda \bar{c} = 0$, and possible poles in G_λ due to massless vector particles will disappear in this projection. A pole due to the massless scalar particle is still present as can clearly be seen from the W-T identity

$$\langle (D_\lambda \bar{c})^a(x), c^b(y) \rangle = i\delta_{ab} \partial_\lambda D_F(x-y), \tag{17}$$

where D_F denotes the free massless propagator. This equation shows that $D_\lambda \bar{c}$ generates a massless scalar particle, and we introduce asymptotic fields corresponding to this massless scalar particle as

$$D_\lambda \bar{c} \to \partial_\lambda \bar{\Gamma}, \quad c \to \Gamma. \tag{18}$$

We then replace $D_\lambda \bar{c}$ by $D_\lambda \bar{c} - \partial_\lambda \bar{\Gamma}$ and write F and $\bar{F}$ for G and $\bar{G}$ in Eqs. (13) and (14). The functions $F_\lambda{}^a(p, q)^{(0)}$ and $\bar{F}_\lambda{}^a(p, q)^{(0)}$ so defined are free of the poles at $(p-q)^2 = 0$ except for the projection operator in Eq. (16).

According to Nakanishi's theorem [5] the asymptotic field $\bar{\Gamma}$ carrying the ghost number (-1) cannot be a BRS singlet, but it must be a member of a BRS doublet. Confinement is realized when $\bar{\Gamma}$ is the second generation of the doublet expressible as

$$\delta \bar{d}(x) = \bar{\Gamma}(x). \tag{19}$$

Then the BRS identity leads to

$$\langle \partial_\lambda \bar{\Gamma}(x), \delta\psi(y), \bar{\psi}(z) \rangle + \langle \partial_\lambda \bar{\Gamma}(x), \psi(y), \delta\bar{\psi}(z) \rangle = 0, \tag{20}$$

and by subtracting Eq. (20) from Eq. (11) we find

$$(p-q)_\lambda \cdot F_\lambda{}^a(p, q)^{(0)} S_F(q) + S_F(p)(p-q)_\lambda \cdot \bar{F}_\lambda{}^a(p, q)^{(0)}$$

$$= iT^a(S_F(p) - S_F(q)). \tag{21}$$

We then put $p - q = \varepsilon P$ with $P^2 \neq 0$, and apply the limiting procedure $\lim_{\varepsilon\to 0} \partial/\partial\varepsilon$ to Eq. (21). Since the individual terms on the ℓ. h. s. of Eq. (21) are of the order of ε because of the absence of poles at $(p - q)^2 = 0$, we obtain

$$P_\lambda \cdot F_\lambda{}^a(p, p: P)^{(0)} S_F(p) + S_F(p) P_\lambda \cdot \bar{F}_\lambda{}^a(p, p: P)^{(0)}$$

$$= iT^a P_\lambda \cdot \frac{\partial}{\partial p_\lambda} S_F(p). \tag{22}$$

F_λ and $\bar{F}_\lambda$ gain a possible dependence on the direction of P through the factor $P_\lambda P_\mu / P^2$ originated from the projection operator in Eq. (16). Eq. (22) shows that F_λ and/or $\bar{F}_\lambda$ must share a pole with $S_F(p)$ at $ip\gamma + m = 0$. For the symmetry reason both must have this pole implying that both $\delta\psi$ and $\delta\bar{\psi}$ generate a pole at the quark mass.

When $\bar{\Gamma}$ is the first generation of the doublet contrary to Eq. (19), however, the BRS identity (20) does not hold. Then we have to go back to Eq. (15) because Eq. (21) does not follow. Since the functions $G_\lambda{}^a$ and $\bar{G}_\lambda{}^a$ are not free of the pole at $(p-q)^2 = 0$, the individual terms on the ℓ. h. s. of Eq. (15) are generally of the order of 1 and only the sum of the two terms is of the order of ε. Then application of the limiting procedure mentioned above leads to an equation in which derivatives of S_F appear not only on the r. h. s. but also on the ℓ. h. s. in a sharp contrast to Eq. (22) in which the S_F on the ℓ. h. s. is not differentiated. In such a case, however, we cannot conclude that G_λ and $\bar{G}_\lambda$ must have a pole at the quark mass. Perturbation theory falls into this category.

Thus, when Eq. (19) holds, $\{\psi^{in}, \delta\psi^{in}\}$ represents a BRS doublet and quarks are confined. A similar argument starting from Eq. (12) shows that gluons are also confined under the same condition.

We shall reexpress the condition (19) in a more convenient form by using the BRS identity.

$$0 = \langle \bar{\Gamma}(x), \Gamma(y) \rangle$$

$$= \langle \delta\bar{d}(x), \Gamma(y) \rangle$$

$$= - \langle \bar{d}(x), \delta\Gamma(y) \rangle. \tag{23}$$

This implies the existence of $\delta\Gamma$. Since Γ is the asymptotic field of c and $\delta c \sim c \times c$, there must exist the asymptotic field of $c \times c$ carrying the same set of quantum numbers as that of c but for the ghost number. Conversely, when $\delta\Gamma$ exists, there must be an asymptotic field $\bar{d}$ for which $\langle \bar{d}, \delta\Gamma \rangle \neq 0$. Then we can replace $D_\lambda \bar{c}$ in Eqs. (13) and (14) by $D_\lambda c - \partial_\lambda \delta\bar{d}$ to introduce the poleless vertex functions F_λ and $\bar{F}_\lambda$. After that we can repeat the same argument leading to the quark confinement.

Thus the existence of the asymptotic field for $c \times c$ is a sufficient condition for color confinement. Quarks and gluons are confined when they form bound states with the ghost c as is clear from the explicit expressions for $\delta\psi$ and δA_μ in Eq. (3). When the ghost c itself forms a bound state with another ghost, the ability of forming a bound state with the ghost is communicated to other colored particles through the BRS identities.

The Bethe-Salpeter equation for the bound states between a pair of Faddeev-Popov ghosts can be solved exactly in the ladder approximation, but the normalization integral is not convergent. It can be shown, however, that introduction of a parameter of the dimension of mass is necessary for the convergence of the normalization integral. A promising way of improving the approximation to make the normalization integral convergent is to exploit the renormalization group method in which a mass parameter enters as a renormalization point.

*) A preliminary version of the present paper will be published in Physics Letters.

[1] K. G. Wilson, Phys. Rev. D10 (1974) 2445.

[2] C. Becchi, A. Rouet and R. Stora, Ann. Phys. (N.Y.) 98 (1976) 287.

[3] T. Kugo and I. Ojima, Phys. Lett. 73B (1978) 459; Suppl. Prog. Theor. Phys. No. 66 (1979).

[4] A Similar relationship has been derived by Hata in a different context.
H. Hata, Kyoto University preprint, KUNS 618 (1981)

[5] N. Nakanishi, Prog. Theor. Phys. 62 (1979) 1396.
T. Kugo and S. Uehara, Prog. Theor. Phys. 64 (1980) 139.

COVARIANT OPERATOR FORMALISM OF GAUGE THEORIES AND ITS EXTENSION TO FINITE TEMPERATURE

Izumi OJIMA
Research Institute for Mathematical Sciences, Kyoto University
Kyoto 606, Japan

On the basis of "thermo field dynamics" allowing the application of the Feynman diagram method to real-time Green's functions at $T\neq 0^\circ K$, a field-theoretical formulation of finite-temperature gauge theory is presented. It is an extension of the covariant operator formalism of gauge theory based upon the BRS invariance: The subsidiary condition specifying physical states, the notion of observables, and the structure of the physical subspace at finite temperatures are clarified together with the key formula characterizing the temperature-dependent "vacuum".

1. Introduction

Although thermodynamic aspects of gauge theory are currently discussed in the so-called imaginary-time formulation of QFT at finite temperatures, which is believed to be the only choice pertaining to the Feynman diagram method, this belief is not correct as shown by Takahashi and Umezawa [1]. They proposed a real-time formulation named "thermo field dynamics" in which statistical averages are expressed in the form of temperature-dependent "vacuum" expectation values and in which the Feynman diagram method can be applied to real-time causal Green's functions at a finite temperature (i.e. statistical averages of time-ordered products). In contrast to the imaginary-time formulation having no time variable, we have here both a temperature and a time variable without being bothered by the cumbersome discrete energy sums over the Matsubara frequencies and the full information about spectral functions is attained without analytic continuation in energy variables.

Thus, this is a formalism to be regarded as a natural extension of QFT (at $T=0^\circ K$) to the case of $T\neq 0^\circ K$. According to [2], we briefly describe here a field-theoretical formulation of gauge theory at $T\neq 0^\circ K$ on the basis of the covariant operator formalism of gauge theory [3] and this "thermo field dynamics".

2. Thermo field dynamics

The point of thermo field dynamics [1] is to introduce fictitious "tilde" operators $\tilde{A}$ corresponding to each of the operators A describing the system and to perform a temperature-dependent Bogoliubov transformation mixing A's with $\tilde{A}^\dagger$'s, which realizes the state space at a finite temperature and the temperature-dependent "vacuum" $|0(\beta)\rangle$ giving statistical averages.

This is seen in the simplest example of a harmonic oscillator defined by

$$H = \varepsilon a^\dagger a, \ [a,a^\dagger] = 1. \tag{2.1}$$

By introducing tilde operators $\tilde{a}$, $\tilde{a}^\dagger [=(\tilde{a})^\dagger = \widetilde{(a^\dagger)}]$ as duplicates of $a, a^\dagger$ commuting with $a, a^\dagger$,

$$[\tilde{a},\tilde{a}^\dagger] = 1, \ [a,\tilde{a}] = [a^\dagger,\tilde{a}] = \ldots = 0, \tag{2.2}$$

the temperature-dependent "vacuum" $|0(\beta)\rangle$ is determined by the Bogoliubov transformation as follows:

$$a(\beta) \equiv a\cosh\theta(\beta) - \tilde{a}^\dagger \sinh\theta(\beta), \ \tilde{a}(\beta) \equiv \tilde{a}\cosh\theta(\beta) - a^\dagger\sinh\theta(\beta); \tag{2.3}$$

$$\cosh\theta(\beta) \equiv 1/(1-e^{-\beta\varepsilon})^{1/2}, \ \sinh\theta(\beta) \equiv e^{-\beta\varepsilon/2}/(1-e^{-\beta\varepsilon})^{1/2}; \tag{2.4}$$

$$a(\beta)|0(\beta)\rangle = \tilde{a}(\beta)|0(\beta)\rangle = 0. \tag{2.5}$$

It can easily be checked by the aid of (2.3)~(2.5) that statistical averages over the Gibbs ensemble are given by the "vacuum" expectation values:

$$\langle A\rangle \equiv \mathrm{Tr}(A\exp(-\beta H))/\mathrm{Tr}(\exp(-\beta H) = \langle 0(\beta)|A|0(\beta)\rangle . \tag{2.6}$$

The origin of the Bogoliubov transformation (2.3)∿(2.5) is traced to the following relations:

$$|0(\beta)\rangle = \sum_{n>0} e^{-\beta n\varepsilon/2}|n\rangle\otimes|n\rangle/(1-e^{-\beta\varepsilon})^{-1/2}$$
$$= \exp[\theta(\beta)(a^{\dagger}\tilde{a}^{\dagger}-\tilde{a}a)]|0\rangle \equiv \exp(-iG)|0\rangle , \tag{2.7}$$

$$a(\beta) = \exp(-iG)a\ \exp(iG),\ \tilde{a}(\beta) = \exp(-iG)a\ \exp(iG). \tag{2.8}$$

Although the unitary operator exp(-iG) has its proper meaning only for the system with finite degrees of freedom, these formulae will be useful heuristically also in the later discussion of gauge theory.

Now, the essence of (2.3)∿(2.5) can be summarized in a formula

$$\exp[\beta(H-\tilde{H})/2]M|0(\beta)\rangle = M^{\dagger}|0(\beta)\rangle , \tag{2.9}$$

which reproduces (2.5) in the case of $M=a,a^{\dagger}$ with the help of (2.1)∿(2.4) and $\tilde{H}=\varepsilon\tilde{a}^{\dagger}\tilde{a}$. By extending the definition of $\tilde{M}$ antilinearly, (2.9) holds for any polynomial M of a and $a^{\dagger}$. Taking account of the commutativity of tilde and non-tilde operators, we can derive from (2.9) the KMS condition [4] characterizing Gibbs states ,

$$\langle AB(t)\rangle = \langle B(t-i\beta)A\rangle . \tag{2.10}$$

The basis for (2.9) can be found [2] in connection with the algebraic formulation of statistical mechanics[5] due to Haag, Hugenholtz and Winnink [6] and Tomita-Takesaki theory [5]. The key concepts there are the modular conjugation operator J and the modular operator $\exp(-\beta\overline{H})$ defined by

$$J\exp(-\beta\overline{H}/2)M|0(\beta)\rangle = M^{\dagger}|0(\beta)\rangle \quad \text{for} \quad M\in\mathfrak{M}, \tag{2.11}$$

where $\mathfrak{M}$ is the algebra of operators desribing the system. J is an antiunitary operator satisfying

$$J^2 = 1,\ J|0(\beta)\rangle = |0(\beta)\rangle , \tag{2.12}$$

$$J\mathfrak{M}J = \mathfrak{M}'(\equiv \text{commutant of } \mathfrak{M}), \tag{2.13}$$

and $\overline{H}$ satisfies

$$\overline{H}|0(\beta)\rangle = 0, \tag{2.14}$$

$$\exp(it\overline{H})\mathfrak{M}\exp(-it\overline{H}) = \mathfrak{M}, \tag{2.15}$$

$$J\overline{H}J = -\overline{H}. \tag{2.16}$$

Thus, by identifying $\tilde{M}$ with JMJ and $\overline{H}$ with $H-\tilde{H}$,

$$\tilde{M} = JMJ, \tag{2.17}$$

$$\overline{H} = H-\tilde{H} = H-JHJ, \tag{2.18}$$

(2.9) can be derived from (2.11) with the help of (2.16) and (2.12).

(2.17) and (2.13) explain the commutativity of tilde operators with nontilde ones. In the case of fermions, however, $\tilde{\psi}$ should anticommute with ψ in order to keep the Bogoliubov transformation meaningful, whence (2.17) for fermions is modified by the Klein transformation [2]:

$$\tilde{\psi} = iJ\psi J\exp(i\pi(F-JFJ)). \tag{2.19}$$

Here F is the fermion number operator. In contrast to (2.9) which requires modification to cover the cases with fermions, (2.11) remains unchanged, and hence, it should be taken as the key formula characterizing the temperature-dependent "vacuum" $|0(\beta)\rangle$ at $T=1/k_B\beta$ independently of a specific model (except for gauge theory discussed later in §4).

Now, corresponding to the "total" Hamiltonian (2.18) which is consistent with (2.16), the "total" Lagrangian $\overline{\mathcal{L}}$ of the "total" system consisting of tilde and non-tilde objects is given by

$$\overline{\mathcal{L}} = \mathcal{L} - \tilde{\mathcal{L}} = \mathcal{L} - J\mathcal{L}J. \tag{2.20}$$

Since the Gell-Mann-Low relation can be verified relative to the splitting of $\overline{\mathcal{L}}$ into the free and interaction parts, $\overline{\mathcal{L}} = \overline{\mathcal{L}}_0 + \overline{\mathcal{L}}_I$, and since the Wick theorem holds (at the operator level in contrast to the Bloch-DeDominicis-Wick theorem in the imaginary time formulation), we can develop the Feynman diagram technique also here on the basis of the propagators such as

$$< 0(\beta)\, T\begin{bmatrix}\phi(x)\\ \tilde{\phi}^\dagger(x)\end{bmatrix}(\phi^\dagger(y),\tilde{\phi}(y))|\, 0(\beta)>$$

$$= i\int\frac{d^4p}{(2\pi)^4}\, e^{-ip(x-y)}U_B(|\vec{p}|,\beta)\begin{pmatrix}1/(p^2-m^2+i0) & 0\\ 0 & -1/(p^2-m^2-i0)\end{pmatrix}U_B^\dagger(|\vec{p}|,\beta); \quad (2.21)$$

$$U_B(|\vec{p}|,\beta) = \begin{bmatrix}\cosh\theta(\beta) & \sinh\theta(\beta)\\ \sinh\theta(\beta) & \cosh\theta(\beta)\end{bmatrix}, \quad (2.22)$$

where $\theta(\beta)$ is given by (2.4) with ε replaced by $(\vec{p}^2+m^2)^{1/2}$.

3. Covariant operator formalism of gauge theory [3]

To apply thermo field dynamics discussed in §2 to the case of gauge fields, we recapitulate the points relevant here of the covariant operator formalism of gauge theory based upon the BRS invariance [3]. The Lagrangian density

$$\mathcal{L} = -F^a_{\mu\nu}F^{a,\mu\nu}/4+\mathcal{L}_{matter}(\varphi,\mathcal{D}_\mu\varphi)-A^a_\mu\partial^\mu B^a+\alpha B^a B^a/2-i\partial^\mu \bar{c}^a(D_\mu c)^a \quad (3.1)$$

is invariant under the BRS transformation whose generator is given by

$$Q_B \equiv \int d^3x[B^a(D_0c)^a-\dot{B}^a c^a+(i/2)g\dot{\bar{c}}^a(c\times c)^a]; \quad (3.2)$$

$$[iQ_B,A_\mu] = D_\mu c,\ [iQ_B,B] = 0,\ [iQ_B,\varphi] = igc^aT^a\varphi,$$

$$\{iQ_B,c\} = -(g/2)c\times c,\ \{iQ_B,\bar{c}\} = iB. \quad (3.3)$$

Although the state space $\mathcal{V}$ necessarily contains unphysical particles with negative norms, they have been shown [3] to decouple from all the physical processes by quite a general norm-cancellation mechanism called quartet mechanism. This is based upon the subsidiary condition specifying the physical subspace $\mathcal{V}_{phys}\equiv\{|phys>\}$ given by

$$Q_B|phys> = 0, \quad (3.4)$$

and the relation for the projection operator P onto the subspace $\mathcal{V}_{phys}$ of states containing physical particles alone,

$$P+\{Q_B,R\} = 1. \quad (3.5)$$

The basic properties of Q_B leading to (3.5) are the following:

$$Q_B^2 = \{Q_B,Q_B\}/2 = 0, \quad (3.6)$$

$$[iQ_c,Q_B] = Q_B, \quad (3.7)$$

where Q_c is the Faddeev-Popov (FP) charge

$$[iQ_c,c] = c,\ [iQ_c,\bar{c}] = -\bar{c},\ \text{otherwise } [iQ_c,\Phi] = 0. \quad (3.8)$$

Since the positive semi-definite space $\mathcal{V}_{phys}$ contains zero-norms, the usual Hilbert space with positive definite metric is obtained by taking the quotient of $\mathcal{V}_{phys}$ with respect to its zero-norm subspace $\mathcal{V}_0 = Q_B\mathcal{V}$,

$$H_{phys} \equiv \mathcal{V}_{phys}/\mathcal{V}_0 = \{|\hat{\Phi}> \equiv |\Phi>+\mathcal{V}_0;|\Phi>\in\mathcal{V}_{phys}\}, \quad (3.9)$$

in which every physical process is described. The trace operation in the statistical average should also be taken in this Hilbert space H_{phys},

$$<A> = Tr(e^{-\beta\hat{H}}\hat{A})/Tr(e^{-\beta\hat{H}}) = \sum<\hat{i}|e^{-\beta\hat{H}}\hat{A}|\hat{i}>/\sum<\hat{i}|e^{-\beta\hat{H}}|\hat{i}>, \quad (3.10)$$

where $\hat{H}$ and $\hat{A}$ are operators in H_{phys} obtained as quotient mappings from the Hamiltonian H and an observable A in $\mathcal{V}$ satisfying the defining relation for observables [7,3],

$$[Q_B,A]\mathcal{V}_{phys} = 0. \quad (3.11)$$

For the observable A satisfying a stronger condition

$$[Q_B,A] = 0, \quad (3.12)$$

(3.10) can be written [8] in terms of the trace in the total space $\mathcal{V}$ as

$$<A> = Tr(A\exp(-\beta H+\pi Q_c))/Z(\beta);\ Z(\beta) = Tr(\exp(-\beta H+\pi Q_c)), \quad (3.13)$$

by using (3.5) and the fact that Q_c vanishes in $\mathcal{H}_{phys}=P\mathcal{V}$. (3.13) applied (in disregard of (3.12)) to FP ghosts leads to the periodic boundary condition of their temperature Green's functions [9,8].

4. Gauge theory at finite temperature [2]

Following the discussion in §2, we first try to express the r.h.s. of (3.13) in a form of a "vacuum" expectation value

$$\mathrm{Tr}(A\exp(-\beta H+\pi Q_c))/Z(\beta) = \langle 0(\beta)|A|0(\beta)\rangle, \qquad (4.1)$$

irrespective of the condition (3.12) to guarantee the equality in (3.13). In this case, the heuristic formula for $|0(\beta)\rangle$ corresponding to (2.7) is

$$|0(\beta)\rangle = \sum_{k,\ell} \exp(-\beta \bar{E}_k/2-iN_k\pi/2)\eta^{-1}_{k\ell}|k\rangle\otimes|\tilde{\ell}\rangle/Z(\beta)^{1/2} \qquad (4.2)$$

with

$$H|k\rangle = E_k|k\rangle,\ iQ_c|k\rangle = N_k|k\rangle; \qquad (4.3)$$

$$\eta_{k\ell} \equiv \langle k|\ell\rangle \propto \delta(N_k,-N_\ell);\ \langle\tilde{k}|\tilde{\ell}\rangle = \langle\ell|k\rangle = \eta_{\ell k} = \eta^*_{k\ell}. \qquad (4.4)$$

Note that the trace in $\mathcal{V}$ with indefinite metric should be understood as

$$\mathrm{Tr}\,\mathcal{O} \equiv \sum_{k,\ell}(\eta^{-1})_{k\ell}\langle\ell|\mathcal{O}|k\rangle. \qquad (4.5)$$

To generalize the relation (2.11) characterizing $|0(\beta)\rangle$ to the case of gauge theory with indefinite metric and to determine the propagators, we work out explicitly the Bogoliubov transformation leading to (4.1) and (4.2) for the free Abelian gauge theory in Feynman gauge (α=1):

$$|0(\beta)\rangle = \exp[\int d\mu(p)\theta(|\vec{p}|,\beta)\{-a^\dagger_\mu(p)\tilde{a}^{\mu\dagger}(p)+\tilde{a}_\mu(p)a^\mu(p)$$
$$-i(c^\dagger(p)\tilde{\bar{c}}^\dagger(p)+\tilde{c}(p)\bar{c}(p)+\bar{c}^\dagger(p)\tilde{c}^\dagger(p)+\tilde{\bar{c}}(p)c(p)\}]|0\rangle; \qquad (4.6)$$

$$a_\mu(p,\beta) = a_\mu(p)\cosh\theta(|\vec{p}|,\beta)-\tilde{a}^\dagger_\mu(p)\sinh\theta(|\vec{p}|,\beta),$$
$$c(p,\beta) = c(p)\cosh\theta(|\vec{p}|,\beta)-\tilde{c}^\dagger(p)\sinh\theta(|\vec{p}|,\beta),$$
$$\bar{c}(p,\beta) = \bar{c}(p)\cosh\theta(|\vec{p}|,\beta)+\tilde{\bar{c}}^\dagger(p)\sinh\theta(|\vec{p}|,\beta), \qquad (4.7)$$

$$a_\mu(p,\beta)|0(\beta)\rangle = c(p,\beta)|0(\beta)\rangle = \bar{c}(p,\beta)|0(\beta)\rangle = 0, \qquad (4.8)$$

with similar equations to (4.7), (4.8) for $\tilde{a}_\mu$, $\tilde{c}$, $\tilde{\bar{c}}$. According to (2.17) for bosons and (2.19) for fermions, tilde operators are defined by

$$\tilde{a}_\mu = Ja_\mu J,\ \tilde{c} = iJcJ\exp(-\pi\bar{Q}_c),\ \tilde{\bar{c}} = iJ\bar{c}J\exp(-\pi\bar{Q}_c), \qquad (4.9)$$

with the antiunitary J satisfying (2.12), (2.13) and the "total" FP charge

$$\bar{Q}_c \equiv Q_c-JQ_cJ. \qquad (4.10)$$

The relations (4.8) as well as the ones with tildes are unified into

$$J\exp[-(\beta\bar{H}-\pi\bar{Q}_c)/2]\mathcal{O}|0(\beta)\rangle = \mathcal{O}^\dagger|0(\beta)\rangle, \qquad (4.11)$$

which is a generalization of (2.11). The propagators to be used in the Feynman diagrams are given in momentum space as

$$\mathrm{F.T.}\langle 0(\beta)|T\begin{pmatrix}A_\mu\\ \tilde{A}_\mu\end{pmatrix}(A_\nu,\tilde{A}_\nu)|0(\beta)\rangle = -iU_B(|\vec{p}|,\beta)g_{\mu\nu}\begin{bmatrix}1/(p^2+i0) & 0\\ 0 & -1/(p^2-i0)\end{bmatrix}U^\dagger_B(|\vec{p}|,\beta)$$

$$\mathrm{F.T.}\langle 0(\beta)|T\begin{pmatrix}c\\ \tilde{c}\end{pmatrix}(\bar{c},-\tilde{\bar{c}})|0(\beta)\rangle=-U_B(|\vec{p}|,\beta)\begin{bmatrix}1/(p^2+i0) & 0\\ 0 & -1/(p^2-i0)\end{bmatrix}U^\dagger_B(|\vec{p}|,\beta), \qquad (4.12)$$

where $U_B(|\vec{p}|,\beta)$ is given by (2.22) with m=0.

In (4.12), we have allowed the appearance of unphysical negative norms by applying (4.1) to <u>non</u>-observable quantities A_μ, c and $\bar{c}$ for the sake of developing the Feynman diagram method. Therefore, we have to treat again at finite temperature an <u>indefinite-metric</u> space $\mathcal{V}(\beta)$ obtained by applying A_μ, $\tilde{A}_\mu$, c, $\tilde{c}$, etc., to the "vacuum" $|0(\beta)\rangle$. This is taken care of by the "total" BRS charge defined by

$$\bar{Q}_B \equiv Q_B-\tilde{Q}_B \equiv Q_B-iJQ_BJ\exp(-\pi\bar{Q}_c) \qquad (4.13)$$

which satisfies $\bar{Q}_B|0(\beta)\rangle=0$. Since $\bar{Q}_c$ defined by (4.10) and $\bar{Q}_B$ satisfy

$$\bar{Q}_B^2 = 0;\ [i\bar{Q}_c,\bar{Q}_B] = \bar{Q}_B, \qquad (4.14)$$

similarly to (3.6) and (3.7), the <u>quartet mechanism</u> discussed in §3 works again here on the basis of the <u>subsidiary condition</u> for physical states

$$\overline{Q}_B|\text{phys}\rangle = 0;\ \mathcal{V}(\beta)_{\text{phys}} \equiv \{|\Phi\rangle \in \mathcal{V}(\beta); \overline{Q}_B|\Phi\rangle = 0\}, \tag{4.15}$$

ensuring the positive semi-definiteness of $\mathcal{V}(\beta)_{\text{phys}}$. Likewise, an observable $\mathcal{O}$ is defined similarly to (3.11) by the condition

$$[\overline{Q}_B, \mathcal{O}]\mathcal{V}(\beta)_{\text{phys}} = 0, \tag{4.16}$$

which allows us to transfer $\mathcal{O}$ into the physical Hilbert space $H(\beta)_{\text{phys}} \equiv \mathcal{V}(\beta)_{\text{phys}}/\mathcal{V}(\beta)_0$ by taking its quotient mapping $\hat{\mathcal{O}}$. Since $\overline{H}$, $\overline{H}-\pi\overline{Q}_c/\beta$, $\overline{Q}_c$ and J also satisfy (4.16), the relation (4.11) for the observable $\mathcal{O}$ satisfying (4.16) can be transferred into $H(\beta)_{\text{phys}}$,

$$\hat{J}\exp(-[\beta\hat{\overline{H}}-\pi\hat{\overline{Q}}_c]/2)\hat{\mathcal{O}}|0(\beta)\rangle = J\exp(-\beta\hat{\overline{H}}/2)\hat{\mathcal{O}}|0(\beta)\rangle = \hat{\mathcal{O}}^{\dagger}|0(\beta)\rangle, \tag{4.17}$$

where we have used $\hat{\overline{Q}}_c=0$ valid in $H(\beta)_{\text{phys}}$. The relation (2.11) for the standard cases with positive metric is thus recovered in $H(\beta)_{\text{phys}}$.

At the end, we add a few comments. Firstly, as for the renormalization of the Feynman diagrams in this formalism, it has been proved [10] that all the UV divergences at $T\neq 0^{\circ}K$ are removed by the counterterms set up at $T=0^{\circ}K$. Secondly, the Lorentz (boost) invariance of relativistic QFT is shown to be broken spontaneously at $T\neq 0^{\circ}K$ without any Goldstone bosons but with continuous zero-energy spectrum due to particle pairs having opposite energy-momenta [11]. Thirdly, in view of the general condition imposed on the well-defined charges, supersymmetry turns out to be unable to be implemented at $T\neq 0^{\circ}K$ [11]. Finally, the application of this formalism to the curved space-time [12] and to quantum gravity will be of interest.

REFERENCES

1. Y. Takahashi and H. Umezawa, Collective Phenom. 2, 55 (1975); H. Umezawa, H. Matsumoto and M. Tachiki, "Thermo Field Dynamics and Condensed States," North-Holland, Amsterdam/New York/Oxford, 1982.
2. I. Ojima, Ann. Phys. 137, 1 (1981).
3. T. Kugo and I. Ojima, Suppl. Prog. Theor. Phys. No.66 (1979).
4. R. Kubo, J. Phys. Soc. Japan 12, 570 (1957); P. C. Martin and J. Schwinger, Phys. Rev. 115, 1342 (1959).
5. O. Bratteli and D. W. Robinson, "Operator Algebras and Quantum Statistical Mechanics", Springer, Berlin/Heidelberg/New York, 1979.
6. R. Haag, N. M. Hugenholtz and M. Winnink, Comm. Math. Phys. 5, 215 (1967).
7. I. Ojima, Nucl. Phys. B143, 340 (1978); Z. Phys. C5, 227 (1980).
8. H. Hata and T. Kugo, Phys. Rev. D21, 3333 (1980).
9. C. Bernard, Phys. Rev. D9, 3312 (1974).
10. H. Matsumoto, I. Ojima and H. Umezawa, in preparation.
11. I. Ojima, unpablished.
12. W. Israel, Phys. Lett. 57A, 107 (1976).

PATH INTEGRATION AT THE CROSSROAD OF STOCHASTIC AND DIFFERENTIAL CALCULUS

Cécile DeWitt-Morette, Department of Astronomy and Center for Relativity,
The University of Texas, Austin, Texas 78712

Ever since Newton differential calculus has been a very successful language for the description of physical systems; stochastic calculus on the other hand has only recently become an instrument of thought but has already been used in challenging problems. For instance the traditional quantization scheme can be summarized in the following chart.

The description of a physical system starts with its configuration space or its phase space	its dynamics being given by a lagrangian or an hamiltonian

The classical lagrangian or the classical hamiltonian is then used to quantize the system

Lagrangian → Hamiltonian - - -→ Hamiltonian operator

Lagrangian - - -→ Path integration ←- - - Hamiltonian

where the dotted arrows indicate that the corresponding construction is not unique.

The stochastic scheme on the other hand proceeds as follows.

The description of a physical system starts with a fibre bundle on its configuration space or on its phase space	its dynamics being given by a stochastic differential equation

Then there is a unique prescription for the following constructions.

stoch. diff. eqn. → Path integral → Hamiltonian operator

↓

Its WKB approximation
(Lagrangian)

Here classical physics is obtained as the WKB limit of quantum physics.

Before presenting some applications of the stochastic scheme, I recall briefly the key concepts of stochastic calculus. Differential calculus is based on the assumption that $dx(t)$ is of order dt; but if $dx(t)$ is for instance of order $(dt)^{1/2}$, as in brownian motion, then $\lim dx(t)/dt$ is undefined and, at best, one can only speak of the probability that a particle which is at x_o at time t_o will be at $x_o + dx(t)$ at time $t_o + dt$. Thus stochastic calculus begins with a probability space $(\Omega,\mathcal{F},w)$ where Ω is a set of points $\omega \in \Omega$, the events are subsets of Ω which make a σ algebra $\mathcal{F}$, and random variables are measurable functions

$\phi\colon (\Omega,\mathcal{F}) \to (\mathbb{R}^n,$ Borel σ-algebra) henceforth abbreviated to $\phi\colon \Omega \to \mathbb{R}^n$.

Random or stochastic processes are sets of random variables indexed, for instance by time T, $\{x(t)\}$ where

$$x\colon T \times \Omega \to \mathbb{R}^n \quad , \quad x(t)\colon \Omega \to \mathbb{R}^n \quad .$$

w is a measure on Ω such that $w(\Omega) = 1$. For quantum physics the concept of measure has to be generalized to the concept of prodistribution - technically a projective family of tempered distributions on a projective system of finite dimensional spaces [1]; a prodistribution is defined by its Fourier transforms $\mathcal{F}w$. The condition $w(\Omega) = 1$ becomes $\mathcal{F}w(0) = 1$. Here I shall simply refer to measures, promeasures, prodistributions as "integrators".

A stochastic differential equation defines a stochastic process x in terms of a known process z, e.g. let $z\colon T \times \Omega \to \mathbb{R}$ be brownian with $z(t_o) = 0$ and let

$$x(t,\omega) = x_o(\omega) + \int_{t_o}^{t} X(x(t,\omega))dz(t,\omega) \quad \text{where X is given.}$$

It is usually abbreviated to $dx = X\,dz \quad , \quad x(t_o) = x_o$.

Stochastic integrals $\int X(x(t))dz(t)$ when $dz(t)$ is not of order dt have been given meaning by Itô and by Stratonovich.

Very few stochastic differential equations can be solved explicitly, but expectation values of functions of stochastic processes

$$\mathbb{E}\phi(x(t)) := \int_{\Omega} dw(\omega)\phi(x(t,\omega)) := \langle\phi(x(t))\rangle$$

are functions of t and the initial point $x_o = x(t_o)$ whose properties can be determined from the stochastic differential equation satisfied by $x(t)$ without solving it.

The prototype of such a situation is the Feynman-Kac formula: given the stochastic differential equation

$$\begin{cases} dx^{\alpha}(t) = X^{\alpha}_{i}(x(t))dz^{i}(t) + A^{\alpha}(x(t))dt & x(t_o) = x_o \\ dv(t) = V(x(t))v(t)dt & v(t_o) = 1 \end{cases} \qquad (1)$$

Then $\Psi(t,x_o) := \int_{\Omega} dw(\omega)\; v(t,\omega)\phi(x(t,\omega))$ is a solution of the diffusion equation

$$\frac{\partial\Psi}{\partial t} = \sum_i X_i^\alpha(x_o)X_i^\beta(x_o)\frac{\partial^2\Psi}{\partial x_o^\alpha \partial x_o^\beta} + A^\alpha(x_o)\frac{\partial\Psi}{\partial x_o^\alpha} + V(x_o)\Psi := \mathfrak{B}\Psi$$

$$\Psi(t_o,x_o) = \phi(x_o) \quad . \tag{2}$$

To obtain a path integral solution of the Schrödinger equation for $H = i\hbar\mathfrak{B}$ with $\mathfrak{B} = \frac{i}{2}\frac{\hbar}{m}\Delta + A^\alpha \frac{\partial}{\partial x_o} + \frac{1}{i\hbar} V$ one must modify the above scheme and work with complex gaussian (technically prodistributions) over the space of paths vanishing at t rather than t_o. Note that the Schrödinger equation obtained from the stochastic process (1) where V is replaced by $V/i\hbar$, is not the equation of a particle in an electromagnetic potential; indeed there should be an A^2 term. One can, of course, modify the stochastic equation (1) to bring out the desired terms but then one is faced with having to choose the order of the factors in the hamiltonian operator. On the other hand if one sets up the problem on the appropriate fibre bundle one has an obvious choice for the stochastic differential equation. In the case of a particle in an electromagnetic field the appropriate fibre bundle is the U(1) bundle over the configuration space. Given a fibre bundle the "obvious" choice of stochastic differential equation is described in the two following examples where (i) the fibre bundle is a principal fibre bundle (ii) the fibre bundle is an associated vector bundle [3], [7].

i) A particle of mass m in a riemannian space M, dimension n, metric g, with the riemannian connection.

Let O(M) be the orthonormal frame bundle, $p \in O(M)$ is a pair (x,u) where $x \in M$ and u is a frame at x

$$u: \mathbb{R}^n \to T_x M$$

Given a path $x: T \to M$, $x(t_o) = x_o$, and a frame u_o at x_o, the connection defines a path $u: T \to O(M)$, $u(t_o) = u_o$ and a map

$$X: O(M) \times \mathbb{R}^n \to T\, O(M)$$

such that

$$\frac{du(t)}{dt} = X(u(t))(u(t))^{-1}\frac{dx(t)}{dt} \quad .$$

Recall that $dx(t)/dt \in T_{x(t)}M$ and $(u(t))^{-1}dx(t)/dt \in \mathbb{R}^n$. If the path x is not differentiable, one can define a stochastic frame by the corresponding stochastic differential equation (in the Stratonovich sense because Itô calculus is not tensorial)

$$du(t) = \mu X(u(t))dz(t) \quad ,$$

where $\mu = (\hbar/m)^{1/2}$ and z is brownian, the brownian path z being multipled by μ, so that μz has the dimension of length. A stochastic path on M is the projection of a path on the bundle: $x(t) = \pi(p(t))$. Given an arbitrary (good) function ϕ: $M \to \mathbb{R}$, then $\int_\Omega dw(\omega)\phi(\pi(p(t,\omega))$ is the path integral representation of the solution $\Psi(t,x_o)$ of the diffusion equation

$$\begin{cases} \dfrac{\partial\Psi}{\partial t} = \dfrac{1}{2}\mu^2 \Delta \Psi \\ \Psi(t_o,x_o) = \phi(x_o) \end{cases}$$

where Δ is the Laplace-Beltrami operator on M defined by the metric g on M. It is an easy matter to write the stochastic differential equation which gives a diffusion equation with a vector and a scalar potential. The corresponding path integral has been computed as an expansion in powers of μ for the terms of order μ^{-2}, μ^{-1}, μ^0 in [1] and more recently for the terms of order μ^1 by a recursion method [2] which in principle can give higher order terms in μ.

ii) A particle in a gauge field in flat space.

Let $\tau_{t_o}^t \phi(x(t))$ be the parallel transport from t to t_o of $\phi(x(t))$ along the stochastic path $x(t) = \mu z(t)$ for $\mu = (\hbar(m))^{1/2}$ and z brownian. Then $\int_\Omega dw(\omega)\tau_{t_o}^t \phi(x(t,\omega))$ is the path integral solution $\Psi(t,x_o)$ of the diffusion equation

$$\begin{cases} \dfrac{\partial\Psi}{\partial t} = \dfrac{1}{2}\mu^2 \Delta \Psi \\ \Psi(t_o,x_o) = \phi(x_o) \end{cases}$$

where Δ is the laplacian constructed from the covariant derivatives defined by the gauge connection.

Both examples can be summarized by saying that a stochastic connection yields a diffusion equation with covariant derivatives. Starting from a stochastic process on a fibre bundle has the following advantages:

i. It is a unifying scheme which applies to a wide class of apparently different problems [3].

ii. It gives simple answers to such problems as parallel transport along a brownian path, short time propagator on riemannian manifolds, canonical relationship between lagrangian function and hamiltonian operator.

iii. It is cast in a framework which guarantees gauge invariance.

The Feynmn-Kac formula and its generalization to stochastic processes on fibre bundles is but a small example of the bartering which goes on at the crossroad of stochastic and differential calculus [4]. Stochastic calculus is also used in quantum field theory [5], [6].

[1] C. DeWitt-Morette, A. Maheshwari and B. Nelson, "Path Integration in Non-Relativistic Quantum Mechanics," Physics Reports 50 (1982), 255-372.

[2] K. D. Elworthy and A. Truman, "The Diffusion Equation and Classical Mechanics: An elementary formula". To appear in Stochastic Processes in Quantum Theory and Statistical Physics: Recent Progress and Applications, S. A. Albeverio, M. Sirugue and M. Sirugue-Collin, eds., (Springer Verlag Lecture Notes in Physics).

[3] C. DeWitt-Morette, K. D. Elworthy, B. L. Nelson and G. S. Sammelmann, "A stochastic scheme for constructing solutions of the Schrödinger equation," Ann. Inst. Henri Poincaré 32 (1980), 327-341.

[4] C. DeWitt-Morette, "Path Integration Quantization" (An expanded version of this talk with a more complete bibliography, presented at the III Marcel Grossmann conference, Shanghai, August 1982).

[5] B. Simon, The $P(\phi)_2$ Euclidean (Quantum) Field Theory, Princeton University Press (1974).

[6] J. Glimm and A. Jaffe, Quantum Physics: A Functional Integral Point of View, Springer Verlag, New York (1981).

[7] K. D. Elworthy, Stochastic Differential Equations on Manifolds, Cambridge University Press (1982).

Credit

Support in part by NSF grant No. PHY81-07381.

Manifestly Covariant Canonical Formalism of Quantum Gravity — A Brief Survey —

Noboru Nakanishi

Research Institute for Mathematical Sciences
Kyoto University, Kyoto 606, Japan

General relativity and quantum field theory are two brilliantly successful fundamental theories of physics. It is, therefore, of very fundamental importance to unify both theories in a consistent and beautiful manner, because if it were impossible to do so then either general relativity or quantum field theory might have to be abandoned in order to achieve the ultimate unified theory. The purpose of this talk is to claim that a quantum field-theoretical formalism of gravity has been formulated in quite a satisfactory way.

Unfortunately, however, there seems to be no consensus about what theory of quantum gravity is to be called satisfactory. Relativists usually expect that quantum gravity should inherit the geometrical concept of general relativity. On the contrary, according to most particle physicists, quantum gravity is nothing more than a quantum field theory of massless spin-two particles, and though the Einstein gravity is unitary and Lorentz invariant, it is not satisfactory because it is not renormalizable. Their primary concern in quantum gravity is thus not the connection between the spacetime structure and gravity at the quantum level but how to remove the ultraviolet divergence in perturbation theory.

I completely disagree to such a standpoint. It is totally nonsense to discuss the divergence problem of quantum gravity in perturbation theory, because Feynman integrals become meaningless when the contributions from the energies greater than the Planck mass are significant. In other words, since the gravitational constant κ is a fundamental constant at the same level as the light velocity c and the Planck constant $\hbar$, the perturbation expansion in quantum gravity is mathematically inadequate just as the expansion in powers of $1/c$ or $\hbar$ is. Accordingly, quantum gravity must be formulated in a non-perturbative way. Of course, it is a very important and extremely difficult problem to invent a divergence-free non-perturbative approximation method, but I emphasize that it is a problem at a stage different from constructing a satisfactory formalism itself. I believe that the invention of an appropriate approximation method is not a prerequisite for the judgement that the formalism is satisfactory.

I make some comments on the path-integral formalism because one might assert that it already provides a satisfactory formalism for quantum gravity. The path-integral formalism is well-formulated for scalar field theories and it yields correct results in perturbation theory. I must emphasize, however, that in gauge theories and in quantum gravity, the non-perturbative approach based on the path-integral formalism has no justification for its validity, because the proof of unitarity has been made only in the perturbative way. Since the non-perturbative functional measure of a path-integral is quite ambiguous, its definition should be made so as to guarantee unitarity, but there is no idea for finding such a definition. I believe that the path-integral formalism may be at best a convenient calculational technique, but it cannot be worth being called a fundamental formalism. Indeed, given an operator formalism, it may be possible to derive the corresponding path-integral, but the converse is generally impossible. This is because the path-integral formalism does not contain complete information; for example, it has no information concerning the hermiticity assignment for field operators in the indefinite-metric case.

In my opinion, a satisfactory operator formalism should be based on the following four principles:

1. Lagrangian and canonical formalism.
2. Manifest covariance.
3. Indefinite-metric Hilbert space with subsidiary conditions.
4. Asymptotic completeness.

I believe that mathematical rigor should not be regarded as a basic principle. The positivity of Hilbert-space metric is mathematically very convenient, but it is rather a source of various pathological features of quantum field theory. As is well known, it is inevitable to use indefinite metric in manifestly covariant gauge theories. Though a consistent positive-metric formalism of quantum electrodynamics is possible in the Coulomb gauge, I conjecture that no consistent non-perturbative formalism is possible in the positive-metric Hilbert space for non-abelian gauge theories nor for quantum gravity.

The celebrated operator formalism of a gauge theory is the Gupta-Bleuler formalism[1] for quantum electrodynamics. It is a satisfactory formalism from my point of view. But I note that it is more reasonable to introduce an auxiliary scalar field $B(x)$ into the fundamental Lagrangian. I believe it natural to have a gauge-fixing condition as an independent field equation. Then $B(x)$ should be regarded as one of fundamental fields. I wish to call such a formalism, in general, "B-field formalism." The Landau gauge can be properly formulated only in the B-field formalism.

It took longer than a quarter of a century to achieve the correct extension of the Gupta-Bleuler formalism to non-abelian gauge theories. This too much long delay has, unfortunately, brought people's blind confidence in the path-integral formalism. The manifestly covariant canonical formalism of non-abelian gauge theories was formulated quite successfully by Kugo and Ojima.[2,3] The basic ingradients of the Kugo-Ojima formalism are as follows:

1. B-field formalism.
2. Correct hermiticity assignment of the Faddeev-Popov (FP) ghosts.
3. Becchi-Rouet-Stora (BRS) invariance.[4]

Then it is crucial to note that there exists the BRS charge Q_B and that it is nilpotent ($Q_B^2=0$) and hermitian ($Q_B^\dagger = Q_B$). The physical states are defined by a subsidiary condition

$$Q_B|\text{phys}\rangle = 0.$$

Then one can quite generally prove the unitarity of the physical S-matrix under the postulate of asymptotic completeness.[3,5] This fact is extremely important because it is the only non-perturbative proof of unitarity in non-abelian gauge theories.

Now, I proceed to the manifestly covariant canonical formalism of quantum gravity.[6-20] As in the Kugo-Ojima formalism, it is natural to introduce the B-field $b_\rho(x)$ and a pair of the hermitian FP ghosts $c^\sigma(x)$ and $\bar{c}_\tau(x)$ and set up the fundamental Lagrangian in such a way that the action integral is BRS invariant. But it must be remarked that there is an important difference from the case of the Yang-Mills theory: The invariance under the general coordinate transformation in general relativity is a spacetime symmetry. Hence the corresponding BRS invariance is also a spacetime symmetry. I defined the "intrinsic" BRS transformation $\underline{\delta}$, as a fermionic derivation satisfying $\underline{\delta}^2=0$, by[6]

$$\underline{\delta}(\Phi^{\mu_1\cdots\mu_k}{}_{\nu_1\cdots\nu_\ell}) = \kappa(\sum_{i=1}^{k}\partial_\lambda c^{\mu_i}\cdot\Phi^{\mu_1\cdots\lambda\cdots\mu_k}{}_{\nu_1\cdots\nu_\ell} - \sum_{j=1}^{\ell}\partial_{\nu_j}c^\lambda\cdot\Phi^{\mu_1\cdots\mu_k}{}_{\nu_1\cdots\lambda\cdots\nu_\ell})$$

for any classical tensor $\Phi^{\mu_1\cdots\mu_k}{}_{\nu_1\cdots\nu_\ell}$ and by (The signs of $\underline{\delta}$ and $\bar{c}_\rho$ are changed for convenience.)

$$\underline{\delta}(x^\mu) = \kappa c^\mu, \quad \text{whence} \quad \underline{\delta}(c^\mu) = 0,$$

$$\underline{\delta}(\bar{c}_\nu) = ib_\nu, \quad \text{whence} \quad \underline{\delta}(b_\nu) = 0,$$

$$[\underline{\delta},\partial_\nu] = -\kappa\partial_\nu c^\lambda\cdot\partial_\lambda.$$

Here $\underline{\delta}(x^\mu) = \kappa c^\mu$ is owing to the basic rule of constructing the BRS transform, namely, the rule that the infinitesimal transformation function is replaced by one of FP ghosts. The vanishing of $\underline{\delta}(c^\mu)$ represents the abelian nature of the translation which is the global version of the general coordinate transformation. The more conventional BRS transformation,[21-25] which I denote here by $\underline{\delta}_*$, is given by

$$\underline{\delta}_*(\varphi) = \underline{\delta}(\varphi) - \kappa c^\lambda\partial_\lambda\varphi$$

for any field $\varphi(\neq b_\nu, \bar{c}_\nu)$. The second term is the "orbital" part of the BRS transformation.

The gauge-fixing term $\mathcal{L}_{GF}$ is chosen so as to be a scalar density under the general linear transformation. I believe that the requirement of general linear invariance for gauge fixing is quite natural, because the introduction of the Minkowski metric $\eta_{\mu\nu}$ into the fundamental Lagrangian is too abrupt and destroys the spirit of general relativity unnecessarily violently. The simplest, most natural expression for $\mathcal{L}_{GF}$ is given by

$$\mathcal{L}_{GF} = -\kappa^{-1}\tilde{g}^{\mu\nu}\partial_\mu b_\nu,$$

where $\tilde{g}^{\mu\nu} \equiv \sqrt{-g}\, g^{\mu\nu}$ with $g \equiv \det g_{\mu\nu}$. It is noteworthy that general linear invariance can be realized only in the framework of the B-field formalism.

The FP-ghost term $\mathcal{L}_{FP}$ is determined by the requirement that

$$\sqrt{-g}^{\,-1}(\mathcal{L}_{GF} + \mathcal{L}_{FP}) = \underline{\delta}(i\kappa^{-1}g^{\mu\nu}\partial_\mu \bar{c}_\nu);$$

then

$$\mathcal{L}_{FP} = -i\tilde{g}^{\mu\nu}\partial_\mu \bar{c}_\rho \cdot \partial_\nu c^\rho .$$

The total Lagrangian is given by

$$\mathcal{L} = \mathcal{L}_E + \mathcal{L}_{GF} + \mathcal{L}_{FP} + \mathcal{L}_M,$$

where $\mathcal{L}_E$ is the Einstein Lagrangian,

$$\mathcal{L}_E = (2\kappa)^{-1}\sqrt{-g}\, R,$$

and $\mathcal{L}_M$ denotes the matter-field Lagrangian.

It is very important to note that $\mathcal{L}_{FP}$ contains simple derivatives only but no covariant derivative. Its form is quite similar to the FP-ghost term of quantum electrodynamics. Though it is widely believed that quantum gravity is similar to the Yang-Mills theory, I emphasize that quantum gravity is much more similar to quantum electrodynamics. This feature is, of course, the consequence of the abelian nature of the translation group. The reason why people could not be aware of this very simple observation is that they did not make clear separation between the intrinsic part and the orbital one.

Since the operator formalism which follows from the above Lagrangian $\mathcal{L}$ is of outstanding beauty, I wish to claim that it is the "correct" formalism of quantum gravity. One might say that since the choice of $\mathcal{L}_{GF}+\mathcal{L}_{FP}$ is rather arbitrary, the fundamental thing is the classical Lagrangian. But I disagree to this opinion. Since quantum theory is to be more fundamental than the classical theory, the quantum Lagrangian must be more fundamental than than the classical one. Hence I believe that the expression for $\mathcal{L}_{GF} + \mathcal{L}_{FP}$ should be uniquely determined in the correct theory. This standpoint is very crucial also in considering quantum field theory in a background curved spacetime. Since it is usually constructed on the basis of classical general relativity, Duff[26] has criticized it by pointing out that the starting Lagrangian is quite ambiguous owing to the freedom of field redefinition. His difficulty is totally resolved if one starts with the full quantum-gravity theory having the uniquely specified gauge fixing.

Now, the field equations which follow from $\mathcal{L}$ are as follows:

$$R_{\mu\nu} - \frac{1}{2} g_{\mu\nu}R - E_{\mu\nu} + \frac{1}{2} g_{\mu\nu}E = -\kappa T_{\mu\nu} \qquad \text{(quantum Einstein equation)},$$

with

$$E_{\mu\nu} \equiv \partial_\mu b_\nu + i\kappa\partial_\mu \bar{c}_\rho \cdot \partial_\nu c^\rho + (\mu \leftrightarrow \nu),$$

$$\partial_\mu \tilde{g}^{\mu\nu} = 0 \qquad \text{(de Donder condition)},$$

$$\partial_\mu(\tilde{g}^{\mu\nu}\partial_\nu c^\rho) = 0,$$

$$\partial_\mu(\tilde{g}^{\mu\nu}\partial_\nu \bar{c}_\rho) = 0.$$

Taking covariant derivative of the quantum Einstein equation, I obtain a remarkably simple equation,[6]

$$\partial_\mu(\tilde{g}^{\mu\nu}\partial_\nu b_\rho) = 0.$$

The four equations other than the quantum Einstein equation can be put together into

$$\partial_\mu(\tilde{g}^{\mu\nu}\partial_\nu X) = 0,$$

where $X = (x^\lambda/\kappa, b_\rho, c^\sigma, \bar{c}_\tau)$. This remark will become very important later.

Canonical quantization can be carried out consistently without employing Dirac's method of quantization[27] (which has serious difficulty in the quantum version of constraints[28]). The second derivatives of $g_{\mu\nu}$ in $\mathcal{L}_E$ and the derivative of b_ρ in $\mathcal{L}_{GF}$ are eliminated by integrating by parts. Canonical fields are $g_{\mu\nu}$, c^σ, $\bar{c}_\tau$, and matter fields; b_ρ is <u>not</u> regarded as a canonical field. Despite this difference from the standard canonical formalism, one can prove the equivalence between field equations and Heisenberg equations.[7]

It is quite remarkable that all equal-time (anti-)commutation relations between any two fields and between any field and the first time-derivative of any field can be calculated explicitly <u>in closed form</u>.[6] For example, I have

$$[g_{\mu\nu}(x), b_\rho(y)]_0 = -i\kappa(\tilde{g}^{00})^{-1}(\delta^0_\mu g_{\rho\nu} + \delta^0_\nu g_{\mu\rho})\delta^3(x-y),$$

$$[g_{\mu\nu}(x), \dot{g}_{\rho\lambda}(y)]_0 = -2i\kappa(\tilde{g}^{00})^{-1}[g_{\mu\nu}g_{\rho\lambda} - g_{\mu\rho}g_{\nu\lambda} - g_{\mu\lambda}g_{\nu\rho}$$

$$+ (g^{00})^{-1}(\delta^0_\mu\delta^0_\rho g_{\nu\lambda} + \delta^0_\mu\delta^0_\lambda g_{\nu\rho} + (\mu\leftrightarrow\nu))]\delta^3(x-y),$$

where a subscript 0 of a commutator indicates to set $x^0 = y^0$.

From the BRS Noether current, I can calculate the expression for the gravitational BRS charge Q_b. By using the quantum Einstein equation and dropping total divergence, I find[6]

$$Q_b = \int d^3x\, \tilde{g}^{0\lambda}(b_\rho \partial_\lambda c^\rho - \partial_\lambda b_\rho \cdot c^\rho),$$

which is nilpotent ($Q_b^2=0$) and satisfies

$$i[Q_b, \varphi]_\mp = \underline{\delta}(\varphi) - \kappa c^\lambda \partial_\lambda \varphi$$

for any field $\varphi(x)$. The subsidiary condition is set up by

$$Q_b|\text{phys}\rangle = 0.$$

Then the unitarity of the physical S-matrix can be proved without recourse to perturbation theory.

The presence of the orbital term in $[Q_b, \varphi]_\mp$ is a very important characteristic of quantum gravity. As its consequence, any local operator $\Phi(x)$ has a non-vanishing (anti-)commutator with Q_b unless $\Phi(x)$ itself is a BRS transform of another operator. Hence, in contrast with the gauge-theory case, $\Phi(x)|0\rangle$ is not a physical state for any non-trivial local operator $\Phi(x)$. Accordingly, Lehmann's spectral function of any non-vanishing two point function can acquire the contribution from negative-norm intermediate states without contradicting unitarity. Thus Lehmann's theorem[29] breaks down in quantum gravity, that is, the exact two-point function may have a milder (perhaps oscillatory) high-energy asymptotic behavior than that of the corresponding free Feynman propagator.[30] This fact supports the old expection that quantum gravity may provide a natural ultraviolet cutoff, without employing any kind of approximation.

The above remark also resolves the Goto-Imamura difficulty[31] for the current-current commutator

$$\langle 0|[j_0(x), j_k(y)]_0|0\rangle.$$

Despite the fact that it must be zero if one relies upon the canonical anticommutation relations, one usually assumes the existence of the Schwinger term[32] in order to avoid the contradiction with the result of general framework. As is well known, however, the Schwinger term is the most pathological beast in quantum field theory. If quantum gravity is taken into account, one cannot prove the non-vanishing of the above commutator because of the presence of negative-norm intermediate states.[30] Thus the Goto-Imamura difficulty is resolved without introducing the Schwinger term.

It is straightforward to define the canonical energy-momentum tensor including gravity. It is interesting to note that the symmetric energy-momentum tensor cannot be defined so as to be a tensor density under the general linear transformation in contrast with the canonical one. By using the quantum Einstein equation and dropping total divergence, I find that the translation generator P_μ is given by a remarkably simple expression[8]

$$P_\mu = \kappa^{-1}\int d^3x\ \tilde{g}^{0\lambda}\partial_\lambda b_\mu$$

independently of the expression for $\mathcal{L}_M$. Of course, the well-defined energy-momentum operator depends on $\mathcal{L}_M$, but the above expression is sensible as a translation generator, because the volume integration should be carried out after commutator is taken.

Likewise, the generator of the general linear transformation is shown to be[8]

$$\hat{M}^\mu{}_\nu = \kappa^{-1}\int d^3x\ \tilde{g}^{0\lambda}[x^\mu\partial_\lambda b_\nu - \delta^\mu_\lambda b_\nu - i\kappa(\bar{c}_\nu\partial_\lambda c^\mu - \partial_\lambda\bar{c}_\nu\cdot c^\mu)].$$

It should be noted that $\hat{M}^\mu{}_\nu$ cannot be a well-defined operator. Indeed, if it were well-defined, one could exponentiate it, that is, one could consider finite general linear transformations. Then the equal-time (anti-)commutation relations imply that any two fields would (anti-)commute at all non-zero spacetime separations. Of course, however, $\hat{M}^\mu{}_\nu$ is sensible as a generator. For example, I have

$$i[\hat{M}^\mu{}_\nu, g_{\sigma\tau}] = \delta^\mu_\sigma g_{\nu\tau} + \delta^\mu_\tau g_{\sigma\nu} + x^\mu\partial_\nu g_{\sigma\tau}.$$

It is very important to note that general linear invariance is necessarily spontaneously broken. Since translational invariance should not be spontaneously broken, the vacuum expectation value of $g_{\sigma\tau}(x)$ must be a constant. Then I can set

$$\langle 0|g_{\sigma\tau}(x)|0\rangle = \eta_{\sigma\tau}$$

without loss of generality just as in the case of the Higgs model in which one can assume without loss of generality that the vacuum expectation value of the complex scalar field is a positive constant. From the above two formulae, I find

$$i\langle 0|[\hat{M}^{\mu}{}_{\nu},g_{\sigma\tau}]|0\rangle = \delta^{\mu}_{\sigma}\eta_{\nu\tau} + \delta^{\mu}_{\tau}\eta_{\sigma\nu} \neq 0.$$

Thus $\hat{M}^{\mu}{}_{\nu}$ is spontaneously broken. But its antisymmetric part

$$\bar{M}_{\alpha\beta} \equiv \eta_{\alpha\mu}\hat{M}^{\mu}{}_{\beta} - \eta_{\beta\mu}\hat{M}^{\mu}{}_{\alpha}$$

is not broken. It is nothing but the Lorentz generator in the absence of spinor fields. The gravitational field is the Goldstone field of the broken ten components of $\hat{M}^{\mu}{}_{\nu}$.[33] Thus the physical graviton mass must vanish exactly.

Now, the most remarkable result of the manifestly covariant canonical formalism of quantum gravity is the existence of a sixteen-dimensional supersymmetry.[14,15] As mentioned already, the (super)current

$$\mathcal{P}^{\mu}(X) \equiv \tilde{g}^{\mu\nu}\partial_{\nu}X$$

is conserved, where $X = (x^{\lambda}/\kappa, b_{\rho}, c^{\sigma}, \bar{c}_{\tau})$, which is natural to be called the sixteen-dimensional supercoordinate. Let X and Y be two sixteen-dimensional supercoordinates. Then $\partial_{\mu}\mathcal{P}^{\mu}(X) = 0$ implies that

$$\mathcal{M}^{\mu}(X,Y) \equiv \sqrt{\epsilon(X,Y)}\,\tilde{g}^{\mu\nu}(X\partial_{\nu}Y - \partial_{\nu}X\cdot Y)$$

is also conserved, where $\epsilon(X,Y) = -1$ if both X and Y are FP ghosts and $\epsilon(X,Y) = +1$ otherwise; $\sqrt{+1} = +1$, $\sqrt{-1} = +i$. From those conservation laws, I obtain conserved (super)charges

$$P(X) \equiv \int d^3x\,\mathcal{P}^0(X),$$

$$M(X,Y) \equiv \int d^3x\,\mathcal{M}^0(X,Y).$$

Here X in $P(X)$ and X and Y in $M(X,Y)$ are not arguments but indices, each of which takes 16 values.

From the definition of $M(X,Y)$, it is evident that

$$M(Y,X) = -\,\epsilon(X,Y)M(X,Y),$$

whence one sees that $M(X,Y)$ has 128 independent components. Since, of course, $P(X)$ has 16 independent components, the theory possesses 144 independent symmetry generators altogether. The previously given generators P_{μ}, $\hat{M}^{\mu}{}_{\nu}$ and Q_b are expressible in terms of $P(X)$ and $M(X,Y)$.

By calculating the (anti-)commutators with field operators, I can determine the symmetry transformation laws corresponding to $P(X)$ and $M(X,Y)$, and verify the invariance of the action under them. It is very important to note that those transformation laws could not be discovered if quantization were made by the path-integral formalism. The Noether (super)currents of those symmetries can be confirmed to reproduce the original (super)currents $\mathcal{P}^{\mu}(X)$ and $\mathcal{M}^{\mu}(X,Y)$ apart from total divergence.[19]

The generators P(X) and M(X,Y) form a superalgebra quite similar to the Poincaré algebra. Hence I call it "sixteen-dimensional Poincaré-like superalgebra" (Some people[34,35] call it "choral symmetry" because it was proposed in the nineth paper of the series.). I define the sixteen-dimensional supermetric $\eta(X,Y)$ by

$$\eta(x^{\lambda}/\kappa, b_{\rho}) = \eta(b_{\rho}, x^{\lambda}/\kappa) = \eta(c^{\lambda}, \bar{c}_{\rho}) = -\eta(\bar{c}_{\rho}, c^{\lambda}) = \delta^{\lambda}_{\rho},$$

$$\eta(X,Y) = 0 \quad \text{otherwise.}$$

Then the (anti-)commutation relations between generators are as follows:[15]

$$[P(X),\ P(Y)]_{\mp} = 0,$$

$$[M(X,Y),\ P(U)]_{\mp} = \sqrt{-\epsilon(XY,U)}\,[\eta(Y,U)P(X) - \epsilon(X,Y)(X\leftrightarrow Y)],$$

$$[M(X,Y),\ M(U,V)]_{\mp} = \sqrt{-\epsilon(XY,UV)}\,\{[\eta(Y,U)M(X,V) - \epsilon(U,V)\eta(Y,V)M(X,U)] - \epsilon(X,Y)[X\leftrightarrow Y]\}.$$

This superalgebra is a natural "super" version of the Poincaré algebra.

The symmetry implied by a generator is a spacetime symmetry if and only if it has the B-field b_{ρ} as an index. Thus the sixteen-dimensional Poincaré-like superalgebra contains <u>both spacetime and internal symmetries in complete harmony</u>. Furthermore, it is quite remarkable that it realizes, in some sense, the <u>democracy between spacetime coordinates and quantum fields</u>.

Some symmetries among the 144 generators are necessarily spontaneously broken. Unbroken ones are 10 Poincaré generators and 74 M(X,Y)'s involving no spacetime as an index. The Ward-Takahashi-type identities

$$\langle 0|[M(X,Y),\ \mathcal{O}]_{\mp}|0\rangle = 0$$

hold for those 74 generators, where $\mathcal{O}$ denotes an arbitrary T-product of field operators. The perturbation-theoretical validity of those Ward-Takahashi-type identities has been confirmed at one-loop level.[18]

Though the sixteen-dimensional Poincaré-like superalgebra includes no particle-physics symmetry other than the Poincaré algebra, in quantum gravi-electrodynamics it is possible to extend this superalgebra so as to include the electromagnetic U(1) symmetry.[17] In this way, therefore, there might be a possibility of unifying all physically relevant symmetries without contradicting the no-go theorem.[36,37]

Now, another very interesting feature of the manifestly covariant canonical formalism of quantum gravity is a revival of general covariance at the purely quantum level; more precisely, in this theory tensor analysis becomes relevant for certain commutation relations. Since the inevitable violation of general covariance in quantization has been quite regretable from the point of view of relativists, this revival of general covariance is quite noteworthy as the evidence showing that the theory is the rightful successor of Einstein's general relativity.

Since the B-field b_ρ is not a canonical field, the equal-time commutator between b_ρ and a canonical field may not necessarily vanish. It is found that the commutator between b_ρ and a local operator which is a tensor at the classical level has, in general, quite remarkable regularity. Let $\Phi^{\mu_1\cdots\mu_k}{}_{\nu_1\cdots\nu_\ell}(x)$ be a tensor generically. Then the general form of the equal-time commutator is[13]

$$[\Phi^{\mu_1\cdots\mu_k}{}_{\nu_1\cdots\nu_\ell}(x), b_\rho(y)]_0$$

$$= i\kappa(\tilde{g}^{00})^{-1}\Big[\sum_{i=1}^{k}\delta_\rho^{\mu_i}\delta^0_{\mu_i'}\Phi^{\mu_1\cdots\mu_i'\cdots\mu_k}{}_{\nu_1\cdots\nu_\ell} - \sum_{j=1}^{\ell}\delta_\rho^{\nu_j'}\delta^0_{\nu_j}\Phi^{\mu_1\cdots\mu_k}{}_{\nu_1\cdots\nu_j'\cdots\nu_\ell}\Big]\delta^3(x-y).$$

This commutation relation is tensorlike in the sense that it is consistent with the rules of tensor analysis, that is, its form is preserved in raising or lowering tensor indices, in constructing tensor product of two tensors, and in contracting upper and lower indices. The validity of the above tensorlike commutation relation has been verified explicitly a large number of examples including the Ricci tensor $R_{\mu\nu}$.

Quite surprisingly, the tensorlike commutation relation can be extended into the four-dimensional form.[20] The four-dimensional commutation relation between b_ρ and a tensor consists of two parts: The main part is tensorlike, consistent with taking covariant derivative, and manifestly affine (i.e., translation and general linear) covariant, while the remaining part is of different character and has the same form as the four-dimensional commutation relation between an FP ghost and that tensor. Hence, in particular, one sees that the equal-time commutation relation between a tensor and b_ρ is tensorlike with no additional terms if and only if that tensor commutes with an FP ghost at the equal time.

In discussing the four-dimensional commutation relation, the quantum-gravity extension of the Pauli-Jordan invariant D-function has been introduced.[20] Since the metric tensor is now an operator, the new invariant D-function, which I denote by $\mathcal{D}(x,y)$, must be a bilocal operator. It is uniquely defined by the following four properties:

(1) $\mathcal{D}(x,y) = -\mathcal{D}(y,x)$.

(2) $\partial^x_\mu[\tilde{g}^{\mu\nu}(x)\partial^x_\nu\mathcal{D}(x,y)] = 0.$

(3) $\mathcal{D}(x,y)|_0 = 0.$

(4) $\partial_0^x \mathcal{D}(x,y)|_0 = -[\tilde{g}^{00}(x)]^{-1}\delta^3(x-y).$

[Here, even if the operator ordering is reversed in (2), the defined $\mathcal{D}(x,y)$ can be shown to be the same.] Then $\mathcal{D}(x,y)$ can be shown to be affine invariant in the sense that

$$i[P_\nu, \mathcal{D}(x,y)] = (\partial_\nu^x + \partial_\nu^y)\mathcal{D}(x,y),$$

$$i[\hat{M}^\mu{}_\nu, \mathcal{D}(x,y)] = (x^\mu\partial_\nu^x + y^\mu\partial_\nu^y)\mathcal{D}(x,y).$$

But, of course, $\mathcal{D}(x,y)$ is not invariant under finite general linear transformations. It has no c-number lightcone singularity, and therefore the short-distance expansion breaks down in quantum gravity. This fact is important in order for quantum gravity to play the role of a natural regulator.

In all the above, the gravitational field has been described by the metric tensor, but when there are Dirac fields, the fundamental gravitational field must be the vierbein (tetrad). Since the six additional degrees of freedom in the vierbein are of local Lorentz transformations, quantum theory can be constructed quite similarly to the Kugo-Ojima formalism for the Yang-Mills field.

It is very crucial that the local-Lorentz gauge-fixing term is chosen to be a scalar density under the general coordinate transformation. The right expression turns out to be[10]

$$\mathcal{L}_{LGF} = -\tilde{g}^{\mu\nu}\hat{\Gamma}_\mu{}^{ab}\partial_\nu s_{ab},$$

where $\hat{\Gamma}_\mu{}^{ab}$ denotes the spin connection and s_{ab} is a new antisymmetric scalar B-field. Since the spin connection contains first derivatives of the vierbein, s_{ab} cannot be regarded as a Lagrange-multiplier field. Owing to this form of $\mathcal{L}_{LGF}$, all components of both the vierbein and the B-field s_{ab} describe dynamical degrees of freedom.

The local-Lorentz FP-ghost term $\mathcal{L}_{LFP}$ is added in such a way that $\mathcal{L}_{LGF} + \mathcal{L}_{LFP}$ becomes a local-Lorentz BRS transform of some quantity. Then the manifestly covariant canonical formalism of quantum gravity can be extended quite beautifully to the vierbein case. Canonical quantization can be carried out consistently without using Dirac's method, and all equal-time (anti-)commutation relations are found explicitly in closed form.[11] Almost all results established in the metric-tensor case, such as the field equations and the equal-time commutation relations for $g_{\mu\nu}$, b_ρ, c^σ, and $\bar{c}_\tau$, various expressions for generators, the sixteen-dimensional Poincaré-like super-algebra, the tensorlike commutation relations, etc., remain intact.[10,11]

An important modification is necessary, however, for the spontaneous breakdown of general linear invariance. In the vierbein case, even the antisymmetric part, $\bar{M}_{\alpha\beta}$, of $\hat{M}^{\mu}{}_{\nu}$ is spontaneously broken. The unbroken one is given by[12]

$$M_{\alpha\beta} \equiv \bar{M}_{\alpha\beta} + \eta_{\alpha a}\eta_{\beta b}M_L{}^{ab},$$

where $M_L{}^{ab}$ is the generator of the global version of the local-Lorentz transformation. The true Lorentz generator $M_{\alpha\beta}$ is thus characterized at the level of the representation of field operators,[19] just as the electromagnetic charge is in the Weinberg-Salam model. This fact is conceptually very important: Lorentz invariance should not be regarded as a first principle determining the fundamental Lagrangian; Lorentz invariance is an S-matrix symmetry rather than a fundamental symmetry. I therefore conjecture that the usual supersymmetry having a spinor charge is not on the right way toward the ultimate unified theory.

Finally, I summarize the main achievements of the manifestly covariant canonical formalism of quantum gravity.

1. The theory is a beautiful and transparent canonical formalism of quantum gravity. Equal-time commutation relations such as $[g_{\mu\nu}, \dot{g}_{\lambda\rho}]$ are explicitly found in closed form.

2. Unitarity is proved in the Heisenberg picture. Since the perturbation series of quantum gravity is unrenormalizable, it is very important to construct the formalism without using perturbation theory.

3. There is a possibility that the ultraviolet divergence difficulty of quantum field theory may be ultimately resolved by taking account of quantum gravity: The theory achieves the evasion of Lehmann's theorem without violating unitarity.

4. The theory is manifestly covariant as in the Gupta-Bleuler formalism of quantum electrodynamics.

5. Though general covariance is broken by gauge fixing, which is necessary for quantization, general linear invariance still remains unbroken at the operator level. It is spontaneously broken up to Lorentz invariance, and the corresponding Goldstone field is nothing but $g_{\mu\nu}$.

6. The theory is invariant under a huge superalgebra, called "sixteen-dimensional Poincaré-like superalgebra", consisting of 144 symmetry generators. It contains both space-time and internal symmetries in complete harmony without contradicting the no-go theorem.

7. The theory has a very interesting property, called "tensor-like commutation relation". General covariance is revived in this way purely at the operator level.

8. All the above establishments are beautifully extended to the case in which vierbein is the fundamental field.

9. The Lorentz invariance of particle physics is characterized by spontaneous breakdown, whence it cannot be a first principle.

References

1. S. N. Gupta, Proc. Phys. Soc. A63, 681 (1950).
2. T. Kugo and I. Ojima, Prog. Theor. Phys. 60, 1869 (1978).
3. T. Kugo and I. Ojima, Prog. Theor. Phys. Suppl. 66, 1 (1979).
4. C. Becchi, A. Rouet and R. Stora. Ann. Phys. 98, 287 (1976).
5. N. Nakanishi, Prog. Theor. Phys. 62, 1396 (1979).
6. N. Nakanishi, Prog. Theor. Phys. 59, 972 (1978).
7. N. Nakanishi, Prog. Theor. Phys. 60, 1190 (1978).
8. N. Nakanishi, Prog. Theor. Phys. 60, 1890 (1978).
9. N. Nakanishi, Prog. Theor. Phys. 61, 1536 (1979).
10. N. Nakanishi, Prog. Theor. Phys. 62, 779 (1979).
11. N. Nakanishi, Prog. Theor. Phys. 62, 1101 (1979).
12. N. Nakanishi, Prog. Theor. Phys. 62, 1385 (1979).
13. N. Nakanishi, Prog. Theor. Phys. 63, 656 (1980).
14. N. Nakanishi, Prog. Theor. Phys. 63, 2078 (1980).
15. N. Nakanishi, Prog. Theor. Phys. 64, 639 (1980).
16. N. Nakanishi and I. Ojima, Prog. Theor. Phys. 65, 728 (1981).
17. N. Nakanishi and I. Ojima, Prog. Theor. Phys. 65, 1041 (1981).
18. N. Nakanishi and K. Yamgishi, Prog. Theor. Phys. 65, 1719 (1981).
19. N. Nakanishi, Prog. Theor. Phys. 66, 1843 (1981).
20. N. Nakanishi, Prog. Theor. Phys. 68, to appear.
21. R. Delbourgo and M. R. Medrano, Nucl. Phys. B110, 476 (1976).
22. K. S. Stelle, Phys. Rev. D16, 953 (1977).
23. P. K. Townsend and P. van Nieuwenhuizen, Nucl. Phys. B120, 301 (1977).
24. K. Nishijima and M. Okawa, Prog. Theor. Phys. 60, 272 (1978).
25. T. Kugo and I. Ojima, Nucl. Phys. B144, 234 (1978).
26. M. J. Duff, in Quantum Gravity 2, A Second Oxford Symposium (ed. by C. J. Isham et al., Clarendon Press, Oxford, 1981), p.81.
27. P. A. M. Dirac, Lectures on Quantum Mechanics (Belfer Graduate School of Science, Yeshiva University, New York, 1964).
28. A. Komar, in Quantum Theory and Gravitation (ed. by A. R. Marlow, Academic Press, 1980), p.127.
29. H. Lehmann, Nuovo Cimento 11, 342 (1954).
30. N. Nakanishi, Prog. Theor. Phys. 63, 1823 (1980).
31. T. Goto and T. Imamura, Prog. Theor. Phys. 14, 396 (1955).
32. J. Schwinger, Phys. Rev. Letters 3, 296 (1959).
33. N. Nakanishi and I. Ojima, Phys. Rev. Letters 43, 91 (1979).
34. P. Pasti and M. Tonin, preprint IFPD 63/81.
35. R. Delbourgo, P. D. Jarvis and G. Thompson, preprint (1982).
36. S. Coleman and J. Mandula, Phys. Rev. 159, 1251 (1967).
37. R. Haag, J. T. Łopuszański and M. Sohnius, Nucl. Phys. B88, 257 (1975).

A GAUGE INVARIANT RESUMMATION OF QUANTUM GRAVITY

Andy Strominger

The Institute For Advanced Study
Princeton, NJ 08540

Abstract

Quantum gravity is expanded in powers of 1/D, where D is the number of dimensions. The extra dimensions are highly compactified. The expansion is gauge invariant. The leading term is equivalent to the iterated one loop matter corrections due to a free, massless scalar field without the $\frac{1}{6} R\phi^2$ term necessary for conformal invariance. The 1/D expansion is renormalizable. Flat space is found to be unstable under small fluctuations.

Despite many valiant efforts, the question of whether or not pure quantum gravity is a consistent theory remains unresolved. The standard expansion in powers of the dimensionless parameter GE = (Newtons constant) x (typical energy) encounters nonrenormalizable high energy divergences. No definitive conclusions can be drawn from this, however, since GE is not small at high energies and we cannot expect an expansion in GE to give a good estimate of high energy corrections.

What is needed is an expansion parameter that is small at high energies. Such a parameter has been suggested by Tomboulis.[1] Tomboulis considers gravity coupled to N matter fields, rescales Newton's constant, and then expands in powers of 1/N. The resulting effective action is, to leading order in 1/N, simply the classical Einstein action with one loop quantum matter corrections. For conformally invariant matter fields and certain choices of renormalization constants, it also turns out to be asymptotically free, renormalizable, and unitary with a Lee-Wick prescription. There are, however, several drawbacks to this approach:

(1) Since no graviton loops are included, it is not clear that we are learning anything about quantized gravity. We may have just swamped out the quantum gravitational effects by dominating the theory with matter fields. If one were to use the same approximation scheme for QCD, for example, one would conclude that it was neither asymptotically free nor confining.

(2) Qualitative features of the expansion depend on the type of matter fields (i.e. scalar or fermion), how they are coupled, and how

it is renormalized. Most choices lead to various types of instabilities. In particular, without conformal invariance, various difficulties arise from the spin zero degrees of freedom. Since conformal invariance is not an observed symmetry of the real world, this somewhat obscures the physical relevance of the expansion.

In view of the above, it would be nice to find a way to resum gravity itself--with no extra matter fields. We could then analyze the internal consistency of quantum gravity and would be spared the ambiguity associated with different choices of matter couplings.

Such a resummation is in fact possible. Pure quantum gravity contains a hidden expansion parameter that is small at all energies. That parameter is 1/D, the inverse number of dimensions. The fundamental fields of gravity are arranged in a DXD matrix. Just as in Yang Mills, Feynman diagrams contain factors of D that arise from traces over this matrix. With an appropriate rescaling of Newton's constant, S matrix elements can be expanded in a series of non-negative powers of 1/D, and the leading term can be explicitly evaluated.

Before proceeding further, however, we must define how the theory is to be extended to D dimensions. There are two inequivalent methods.

The first method is to simply take the standard D dimensional Einstein action on a D dimensional manifold. This theory is invariant under the full D dimensional diffeomorphism group. Extraction of the leading term in the 1/D expansion requires analysis of the D dependence of both the trace factors and of the phase space factors in the D dimensional Feynman integrations. This analysis can be found in Reference [2] and will not be discussed further here.

In this talk we consider a different approach. The extra dimensions are compactified to very small circles. Excitations of the metric along those dimensions are necessarily very short wavelength and very high energy. If the circles are made small enough, such excitations are negligible. The effective theory then consists of a D dimensional matrix of fields on a four dimensional manifold.

This theory is equivalent, via the Kaluza-Klein mechanism, to gravity coupled to (D-4) massless U(1) fields and (D-4)(D-3)/2 scalar fields. For the purpose of analyzing the large D behavior, however, it is not convenient to reexpress the theory in terms of these fields.

The Feynman rules for this theory are determined from the standard Einstein action with gauge fixing and ghost terms:

$$S = -\frac{1}{2K^2}\int d^4x\, \det{}^{-1/2}[G]\Big\{+\frac{1}{2}g^{\alpha\beta}\Big[\mathrm{tr}[G,_\beta G^{-1},_\alpha]$$
$$+\,\mathrm{tr}[G,_\beta G^{-1}]\,\mathrm{tr}[G,_\alpha G^{-1}]\Big]$$
$$+\,3g^{\mu\nu},_\nu \mathrm{tr}[G^{-1},_\mu G] + \frac{1}{2}g^{A\beta},_\nu g^{B\nu},_\beta g_{AB}$$
$$+\,\frac{1}{2\alpha}F_A F^A + \bar{\varepsilon}_A M^{AB}\varepsilon_A\Big\} \qquad (1)$$

where

$$(G)_{AB} = g_{AB}$$
$$A,B = 1,2,\ldots D$$
$$\alpha,\beta,\mu,\nu = 1,2,3,4.$$

$\frac{1}{2\alpha}F_A F^A$ is a gauge fixing term and $\bar{\varepsilon}_A M^{AB}\varepsilon_B$ the corresponding ghost action. K is the gravitational coupling. Where the indices label derivatives, they only run from 1 to 4 since the arguments of the fields are four dimensional. This has been indicated by the use of Greek indices. The invariance group of this action is

$$x_A \to x_A + \varepsilon_A(x_\mu)$$

under which

$$g_{AB}(x_\mu) \to g_{AB}(x_\mu) + \varepsilon_A(x_\mu);_B + \varepsilon_B(x_\mu);_A. \qquad (2)$$

This contains four dimensional coordinate transformations and constant translations in D dimensions. The expansion presented here is in the inverse dimensionality of this latter invariance group.

Isolating the large D behavior in terms of $h^{\mu\nu} = g^{\mu\nu} - \eta^{\mu\nu}$ is awkward because of the double trace term in (1). It is not described by any simple set of diagrams, and in fact gets a contribution from every diagram.

This difficulty is circumvented by using an exponential parametrization:

$$g^{AB} = e^{2K\Omega/D}(e^{K\phi})^{AB}. \qquad (3)$$

ϕ is a traceless DxD matrix and Ω is a scalar. The factor of 1/D in front of Ω is necessary to ensure that Ω has a D independent propagator. The action may now be written:

$$S = \int d^4x\, e^{-K\Omega(1-2/D)}\Big\{\frac{1}{4}\mathrm{tr}[\phi,_\alpha\phi,_\beta](e^{K\phi})^{\alpha\beta} + \Omega,_\alpha\Omega,_\beta(e^{K\phi})^{\alpha\beta}(-1+\frac{7}{D}+\frac{2}{D^2})$$
$$+\,\frac{1}{K}(e^{K\phi})^{\alpha\beta},_\alpha\Omega,_\beta(6-\frac{1}{D}) - \frac{1}{2K^2}(\frac{1}{2}(e^{K\phi})^{A\beta},_\mu(e^{K\phi})^{\mu B},_\beta(e^{K\phi})_{AB}$$
$$+\,\frac{1}{2\alpha}F_A F^A + \bar{\varepsilon}_A M^{AB}\varepsilon_B)\Big\}$$

The large D behavior can now be examined by rescaling the coupling:

$$K \to K/D \tag{5}$$

and counting powers of D in Feynman diagrams. Alternately, one may note that, because of the trace, the first term in (4) is order D while the subsequent terms are order one. The ghosts are the fermions (fundamental multiplet) of this theory and can't contribute to the large D limit. This has the pleasant consequence that <u>the large D limit is gauge invariant</u>.

The quantum contribution to the large D limit thus comes from fluctuations of the first, trace term of (4). Since

$$\frac{1}{4} e^{-K\Omega(1-2/D)} (e^{K\phi})^{\alpha\beta} \mathrm{tr}[\phi,_{\alpha}\phi,_{\beta}] = \frac{1}{4}\sqrt{-g}\, g^{\alpha\beta}\, \mathrm{tr}[\phi,_{\alpha}\phi,_{\beta}] \tag{6}$$

we arrive at the following conclusion: <u>The 1/D expansion of quantum gravity is equivalent to the 1/N expansion of gravity coupled to N free, massless scalars, where $N = D^2$.</u>

This result might seem obvious in view of the fact that the theory is equivalent, via the Kaluza-Klein mechanism, to gravity coupled to (D-4)(D-3)/2 scalars and (D-4) U(1) fields. As we have seen here, however, the 1/D expansion arranges the fields in such a way that it is not just the scalar fields that contribute to the large D limit. This analysis shows that contributions remain when D=4 and the "extra" scalar fields are not there. The expansion should remain valid at D=4, where it describes pure quantum gravity.

The leading 1/N corrections for quantum gravity coupled to N massless, free scalar fields has been discussed in various forms in the literature. The resummed propagator has a $1/p^4$ behavior at high energies which allows the theory to be renormalized with an R^2 type counterterm. Because there is no conformal invariance, however, difficulties arise from the spin zero modes. This can be seen by evaluating the energy of static, spatially varying perturbations of flat space. This energy is equal, by standard arguments, to minus the effective action and can be obtained from general formulae computed by Hartle and Horowitz.[3] For our case, it is given by:

$$E[h_{\alpha\beta}] = \frac{K^{-2}}{4}\int \frac{d^3p}{(2\pi)^3}\left[\vec{p}^{\,2}|h^{TT}_{\alpha\beta}|^2 - \frac{45}{32}\left(\frac{768\pi^2}{K^2 \ln \vec{p}^{\,2}/\mu^2} - \vec{p}^{\,2}\right)|h^T|^2\right]\left[1+\frac{K^2\vec{p}^{\,2}\ln\vec{p}^{\,2}/\mu^2}{1920\pi^2}\right]$$

where $h^T = P^{\alpha\beta}h_{\alpha\beta}$ is the trace part of $h_{\alpha\beta}$ and $h^{TT}_{\alpha\beta} = P_{\alpha}{}^{\gamma}h_{\gamma\delta}P^{\delta}{}_{\beta} - \frac{1}{2}h^T_{\alpha\beta}$ is the traceless part of $h_{\alpha\beta}$ in momentum space. It is readily seen that the energy can be decreased by fluctuations in h^T. This means that flat space is unstable and is not the ground state expectation value of the metric.

Several different conclusions may be inferred from this:

(1) The Einstein action is not a fundamental action but an effective action and should not be quantized. The sickness we found is a result of incorrect quantization.

(2) Quantum gravity needs matter fields for consistency (e.g. supergravity).

(3) Quantum gravity is a good theory. The instabilities are just telling us that we have the wrong ground state.

(4) Quantum gravity is a good theory but the 1/D expansion is bad at D=4.

A final conclusion awaits further analysis.

Acknowledgements

I am grateful to G. Horowitz, E. Tomboulis, and E. Witten for useful conversations. I would also like to thank H. Georgi, P. Ginsparg and the Harvard theory group for their hospitality.

References

1. E. Tomboulis, Phys. Lett. 70B (1977) 361, and Phys. Lett. 97B (1980) 77.
2. A. Strominger, Phys. Rev. D24 (1981) 3082.
3. J.B. Hartle and G. Horowitz, Phys. Rev. D24 (1981) 257.

"The Gauge Invariant Effective Action for Quantum Gravity and Its Semi-Quantitative Approximation"

Bryce S. DeWitt
University of Texas at Austin

Introduction

The configuration space of quantum gravity is PR(M), the set of all pseudo-Riemannian metrics on spacetime M. The gauge group of quantum gravity is $\mathrm{Diff}^*(M)$. A gauge transformation ξ acts on PR(M):

$$\begin{aligned} \xi : PR(M) &\to PR(M) & \xi &\in \mathrm{Diff}^*(M) \\ \varphi &\to \xi(\varphi) & \varphi &\in PR(M) \end{aligned} \tag{1}$$

Gauge transformations divide PR(M) into orbits. The orbits can be shown to comprise an infinite dimensional manifold $PR(M)/\mathrm{Diff}^*(M)$. This manifold is the space of physically distinct fields.

Infinitesimal gauge transformations take the form[1]

$$\varphi \to \varphi + \delta\varphi \qquad \delta\varphi^i = Q^i{}_\alpha[\varphi]\,\delta\xi^\alpha \tag{2}$$

The $Q^i{}_\alpha$ are components of a set of vector fields on PR(M), the Killing flows. The Lie brackets of the Killing flows define the structure constants of $\mathrm{Diff}^*(M)$

$$[\underset{\sim}{Q}_\alpha, \underset{\sim}{Q}_\beta] = \underset{\sim}{Q}_\gamma c^\gamma{}_{\alpha\beta}. \tag{3}$$

PR(M) may be endowed with a gauge invariant metric $\underset{\sim}{\gamma}$. Gauge invariance is expressed by

$$\mathcal{L}_{\underset{\sim}{Q}_\alpha} \underset{\sim}{\gamma} = 0 \tag{4}$$

Any metric that satisfies this equation projects to a metric on $PR(M)/\mathrm{Diff}^*(M)$. In the case of quantum gravity Eq.4 has a unique one parameter family of local solutions,

[1] The dynamical variables (which in gravity theory are the metric components $g_{\mu\nu}$) are denoted by φ^i. The index i is to be understood as a combined discrete-continuous label. The implicit summation convention involves integrals as well as sums.

$$\gamma^{\mu\nu\sigma'\tau'} = g^{\frac{1}{2}}(g^{\mu\sigma}g^{\nu\tau}+g^{\mu\tau}g^{\nu\sigma}+\lambda g^{\mu\nu}g^{\sigma\tau})\delta(x,x')$$

$$\lambda \neq -\frac{1}{2} \qquad g = -\det(g_{\mu\nu}) \tag{5}$$

The dynamics of the gravitational field is described by the classical action S, which is a real valued scalar function on PR(M) [2]

$$S[\varphi] = \mu_P{}^2 \int g^{\frac{1}{2}} R d^4x \qquad 16\pi G = \mu_P{}^{-2} \tag{6}$$

The classical action is gauge invariant:

$$\underset{\sim}{Q}_\alpha S = 0 \tag{7}$$

With $g_{\mu\nu}$ chosen for the basic dynamical variables φ^i the action of the gauge group on PR(M) is linear. Linearity may be expressed by

$$Q^i{}_{\alpha,jk} \equiv 0 \tag{8}$$

where the comma denotes functional differentiation. By repeatedly functionally differentiating Eq.7 and making use of Eq.8 one obtains the infinite sequence of equations

$$S_{,i}Q^i{}_\alpha \equiv 0$$

$$S_{,ij}Q^j{}_\alpha \equiv -S_{,j}Q^j{}_{\alpha,i}$$

$$S_{,ijk}Q^k{}_\alpha \equiv -S_{,kj}Q^k{}_{\alpha,i} - S_{,ik}Q^k{}_{\alpha,j}$$

$$S_{,ijk\ell}Q^\ell{}_\alpha \equiv -S_{,\ell jk}Q^\ell{}_{\alpha,i} - S_{,i\ell k}Q^\ell{}_{\alpha,j} - S_{,ij\ell}Q^\ell{}_{\alpha,k}, \text{ etc.} \tag{9}$$

These are the bare Ward-Takahashi identities of the theory.

The classical field equations are

$$S_{,i}[\varphi] = 0 \tag{10}$$

Given a solution φ of these equations one is often interested in a solution $\varphi + \delta\varphi$ which differs infinitesimally from φ . $\delta\varphi$ satisfies the equation of small disturbances

---

[2] We use units for which $\hbar = c = 1$ and a spacetime signature $-{+}{+}{+}$.

$$0 = S_{,i}[\varphi+\delta\varphi] = S_{,i}[\varphi]+S_{,ij}[\varphi]\delta\varphi^j = S_{,ij}[\varphi]\delta\varphi^j \tag{11}$$

The second functional derivative $S_{,ij}[\varphi]$ appearing in this equation is effectively a linear differential operator. Because of the gauge invariance of the theory this operator is singular. Equation 11 has a well defined solution for a given set of boundary conditions only if one imposes a supplementary condition

$$P^{\alpha}{}_{i}[\varphi]\delta\varphi^j = 0 \tag{12}$$

When condition 12 is satisfied $\delta\varphi$ satisfies

$$F_{ij}[\varphi]\delta\varphi^j = 0 \tag{13}$$

where

$$F_{ij} \overset{\text{def}}{=} S_{,ij} + \eta_{\alpha\beta} P^{\alpha}{}_{i} P^{\beta}{}_{j} \tag{14}$$

The functions $P^{\alpha}{}_{i}$ appearing in Equation 12 are often chosen so that small disturbances are $\underset{\sim}{\gamma}$-orthogonal to gauge variations:

$$P^{\alpha}{}_{i} = \eta^{-1\alpha\beta} Q^{j}{}_{\beta} \gamma_{ji} \longrightarrow Q^{j}{}_{\beta} \gamma_{ji}\delta\varphi^{i} = 0 \tag{15}$$

It is convenient to impose the following gauge covariance condition on the continuous matrix $\eta_{\alpha\beta}$:

$$\eta_{\alpha\beta,i}Q^{i}{}_{\gamma} \equiv -\eta_{\delta\beta}c^{\delta}{}_{\gamma\alpha} - \eta_{\alpha\delta}c^{\delta}{}_{\gamma\beta} \tag{16}$$

The effective action

Suppose spacetime is such that we can introduce coherent "in" and "out" states, $|in,vac>$, $|out,vac>$. These states are sometimes known as relative vacua, i.e., they are vacua relative to a given background. The effective action Γ is defined by

$$<out,vac|in,vac> = e^{i\Gamma[\varphi]} \tag{17}$$

Γ is a complex valued scalar field on the space CR(M) of complex metrics on spacetime M. The field φ appearing on the right hand side of Eq.17 is arbitrary. It does not need to be a classical background.

In what follows it will be convenient to define

$$\begin{aligned} \mathfrak{F}^{\alpha}{}_{\beta}[\varphi] &= P^{\alpha}{}_{i}[\varphi]Q^{i}{}_{\beta}[\varphi] \\ \mathfrak{F}^{\alpha}{}_{\gamma}[\varphi]\mathfrak{G}^{\gamma}{}_{\beta}[\varphi] &= -\delta^{\alpha}{}_{\beta} \\ V^{\alpha}{}_{\beta i}[\varphi] &= P^{\alpha}{}_{j}[\varphi]Q^{j}{}_{\alpha,i} \end{aligned} \tag{18}$$

and to extend the domain of all functionals to CR(M).

$\mathfrak{G}^{\alpha}{}_{\beta}$ is known as the <u>ghost propagator</u> and $V^{\alpha}{}_{\beta i}$ is known as the <u>ghost vertex</u>. The vacuum-to-vacuum amplitude (Eq.17) may be expressed by the following functional integral

$$\begin{aligned} e^{i\Gamma[\varphi]} &= \text{const.} \times \int e^{iS[\varphi+\phi]}d\phi \\ &= \text{const.} \times (\det\eta[\varphi])^{\frac{1}{2}}(\det\mathfrak{G}[\varphi])^{-1} \\ &\quad \times \int e^{i(S[\varphi+\phi]+\frac{1}{2}\eta_{\alpha\beta}[\varphi]P^{\alpha}{}_{i}[\varphi]P^{\beta}{}_{j}[\varphi]\phi^{i}\phi^{j})} \\ &\quad \times \det(1-\mathfrak{G}[\varphi]V[\varphi]\phi)^{-1}\, d\phi \end{aligned} \tag{19}$$

The <u>value</u> (although not in the explicit functional form) of the second integral is independent of the choice of $P^{\alpha}{}_{i}$ and $\eta_{\alpha\beta}$ provided a regularization is adopted that yields

$$c^{\beta}{}_{\alpha\beta} = 0 \quad , \qquad Q^{i}{}_{\alpha,i} = 0 \tag{20}$$

If the P's and η's are chosen covariantly then the effective action is gauge invariant ($\underset{\sim}{Q}_\alpha \Gamma \equiv 0$) and satisfies simple Ward-Takahashi identities:

$$\Gamma_{,i}Q^i{}_\alpha \equiv 0$$

$$\Gamma_{,ij}Q^j{}_\alpha \equiv -\Gamma_{,j}Q^j{}_{\alpha,i}$$

$$\Gamma_{,ijk}Q^k{}_\alpha \equiv -\Gamma_{,kj}Q^k{}_{\alpha,i} - \Gamma_{,ik}Q^k{}_{\alpha,i}$$

$$\Gamma_{,ijk\ell}Q^\ell{}_\alpha \equiv -\Gamma_{,\ell jk}Q^\ell{}_{\alpha,i} - \Gamma_{,i\ell k}Q^\ell{}_{\alpha,j} - \Gamma_{,ij\ell}Q^\ell{}_{\alpha,k} \qquad (21)$$

The equations

$$\Gamma_{,i}[\varphi] = 0 \qquad (22)$$

are called the <u>effective field equations</u>. The solution of these non-local equations satisfying the given boundary conditions is called the effective field. If the background is chosen to be the effective field the 1-particle reducible graphs may be omitted from the <u>loop expansion</u> of Γ:

$$\Gamma[\varphi] = S[\varphi] + \Sigma[\varphi]$$

$\Sigma[\varphi] = -\frac{i}{2}\ \ell n \det\eta[\varphi] - \frac{i}{2}$ $+\ i$

$-\frac{1}{12}$  $+\ \frac{1}{2}$  $-\ \frac{1}{8}$ 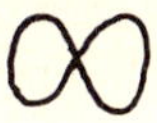 $+\ \cdots$ (23)

In these graphs solid lines denote the Feynman propagator for $F_{ij}[\varphi]$, written $G^{ij}[\varphi]$. A dotted line represents $\mathfrak{G}^\alpha{}_\beta[\varphi]$. A vertex at which a solid line meets two dotted lines represents the vertex function $V^\alpha{}_{\beta i}[\varphi]$. Vertices at which three or more solid lines come together represent functional derivatives of the classical action (the $S_{,ijk}$, $S_{,ijk\ell}$, etc.). The solid circle denotes $\ell n \det G[\varphi]$. The dotted circle denotes $\ell n \det \mathfrak{G}[\varphi]$.

The effective action has the following important properties:

1. The functional form of Γ depends on the P's and η's but the effective field and the value of Γ do not.

2. The tree functions built out of Γ yield the exact scattering amplitudes.

3. The effective field is an operator average:

$$\varphi^i = \frac{\langle \text{out,vac} | \underset{\sim}{\varphi}^i | \text{in,vac} \rangle}{\langle \text{out,vac} | \text{in,vac} \rangle} \qquad (24)$$

4. If $\mathrm{Im}\Gamma \approx 0$ then the "in" and "out" states are nearly identical and φ^i becomes (approximately) an expectation value.

5. Γ , not S, governs the dynamics of quantized spacetime (e.g., near the Big Bang or near classical singularities).

6. In Yang-Mills theory the use of Γ simplifies the renormalization program. (See Abbot 1981, CERN preprints TH. 2973 and 3113,) and (Hart 1981, Ph.D. dissertation, University of Texas).

The general structure of Γ

In this section I shall attempt to adduce some plausibility arguments for the general structure of Γ . Begin by considering the functional $\Sigma[\varphi]$ which represents the difference between Γ and the classical action, i.e. the radiative corrections to the classical action (Eq.23). The functional Σ, like S and Γ , is gauge invariant. In quantum gravity this is expressed by:

$$(\delta\Sigma/\delta g_{\mu\nu})_{;\nu} \equiv 0 \tag{25}$$

Let us assume that the Minkowski metric $\eta_{\mu\nu}$ is a stable solution of the effective field equation just as it is of the classical field equation $\delta S/\delta g_{\mu\nu} = 0$. That is, let us assume that Poincaré group,which is relevant for asymetrically flat spacetimes, is not dynamically broken.

$$(\delta\Sigma/\delta g_{\mu\nu})_{\underset{\sim}{g} = \underset{\sim}{\eta}} = 0 \tag{26}$$

whence in virtue of 25,

$$\left[\left(\frac{\delta^2\Sigma}{\delta g_{\mu\nu}\,\delta g_{\sigma'\tau'}}\right)_{;\nu}\right]_{\underset{\sim}{g} = \underset{\sim}{\eta}} = 0 \tag{27}$$

Denote by $\Sigma^{\mu\nu\sigma\tau}(p)$ the Fourier transform of $\left(\dfrac{\delta^2\Sigma}{\delta g_{\mu\nu}\delta g_{\sigma'\tau'}}\right)_{\underset{\sim}{g} = \underset{\sim}{\eta}}$,

with the δ function expressing momentum conservation removed. Equation 25 is equivalent to

$$\Sigma^{\mu\nu\sigma\tau}(p)p_\nu = 0 \tag{28}$$

of which the general solution, respecting Lorentz invariance and the index symmetries of $\Sigma^{\mu\nu\sigma\tau}$, is:

$$\begin{aligned}
&\Sigma^{\mu\nu\sigma\tau}(p)\\
&= \tfrac{1}{4}[(\eta^{\mu\sigma}\eta^{\nu\tau}+\eta^{\mu\tau}\eta^{\nu\sigma}-\tfrac{2}{3}\eta^{\mu\nu}\eta^{\sigma\tau})p^4\\
&\quad -(\eta^{\mu\sigma}p^\nu p^\tau+\eta^{\nu\tau}p^\mu p^\sigma+\eta^{\mu\tau}p^\nu p^\sigma+\eta^{\nu\sigma}p^\mu p^\tau-\tfrac{2}{3}\eta^{\mu\nu}p^\sigma p^\tau-\tfrac{2}{3}\eta^{\sigma\tau}p^\mu p^\nu)p^2\\
&\quad +\tfrac{4}{3}p^\mu p^\nu p^\sigma p^\tau]\Sigma_1(p^2)\\
&\quad -\tfrac{1}{3}[\eta^{\mu\nu}\eta^{\sigma\tau}p^4-(\eta^{\mu\nu}p^\sigma p^\tau+\eta^{\sigma\tau}p^\mu p^\nu)p^2+p^\mu p^\nu p^\sigma p^\tau]\Sigma_2(p^2)
\end{aligned} \tag{29}$$

If λ in Equation 5 is chosen to be -1, then the Fourier transform of $(F_{ij})_{g=\eta}$ is:

$$\frac{1}{4}\mu_P^2(\eta^{\mu\sigma}\eta^{\nu\tau}+\eta^{\mu\tau}\eta^{\nu\sigma}-\eta^{\mu\nu}\eta^{\sigma\tau})p^2+\Sigma^{\mu\nu\sigma\tau}(p) \quad (30)$$

The full graviton propagator is the 10 x 10 matrix inverse of this:

$$\begin{aligned}\Gamma_{\mu\nu\sigma\tau}(p) = \mu_P^{-2}[&\eta_{\mu\sigma}\eta_{\nu\tau}+\eta_{\mu\tau}\eta_{\nu\sigma}-\frac{2}{3}\eta_{\mu\nu}\eta_{\sigma\tau}\\ &+\mu_P^{-2}(\eta_{\mu\sigma}p_\nu p_\tau+\eta_{\nu\tau}p_\mu p_\sigma+\eta_{\mu\tau}p_\nu p_\sigma+\eta_{\nu\sigma}p_\mu p_\tau)\Sigma_1(p^2)]\\ &\times[p^2+\mu_P^{-2}p^4\Sigma_1(p^2)]^{-1}\\ &-\frac{1}{3}\mu_P^{-2}\eta_{\mu\nu}\eta_{\sigma\tau}[p^2+\mu_P^{-2}p^4\Sigma_2(p^2)]^{-1}\\ &-\frac{2}{3}\mu_P^{-4}[(\eta_{\mu\nu}p_\sigma p_\tau+\eta_{\nu\tau}p_\mu p_\sigma)p^2+2p_\mu p_\nu p_\sigma p_\tau]\\ &\times[\Sigma_1(p^2)-\Sigma_2(p^2)][p^2+\mu_P^{-2}p^4\Sigma_1(p^2)]^{-1}\\ &\times[p^2+\mu_P^{-2}p^4\Sigma_2(p^2)]^{-1} \end{aligned} \quad (31)$$

As can be seen from the form of this expression the particle spectrum is determined by the zeros of the functions $[p^2+\mu_P^{-2}p^4\Sigma_1(p^2)]$ and $[p^2+\mu_P^{-2}p^4\Sigma_2(p^2)]$. It is not difficult to show that if Σ is expanded as a functional power series in $\varphi_{\mu\nu}=g_{\mu\nu}-\eta_{\mu\nu}$ then the term of lowest order is quadratic in $\varphi_{\mu\nu}$ and is uniquely determined by Eq.29 to have the form

$$\Sigma^{(2)} = \int d^4x\int d^4x'\left[\frac{1}{2}\tilde{\Sigma}_1((x-x')^2)C_{\mu\nu\sigma\tau}(x)C^{\mu\nu\sigma\tau}(x') -\frac{1}{6}\tilde{\Sigma}_2((x-x')^2)R(x)R(x')\right] \quad (32)$$

where $C_{\mu\nu\sigma\tau}$ is the linearized Weyl tensor, R is the linearized curvature scalar and $\tilde{\Sigma}_1$ and $\tilde{\Sigma}_2$ are the Fourier transforms of Σ_1 and Σ_2 respectively.

In the one-loop approximation <u>without</u> subtraction, the dominant singularities of both $\tilde{\Sigma}_1$ and $\tilde{\Sigma}_2$ are proportional to $i/(x-x')^4$. This singularity structure, which renders expression (32) logarithmically divergent, arises from products of pairs of Green's functions $i/(x-x')^2$, together with loop factors $-i$, in typical self-energy graphs.

How does it get modified in the exact theory?

A partial answer to this question is known [3,4] in the case of ladder graphs in which the free ends at the top of each ladder are joined together to make a single line, leaving only the two free ends at the bottom. The dominant high-energy contribution to the infinite sum of all such graphs can be expressed as the solution of the following simple integral equation

$$X(p) = \frac{1}{p^2-i0} - \frac{1}{(2\pi)^4\mu_P{}^2}\int \frac{X(k)}{(p-k)^2-i0}\, d^4k \tag{33}$$

Since the integral in this equation is a convolution integral the equation is easily solved by taking the Fourier transform:

$$\tilde{X}(x) = G(x)[1-i\mu_P{}^{-2}\tilde{X}(x)] \tag{34}$$

where G(x) is the standard scalar propagator,

$$G(x) = \frac{1}{(2\pi)^4}\int \frac{e^{ip\cdot x}}{p^2-i0}\, d^4p = \frac{i}{(2\pi)^2}\;\frac{1}{x^2+i0} \tag{35}$$

yielding

$$\tilde{X}(x) = \frac{G(x)}{1+i\mu_P{}^{-2}G(x)} = \frac{i}{(2\pi)^2}\;\frac{1}{x^2-\lambda_P{}^2+i0} \tag{36}$$

with

$$\lambda_P = \frac{1}{2\pi\mu_P} \tag{37}$$

The line at the top of the ladder graph contributes a factor $1/(x-x')^2$ as always, but the rungs when summed to all orders, contribute expression 36 as a factor. The singularity of the rung factor lies on a hyperboloid at a distance λ_P outside the Minkowski light cone and implies noncausal propagation relative to Minkowski space-time. This is neither surprising nor alarming. When the metric itself undergoes quantum fluctuations "real" space-time is Minkowskian only in an average sense.

These results suggest that $\tilde{\Sigma}_1$ and $\tilde{\Sigma}_2$ may be well approximated by choosing each to be proportional to

---

[3] B. S. DeWitt, Phys. Rev. Lett. 13, 114 (1964).

[4] I. B. Khriplovich, Yad. Fiz. 2, 950 (1965) [Sov. J. Nucl. Phys. 3, 415 (1966)].

$$-iG(x-x')\tilde{X}(x-x') = \frac{i}{(2\pi)^4}\frac{1}{(x-x')^2[(x-x')^2-\lambda_p{}^2]}$$

$$= \frac{i}{(2\pi)^4\lambda_p{}^2}\left[\frac{1}{(x-x')^2-\lambda_p{}^2+i0} - \frac{1}{(x-x')^2+i0}\right]$$

$$= \frac{i}{(2\pi)^2}\,\mu_p{}^2\int_0^1 \frac{d\xi}{[(x-x')^2-\xi\lambda_p{}^2+i0]^2} \qquad (38)$$

The final integral gives concrete expression to the old idea that quantum gravity smears the light cone. A more complete theory, which sums other graphs besides ladder graphs, would presumably insert a smearing function $w(\xi)$ in the integrand.

If $\tilde{\Sigma}_1$ and $\tilde{\Sigma}_2$ have the form 38 then their Fourier transforms are given by

$$\Sigma_{1,2}(p^2) \approx A_{1,2}\left[-\frac{\pi i}{2}\,\frac{H_1^{(2)}((\lambda_p{}^2p^2-i0)^{\frac{1}{2}})}{(\lambda_p{}^2p^2-i0)^{\frac{1}{2}}} - \frac{1}{\lambda_p{}^2p^2-i0}\right] \qquad (39)$$

This function is complex for space-like momenta and real for time-like momenta. In both cases it tends to zero for $|p^2|\gg\mu^2$. With this approximation the functions $p^2+\mu_p{}^{-2}p^4\Sigma_{1,2}(p^2)$ have no zeros other than $p^2=0$ on the real p^2 axis, provided $a_{1,2}$ avoid values lying between approximately .024 and .054 as well as an infinity of isolated points clustering about the origin between -.011 and .009. If $a_{1,2}$ lies between .024 and .054 then the graviton propagator has timelike ghosts. If $a_{1,2}$ = one of the discrete values then there are tachyon ghosts. The functions $p^2 + \mu_p{}^{-2}p^4\Sigma_{1,2}(p^2)$ have an infinity of zeros in the lower half p^2 plane, i.e., in the upper half $(p^0)^2$ plane. In the p^0 plane these zeros are in the first and third quadrants. Let $E=\omega+i\gamma\,(\omega>0,\gamma>0)$ be one of the first-quadrant zeros. The corresponding "instability" modes have time dependence $e^{\pm iEt}$. The mode that propagates positive frequencies into the future is

$$e^{-iEt} = e^{-i\omega t + \gamma t} \qquad (40)$$

The mode that propagates negative frequencies into the past is

$$e^{iEt} = e^{i\omega t - \gamma t} \qquad (41)$$

Both modes are eliminated by the "in" and "out" boundary conditions. Therefore they do not lead to real instabilities. However, they make Wick rotation impossible. This means that if the above approximation has any validity whatever, Euclideanization is not permitted in quantum gravity.

Equations 32 and 38 admit of immediate generalization to an approximation for Σ , and hence for Γ , that is invariant under the full diffeomorphism group:

$$\Gamma \approx \mu_P{}^2 \int g^{\frac{1}{2}} R d^4x$$
$$+\mu_P{}^2 \int d^4x \int d^4x' g^{\frac{1}{2}} g'^{\frac{1}{2}} \left[\frac{i}{\sigma(x,x') - \frac{1}{2}\lambda_P{}^2 + i0} - \frac{i}{\sigma(x,x') + i0} \right]$$
$$\times \left[\tfrac{1}{4} A_1 g^{\mu\alpha'} g^{\nu\beta'} g^{\sigma\gamma'} g^{\tau\delta'} C_{\mu\nu\sigma\tau} C_{\alpha'\beta'\gamma'\delta'} - \frac{1}{12} A_2 R R' \right] \qquad (42)$$

Here g is $-\det(g_{\mu\nu})$, $g^{\mu\alpha'}$ is the parallel displacement bivector,[5] $\sigma(x,x')$ is half the square of the geodetic distance between x and x',[5] and $C_{\mu\nu\sigma\tau}$ and R are the Weyl tensor and curvature scalar of the full nonlinear theory. A_1 and A_2 are numerical coefficients whose precise values depend on the numbers and kinds of matter fields included, but whose magnitudes are not vastly different from unity. The i0 in the "propagators" specifies how the poles are to be skirted in the double integral, and the other factors i remind us that both Γ and the effective field, which is an "in-out" average, are generally complex valued.

Although expression (42) has been derived by arguments starting from flat space-time, I propose that it be taken seriously even under conditions of strong curvature ($R_{\mu\nu\sigma\tau} > \mu_P{}^2$) and with topologies other than $\mathbb{R}^4$. Efforts are currently underway at the University of Texas to test it on Friedmann-Robertson-Walker universes to see whether, under generic realistic conditions, it will suppress the initial curvature singularity. Among the properties of Friedmann-Robertson-Walker models that simplify this investigation is conformal flatness. The Weyl tensor disappears from expression (42) taking with it the parallel displacement bivectors, leaving $\sigma(x,x')$ as the only difficult geometrical quantity to compute and A_2 as the only adjustable constant.

Before describing these efforts, I wish to make a few comments on the reasonableness of expression (42) as an approximation to the true effective action. Expression (42), based as it is on a quadratic approximation to Σ that is determined solely by the graviton propagator, cannot be expected to yield accurate vertex functions (third functional derivatives and higher). Nevertheless it is well known that in regions of momentum space where $\Sigma_1(p^2)$ and $\Sigma_2(p^2)$ are slowly varying, e.g., in

---

[5] B. S. DeWitt, Dynamical Theory of Groups and Fields (Gordon and Breach, New York, 1965), Chap. 17.

the ultrahigh-energy region $|p^2| >> \mu p^2$ (see comments following Eq.(39)) the vertex functions are fully determined by the graviton propagator in virtue of the gauge-invariance condition (25). Therefore, if $\tilde{\Sigma}_1$ and $\tilde{\Sigma}_2$ are well approximated by (38) then expression (42) has the correct structure as $x' \to x$ and will yield qualitatively correct dynamical behavior. More accurate vertex functions at lower energies could in principle be obtained by adding to expression (42) higher multiple integrals in which the curvature appears cubically, quartically, etc., along with factors involving $g^{\mu\alpha'}$ and $\sigma(x,x')$.

Numerical Work

The effort to solve the effective field equations based on the effective action (42) is being carried out by Richard Rohwer at the University of Texas. In the case of Friedmann-Robertson-Walker universes the line element may be written in the form

$$ds^2 = -\alpha^2(t)dt^2 + a^2(t)d\underset{\sim}{r}^2 \tag{43}$$

Here we are specializing to the case in which the spatial sections t = constant are flat. This is because we are primarily interested in the neighborhood of the Big Bang and in our actual universe the curvature in time at this epoch is much more important than the curvature in space. With this line element Eq.42 takes the form

$$\begin{aligned}\Gamma \approx\ & 6V\mu_P^{\,2}\int_{-\infty}^{\infty}\alpha^{-1}a\,\dot{a}^2dt\\ & -12\pi V\mu_P^{\,2}A_2\int_{-\infty}^{\infty}dt\int_{-\infty}^{\infty}dt'\int_0^{\infty}s^2ds\,\alpha^{-2}\alpha'^{-2}(\alpha a^2\ddot{a}+\alpha a\dot{a}^2-a^2\dot{a}\dot{\alpha})\\ & \times(\alpha' a'^2\ddot{a}'+\alpha' a'\dot{a}'^2-a'^2\dot{a}'\dot{\alpha}')\left[\frac{i}{\sigma(t,t',s)-\frac{1}{2}\lambda_P^{\,2}+i0}\right.\\ & \qquad\left.-\frac{i}{\sigma(t,t',s)+i0}\right]\\ & +6V\mu_P^{\,4}\int_{\infty}^{\infty}\alpha a^{-1}dt\end{aligned} \tag{44}$$

where V is the volume of space and the last term expresses the effect of the radiation which is assumed to fill the universe, a choice of scale being made so that the energy density is equal to $6\mu_P^{\,4}$ when a = 1.

There are two effective field equations for this effective action

$$\frac{\delta\Gamma}{\delta\alpha} = 0$$

$$\frac{\delta\Gamma}{\delta a} = 0 \tag{45}$$

Because of gauge invariance these two equations are not independent but satisfy the identity

$$\alpha\,\frac{d}{dt}\,\frac{\delta\Gamma}{\delta\alpha} \equiv \dot{a}\,\frac{\delta\Gamma}{\delta a} \tag{46}$$

It evidently suffices to work with $\frac{\delta\Gamma}{\delta\alpha} = 0$. Explicitly one finds

$$0 = -(6V\mu_P{}^2)^{-1}(\delta\Gamma/\delta\alpha)_{\alpha=1}$$

$$= a\dot{a}^2 - \mu_P{}^2 a^{-1}$$

$$+4\pi A_2\int_{-\infty}^{\infty} dt'\int_0^{\infty} s^2 ds \left\{ a\dot{a}^2\left[\frac{i}{\sigma(t,t',s)-\tfrac{1}{2}\lambda_P{}^2+i0} - \frac{i}{\sigma(t,t',s)+i0}\right]\right.$$

$$\left. -a^2\dot{a}\sigma,_t(t,t',s)\left[\frac{i}{[\sigma(t,t',s)-\tfrac{1}{2}\lambda_P{}^2+i0]^2} - \frac{i}{[\sigma(t,t',s)+i0]^2}\right]\right\}$$

$$\times(a'^2 a' + a'\dot{a}'^2)$$

$$+4\pi A_2\int_{-\infty}^{\infty} dt'\int_{-\infty}^{\infty} dt''\int_0^{\infty} s^2 ds\,(a'^2\ddot{a}' + a'\dot{a}'^2)(a''^2 a'' + a''\dot{a}''^2)$$

$$\times\left\{-a^{-2}(t)[\sigma,_s(t',t'',s)]^2 - 2\sigma(t',t'',s)\right\}^{\frac{1}{2}}$$

$$\times\left[\frac{i}{\sigma(t',t'',s)-\tfrac{1}{2}\lambda_P{}^2+i0} - \frac{i}{\sigma(t',t'',s)+i0}\right] \qquad (47)$$

The idea of the computation is the following. Begin with a Big Crunch followed by a Big Bang, having the form $a \sim |t|^{\frac{1}{2}}$ (which is a solution of the classical field equation) but with the Crunch at $t = 0$ smoothed out (by hand) over a region of the order of the Planck time. Substitute this value of $\underline{a}$ into the integrals appearing in Eq.47 and obtain new values for $\underline{a}$ by putting the terms involving no integrals on the left hand side of the equation. Then iterate this procedure, hoping that the sequence will converge. Unfortunately we have been unable as yet to get the program to this stage because we have been encountering unforseen difficulties in evaluating the bilinear σ . We began by attempting to compute it from the Hamilton-Jacobi equation,

$$2\sigma = g^{\mu\nu}\sigma_{,\mu}\sigma_{,\nu} = -(\sigma_{,t})^2 + a^{-2}(\sigma_{,s})^2 \qquad (48)$$

but discovered subsequently that every reasonable way for converting this equation to a set of difference equations leads to unconditional instabilities. Another possibility is to solve directly the geodesic equations. However a new problem arises, namly that of caustics, which indeed occur for these metrics. We are now thinking in terms of computing the scalar propagator in the given metric and inverting it to obtain an approximation to σ . Note that the presence of the factors i and i0 in Eq.47 causes a to become complex. This in turn causes σ to become complex. Complex functions can be handled on the computer in a straightforward way but it is important to call attention to the added complication.

Background Field Method of Gauge Theory and the Renormalization Problem

S. Ichinose and M. Omote
Institute of Physics, University of Tsukuba, Ibaraki 305, Japan

§1. Introduction

The background field method was proposed by DeWitt[1] and has been discussed later in many papers.[2] This method has an interesting feature that we can quantize gauge theories without losing gauge invariance. As a result, in the renormalization procedure gauge invariant quantities need to be considered. This is a very important advantage in discussing the renormalization problem of non-Abelian gauge theories (especially of gravitation).

In this paper we will discuss a systematic renormalization procedure by using the background field method and by generalizing the counter-term formula of 't Hooft[3] to two loop processes.

§2. Background Field Method

In the background field method the generating functional of the S-metrix is given by

$$W(\tilde{\phi}) = \int D\phi e^{iS(\tilde{\phi}+\phi)} = \int D\phi \exp i\{S(\tilde{\phi}) + S_{,i}\phi_i + \frac{1}{2!} S_{,ij}\phi_i\phi_j + \cdots\}, \tag{2.1}$$

where $S_{,i}$, $S_{,ij}$ $\cdots$ denote functional derivatives of the action $S(\tilde{\phi})$ with respect to $\tilde{\phi}_i(x)$, $\tilde{\phi}_j(y)$, $\cdots$, and $S_{,i}\phi_i$, $S_{,ij}\phi_i\phi_j$ are abbreviations of

$$S_{,i}\phi_i = \int d^4x \frac{\delta S(\tilde{\phi})}{\delta\tilde{\phi}_i(x)} \phi_i(x) , \quad S_{,ij}\phi_i\phi_j = \int d^4x d^4y \frac{\delta^2 S(\tilde{\phi})}{\delta\tilde{\phi}_i(x)\delta\tilde{\phi}_j(y)} \phi_i(x)\phi_j(y). \tag{2.2}$$

In (2.1) the background field $\tilde{\phi}$ is defined as a solution of the field equation $S_{,i}(\tilde{\phi}) = 0$.

If we consider a system which includes gauge fields in this formalism we can introduce two kinds of gauge transformations. By taking the Yang-Mills fields as an example we define such transformations as follows:

c-type gauge transformations

$$\begin{aligned} \tilde{\phi}^{a\prime}_{\mu} &= \tilde{\phi}_{\mu}{}^{a} + gf^{abc}\omega^{b}\tilde{\phi}_{\mu}{}^{c} - \partial_{\mu}\omega^{a} , \\ \phi'^{a}_{\mu} &= \phi_{\mu}{}^{a} + gf^{abc}\omega^{b}\phi_{\mu}{}^{c} , \end{aligned} \tag{2.3}$$

q-type gauge transformations

$$\tilde{\phi}'^{a}_{\mu} = \tilde{\phi}^{a}_{\mu} ,$$
$$\phi'^{a}_{\mu} = \phi^{a}_{\mu} + gf^{abc}\omega^{b}(\phi^{c}_{\mu} + \tilde{\phi}^{c}_{\mu}) - \partial_{\mu}\omega^{a} . \qquad (2.4)$$

Under the both transformations (2.3) and (2.4) the total field $A = (\tilde{\phi}+\phi)$ has ordinary gauge transformation properties

$$A'^{a}_{\mu} = A^{a}_{\mu} + gf^{abc}\omega^{b}A^{c}_{\mu} - \partial_{\mu}\omega^{a} . \qquad (2.5)$$

The important point is the fact that only $\tilde{\phi}$ field (ϕ field) has the transformation properties of gauge field under c-type (q-type) gauge transformations.

As concerning gauge fixing we have to fix the q-type gauge only in (2.1). Then we can choose a gauge fixing condition (the background gauge) such as

$$D_{\mu}\phi^{a}_{\mu} \equiv (\partial_{\mu}\phi^{a}_{\mu} + gf^{abc}\tilde{\phi}^{b}_{\mu}\phi^{c}_{\mu}) = 0 . \qquad (2.6)$$

Since (2.6) transforms covariantly under (2.3) it is evident that $W(\tilde{\phi})$ is invariant under (2.3) even if the gauge fixing has been performed. This important fact simplifies the discussion of the renormalization problem of the gauge theories in the point that the counter Lagrangian can be expressed in term of gauge invariant combinations such as $(F^{a}_{\mu\nu})^2$.

§3. Counter-term Formula

Without loss of generality we can expand the action $S(\tilde{\phi}+\phi)$ around $\tilde{\phi}$ as

$$\begin{aligned} &S(\tilde{\phi}+\phi) - S(\tilde{\phi}) - S_{,i}\phi_i \\ &= \int d^4x\{\tfrac{1}{2}\,\partial_\mu\phi_i W^{ij}\partial_\mu\phi_j + \phi_i N^{\mu}_{ij}\partial_\mu\phi_j + \tfrac{1}{2}\,\phi_i M^{ij}\phi_j + \Xi^{ijk}_{\mu\nu}\phi_i\partial_\mu\phi_j\partial_\nu\phi_k \\ &\quad + \Omega^{ijk}_{\mu}\phi_i\phi_j\partial_\mu\phi_k + \Lambda^{ijk}\phi_i\phi_j\phi_k + \Gamma^{ijk\ell}_{\mu\nu}\phi_i\phi_j\partial_\mu\phi_k\partial_\nu\phi_\ell \\ &\quad + \Sigma^{ijk\ell}_{\mu}\phi_i\phi_j\phi_k\partial_\mu\phi_\ell + \Theta^{ijk\ell}\phi_i\phi_j\phi_k\phi_\ell + \cdots\} , \end{aligned} \qquad (3.1)$$

where coefficients W, N_μ, M, Ξ, Ω_μ, Λ, $\Gamma_{\mu\nu}$, Σ_μ and Θ are functions of the $\tilde{\phi}$. In (3.1) we assumed that $S(\tilde{\phi}+\phi)$ has derivative couplings up to second order. This is satisfied both for Einstein gravity and for the Yang-Mills type gauge theories.

In this paper we will restrict ourselves to the case $\Xi = \Gamma = \Sigma = 0$, $W^{ij} = -\delta^{ij}$. By introducing a kind of covariant derivatives $\nabla_\mu\phi_i = \partial_\mu\phi_i + N^{\mu}_{ij}\phi_j$ under some transformations which will be mentioned later (3.1) can be rewritten by

$$S(\tilde{\phi}+\phi) - S(\tilde{\phi}) - S_{,i}\phi_i$$

$$= \int d^4x\{-\frac{1}{2}\nabla_\mu\phi_i\nabla_\mu\phi_i + \frac{1}{2}\phi_i X_{ij}\phi_j + \Omega_{ijk}\phi_i\phi_j\nabla_\mu\phi_k + \tilde{\Lambda}_{ijk}\phi_i\phi_j\phi_k + \Theta_{ijk\ell}\phi_i\phi_j\phi_k\phi_\ell\}, \quad (3.2)$$

where $X_{ij} = M_{ij} - (N^\mu N^\mu)_{ij}$,

$$\tilde{\Lambda}_{ijk} = \Lambda_{ijk} + \frac{1}{3}(N^\mu_{i\ell}\Omega^\mu_{jk\ell} + N^\mu_{j\ell}\Omega^\mu_{ik\ell} + N^\mu_{k\ell}\Omega^\mu_{ij\ell}) \; . \quad (3.3)$$

The action (3.1) is invariant under the following transformations

$$\phi'_i = \phi_i + \lambda_{ij}\phi_j \;, \quad N'^\mu_{ij} = N^\mu_{ij} + \lambda_{ik}N^\mu_{kj} - N^\mu_{ik}\lambda_{kj} - \partial_\mu\lambda_{ij} \;,$$

$$M'_{ij} = M_{ij} + \lambda_{ik}M_{kj} - M_{ik}\lambda_{kj} - N^\mu_{ik}\partial_\mu\lambda_{kj} - \partial_\mu\lambda_{ik}N^\mu_{kj} \;,$$

$$\Omega'^\mu_{ijk} = \Omega^\mu_{ijk} + \lambda_{i\ell}\Omega^\mu_{\ell jk} + \lambda_{j\ell}\Omega^\mu_{i\ell k} + \lambda_{k\ell}\Omega^\mu_{ij\ell} \;,$$

$$\Lambda'_{ijk} = \lambda_{i\ell}\Lambda_{\ell jk} + \frac{1}{3}\partial_\mu\lambda_{i\ell}\Omega^\mu_{jk\ell} + \text{cyclic}(i,j,k) \;,$$

$$\Theta'_{ijk\ell} = \lambda_{im}\Theta_{mjk\ell} + \text{cyclic}(i,j,k,\ell) \; . \quad (3.4)$$

By considering dimensions of the coefficient functions and their transformation properties under (3.4) which include c-type gauge transformations as a special case, one-loop counter-terms for (3.2) are given by[3]

$$\Delta L^{\text{one-loop}}(\tilde{\phi}) = -\frac{1}{8\pi^2\varepsilon}(\frac{1}{4}X_{ij}X_{ji} + \frac{1}{24}Y^{\mu\nu}_{ij}Y^{\mu\nu}_{ji}) \;, \quad (3.5)$$

where $Y^{\mu\nu}_{ij} = \partial_\mu N^\nu_{ij} - \partial_\nu N^\mu_{ij} + N^\mu_{ik}N^\nu_{kj} - N^\nu_{ik}N^\mu_{kj}$ and $\varepsilon = 4 - n$. Similarly we can obtain two-loop counter-terms which were given explicitly in our paper[4].

When we apply these counter-term formula to the pure Yang-Mills type theory as an example we find

$$\Delta L^{\text{one-loop}} + \Delta L^{\text{two-loop}} = -\{\frac{11}{96\varepsilon}\frac{g^2C_2}{\pi^2} + \frac{17}{3\cdot 29\varepsilon}\frac{(g^2C_2)^2}{\pi^4}\}(F_{\mu\nu}{}^a)^2 \;, \quad (3.6)$$

from which the renormalization group function $\beta(g)$ can be found to be

$$\beta = -\frac{11}{3\cdot 2^4}\frac{g^3C_2}{\pi^2} - \frac{17}{3\cdot 2^7}\frac{g^5C_2^2}{\pi^4} + \cdots \; . \quad (3.7)$$

In the calculation of (3.6) we used the background field gauge condition (2.6)

$$L_{\text{gauge}}(\alpha) = -\frac{1}{2\alpha}(D_\mu\phi_\mu{}^a)^2 \;, \quad (3.8)$$

where the gauge parameter α is fixed to be $\alpha = 1$. In order to cancel all the subdivergence it is necessary to be renormalized such that

$$\alpha = 1 - \frac{g^2}{4\pi^2\varepsilon} C_2 + \cdots . \tag{3.9}$$

§4. Conclusions and Discussions

We obtained the counter-term formula up to two-loop by using the background field method. Although we did not discuss the cancellation mechanism of subdivergences in this formalism, it was examined in detail in our previous paper(4). The generalization of this formula to the gravitational theory is in progress now.

References

1) B. S. DeWitt, Phys. Rev. 162 (1967) 1195, 1239.

2) I. Ya. Aref'eva, A. A. Slavnov and L. D. Faddeev, Theor. Mat. Fiz. 21 (1974) 311,
G. 't Hooft, in Acta Universitateis Wratislavensis no.38, XII Winter School of Theoretical Physics in Karpacz,
D. Boulware, Phys. Rev. D23 (1981) 389,
B. S. DeWitt, in Quantum Gravity II, ed. C. Isham, R. Penrose and D. Sciama,
L. F. Abbott, Nucl. Phys. B185 (1981) 189.

3) G. 't Hooft, Nucl. Phys. B62 (1973) 444.

4) S. Ichinose and M. Omote, Nucl. Phys. B203 (1982) 221.

SUPERSYMMETRIC GRAND UNIFICATION

Norisuke Sakai
National Laboratory for High Energy Physics
Tsukuba, Ibaraki, 305 Japan

Abstract:

After a brief discussion of motivations and prototype models, we review recent studies of model building and of proton decay in supersymmetric grand unification. A new effect is mentioned for $\Delta B \neq 0$ four-scalar interactions induced by an intermediate scale ($10^{10} \sim 10^{12}$ GeV) supersymmetry breaking.

I. Gauge hierarchy and naturalness

Up to the highest accelerator energies electroweak interactions are now adequately described by the $SU(2) \times U(1)$ gauge model. Color SU(3) for strong interactions seems to be supported by all the available data too. This $SU(3) \times SU(2) \times U(1)$ gauge model has been beautifully unified into grand unified theories (GUT) such as SU(5) [1]. GUT has achieved several nice points:

i) The unique gauge coupling provides a true unification of electromagnetic, weak, and strong interactions.

ii) Quantization of charge. The equality of charges of proton and positron is a mysterious accident in the $SU(3) \times SU(2) \times U(1)$ gauge model, but it is explained by symmetry reasons in grand unified theories.

iii) A number of quantitative successes such as $\sin^2\theta_W(M_W)$ and m_b/m_τ.

iv) Possibility of proton decay and of explaining the origin of the baryon number in the universe.

On the other hand GUT has left several important problems still unanswered:

i) There are vastly different mass scales of gauge symmetry breaking for $SU(2) \times U(1)$ and GUT (gauge hierarchy), e.g. $M_W^2/M_{GUT}^2 \simeq 10^{-26}$ in SU(5).

ii) How to explain fermion masses and generations?

iii) How to incorporate gravity?

and so forth. Among them we now have some hope for a natural solution of the gauge hierarchy problem.

Some time ago 'tHooft has clarified the concept of naturalness: If a theory contains small parameters, it should acquire a larger symmetry for vanishing values of the parameters[2]. Namely these parameters are protected from getting large values by the approximate symmetry. In that view gauge hierarchy becomes natural, if there is a larger symmetry in the limit of vanishing mass-squared for Higgs scalar which is responcible for the $SU(2) \times U(1)$ symmetry breaking. For particles with spin one-half (one), chiral (local gauge) symmetry protects their masslessness. Since no symmetry is known which directly guarantees masslessness for spinless particles, we are led to two alternatives:

i) Theories without elementary scalar particles (technicolor models)[3].

ii) Supersymmetry (SUSY)[4]. Higgs scalar can be guaranteed to be massless if SUSY relates it to a spin 1/2 fermion which is massless because of chiral symmetry.

II. Supersymmetric grand unified models

1. SU(5) with explicit soft breaking of SUSY[5],[6]

i) The standard SU(5) GUT has been successfully made supersymmetric. One fine tuning at the tree level is needed for light Higgs doublet, but is not disturbed by radiative corrections (nonrenormalization theorem[7]).

ii) Since the naturalness is not spoiled by explicit soft SUSY breaking of order $\Delta m <$ TeV, superpartners of quarks and leptons can be given small masses ($<$ TeV).

iii) Renormalization-group analysis including many new particles showed phenomenologically acceptable values for $\sin^2\theta_W(M_W)$ and m_b/m_τ [8], but grand unification scale M_{GUT} tends to be larger than the nonsupersymmetric model.

2. SU(3) × SU(2) × U(1) × $\tilde{U}$(1) models

i) Experimentally scalar partners of charged leptons are found to be heavier than 16 GeV[9]. On the other hand spontaneous breakdown of SUSY gives a mass sum rule at the tree level[10]

$$\Sigma m^2_{\text{Boson}} - \Sigma m^2_{\text{Fermion}} = \sum_{U(1)} g\langle D\rangle \qquad (1)$$

where the right-hand side is a measure of SUSY breaking, so-called D-terms associated to broken U(1) subgroups. There are only two such U(1) generators in the SU(3) × SU(2) × U(1) model: weak hypercharge Y and the third component I_3 of SU(2). Unfortunately both Y and I_3 vanishes by summing over quarks (leptons).

$$\Sigma m^2_{\text{scalar quark}} - \Sigma m^2_{\text{quark}} = 0 \qquad (2)$$

Therefore we are forced to enlarge the gauge group in order to have a scalar partner heavier than quarks and leptons[6]. The simplest of such possibilities is the SU(3) × SU(2) × U(1) × $\tilde{U}$(1) models[11],[12].

ii) Irrespective of details of grand unification, one may look for a low energy SUSY model with SU(3) × SU(2) × U(1) × $\tilde{U}$(1) gauge symmetry. The model must satisfy:

a) Anomaly cancellation for renormalizability.

b) U(1) and $\tilde{U}$(1) be traceless for the absence of quadratic divergences (this might be unnecessary according to ref. 7).

c) Spontaneous breaking of supersymmetry.

Much efforts have been devoted to build such a model[11]-[13], but so far their models did not satisfy either one of the above three requirements. Recently we succeeded to construct a model with all three properties[14]. The result shows, however, a few annoying features too: a) Asymptotic nonfree (β-function for SU(3) is zero at one loop). b) Too many fields with the same quantum number. Therefore it appears difficult to embed the model into a simple group.

iii) A different viewpoint was proposed by a CERN group[15]. They take a mass scale in the Lagrangian ($\tilde{U}$(1) D-term) to be of the order of the Planck mass M_{Pl}. Because of that they argued to disregard the nonrenormalizability due to anomalies. Their model contains a SUSY breaking mass scale μ much larger than the electroweak mass scale M_W. However the effect almost decouples from our low energy world of quarks, leptons and Higgs doublet except effects of order μ^2/M_{Pl}, which is identified as M_W.

3. Intermediate scale SUSY breaking

i) The mass sum rule (1) was the stumbling block of SUSY model building. Recently several people realized that radiative corrections violate the sum rule to give a desirable mass pattern at two loop order[16]. In this picture the electroweak mass scale arises as a radiative correction to the SUSY breaking at higher energy scale. In particular if particles affected by the SUSY breaking μ at the tree level are themselves extremely heavy (of order $M \sim M_{GUT}$), they approximately decouple from the low energy world. Therefore the effective SUSY breaking in quarks, leptons and Higgs doublet supermultiplets is of order $\alpha\mu^2/M$. Identifying $\alpha\mu^2/M \sim M_W$ and $M \sim M_{GUT}$ or M_{Pl}, one obtains the SUSY breaking mass scale μ as $10^{10} \sim 10^{12}$ GeV, intermediate between M and M_W (geometric hierarchy[17]).

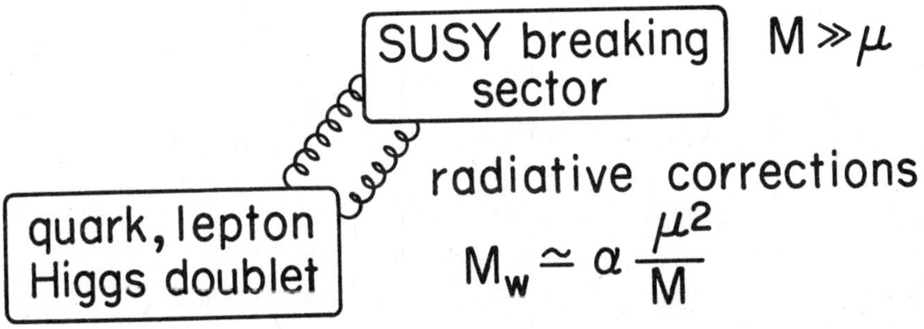

Fig. 1. Particles in the SUSY breaking sector are superheavy. Effective SUSY breaking in quark, lepton, and Higgs doublet is induced by radiative corrections.

ii) A realistic SU(5) model based on the above idea was constructed by Dine and Fischler[18] with $M \sim M_{GUT}$. Witten's mechanism[19] of generating large mass scale as a radiative correction was also implemented in a SU(5) model[17], but was found to have several severe problems[20]. Polchinski and Susskind studied decoupling and worked out a systematic way to extract the effective low energy theory[21]. The resulting picture is a theory with explicit soft breakings of SUSY, which are derivable from and constrained by the underlying high energy theory.

iii) Most recently a very interesting type of models were proposed where an explicit SUSY breaking arises as an effect of embedding SUSY GUT models into supergravity. This will be discussed by R. Arnowitt in this conference.

III. Proton decay in supersymmetric models

1. Model independent analysis in SUSY models

i) Proton decay offers the most spectacular and important information on the grand unification. Possible baryon-number violating effective interactions are known to be constrained by low-energy symmetries such as SU(3) × SU(2) × U(1)[22]. Model-independent operator-analysis has also been done for SUSY and SU(3) × SU(2) × U(1) as the low-energy symmetry[23],[12]. One finds a dimension-four $\Delta B \neq 0$ operator, but one can easily forbid it, for instance, by imposing a discrete symmetry, "matter parity", namely a sign change of quark and lepton superfields. On the other hand in SUSY models, dimension-five operators generally exist such as

$$\frac{1}{M}[q_- q_- q_- \ell_-]_F = \frac{1}{M} A_{q_-} A_{q_-} \psi_{q_-} \psi_{\ell_-} + \ . \ . \ . \qquad (3)$$

where q_- and ℓ_- are quark and lepton superfields and A_{q_-} and ψ_{q_-} are scalarquark and quark. The GUT mass scale is denoted by M. A typical term arises as an interaction between two scalars and two fermions (of quarks and a lepton) due to the baryon-number violating Higgs fermion exchange in SU(5) model for instance. If there are SUSY breaking Majorana masses for gauge fermions, the dominant contribution to proton decay comes from a loop diagram containing the dimension-five operator[24].

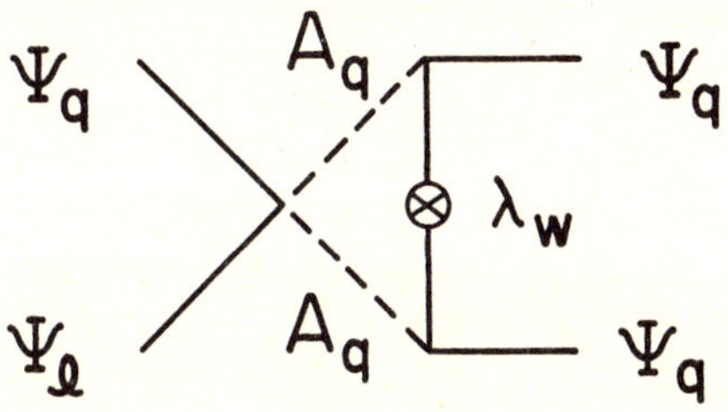

Fig. 2. A loop diagram with the dimension-five operator and the SUSY breaking Majorana mass for gauge fermion λ_W.

A detailed analysis gave the rate consistent with the present experimental bound and predicted the dominance of decay modes with neutrinos and higher generations[25].

Since proton decay due to the dimension-five operators is rather close to the experimental bound, one may wish to construct models which forbid these operators. Discrete symmetries, R-invariance (a global symmetry)[23], and local U(1) symmetry[12] were proposed for such a purpose. In this case SUSY restricts the dimension-six four-fermion operators to those of mixed chirality which can be experimentally verified from lepton-polarization measurement.

2. $\Delta B \neq 0$ four-scalar interaction induced by intermediate scale SUSY breaking

In the case of intermediate scale SUSY breaking, there are additional important operators for proton decay. Using the method of ref. 21, we performed a systematic operator-analysis with a new superfield $c_\pm$ responcible for the SUSY breaking[26]. More precisely the F-component of $c_\pm$ develops a vacuum expectation value μ^2 intermediate between $M \sim M_{GUT}$ and M_W, and hence $c_\pm$ must be singlet of SU(3) × SU(2) × U(1). For instance we obtain new operators of dimension six which give four-scalar interactions after SUSY breaking such as

$$\frac{1}{M^2}[q_-q_-q_-\ell_-c_-]_F = \frac{\mu^2}{M^2} A_{q_-}A_{q_-}A_{q_-}A_{\ell_-} \qquad (4)$$

where q_- (ℓ_-) and A_{q_-} (A_{ℓ_-}) denote left-handed quark (lepton) superfield and its scalar component. It contributes to proton decay through two-loop diagrams with Majorana masses of order M_W for gauge fermions.

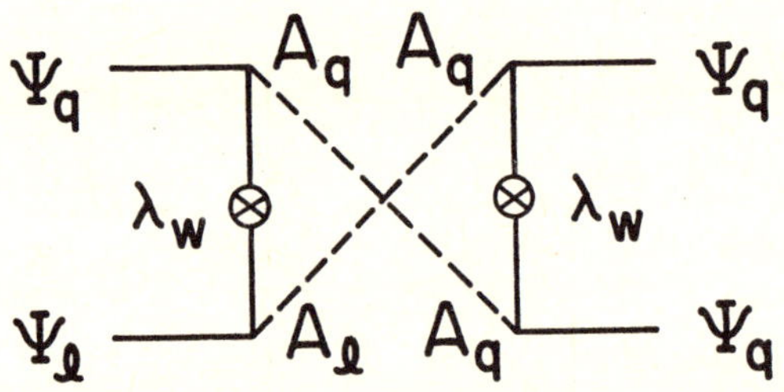

Fig. 3. A two-loop diagram with the SUSY breaking four-scalar interaction and the SUSY breaking Majorana mass for gauge fermion λ_W.

Since $M_W \simeq \mu^2/M$ in the geometric hierarchy picture, the contribution of the four-scalar interaction to proton decay is of the same order as the dimension-five SUSY operator (apart from powers of couplings etc.): $(\mu/M)^2 \cdot (1/M_W^2) \simeq (1/M) \cdot (1/M_W) \simeq 1/\mu^2$. In certain models SUSY breaking Majorana masses for gauge fermions are negligible[27] and make the above four-scalar interactions and the dimension-five SUSY operators unimportant for proton decay. In that case more important operator is another type of four-scalar interactions with mixed chirality such as

$$\frac{1}{M^4}[q_- q_- u_+ e_+ c_+ c_-]_D = \left(\frac{\mu}{M}\right)^4 A_{q_-} A_{q_-} A_{u_+} A_{e_+} \qquad (5)$$

where u_+ (e_+) and A_{u_+} (A_{e_+}) are right-handed u-quark (electron) superfield and its scalar component. The operator can contribute to proton decay through two-loop diagrams without the Majorana masses for gauge fermions. In the geometric hierarchy picture its contribution to the proton decay is of the same order as the dimension six SUSY operator $(\mu/M)^4 \cdot (1/M_W^4) \simeq 1/M^2$. Therefore the new $\Delta B \neq 0$ four-scalar interactions can contribute to proton decay with comparable order of magnitude as the supersymmetric $\Delta B \neq 0$ operators for both cases with or without significant Majorana mass for gauge fermions.

Existence of a type of four-scalar interactions in a model was also noted recently in ref. 28, but a systematic operator-analysis of proton decay in intermediate scale SUSY breaking is found in ref. 26.

References

1. H. Georgi and S.L. Glashow, Phys. Rev. Lett. 32(1974)438.
2. G. 'tHooft, Recent Developments in Gauge Theories, Cargèse summer school 1979, p.135;
   M. Veltman, Acta Phys. Pol. B12(1981)437.
3. S. Weinberg, Phys. Rev. D13(1976)974 and D19(1979)1277;
   L. Susskind, Phys. Re. D20(1979)2619;
   S. Dimopoulos and L. Susskind, Nucl. Phys. B155(1979)237;
   E. Eichten and K. Lane, Phys. Lett. 90B(1980)125.
4. For a review, see P. Fayet and S. Ferrara, Phys. Rep. 32C(1977) 249;
   A. Salam and J. Strathdee, Fortschr. Phys. 26(1978)57.
5. N. Sakai, Z. f. Phys. C11(1981)153.
6. S. Dimopoulos and H. Georgi, Nucl. Phys. B193(1981)150.
7. P.K. Townsend, talk at this conference;
   M.T. Grisaru, W. Siegel and M. Roček, Nucl. Phys. B159(1979)429.
8. M. Einhorn and D.R.T. Jones, Nucl. Phys. B196(1982)475.
9. D. Cords, in Proc. XX Int. Conf. High Energy Phys. Wisconsin, 1980, p.590.
10. S. Ferrara, L. Girardello, and F. Palumbo, Phys. Rev. D20(1979) 403.
11. P. Fayet, Phys. Lett. 69B(1977)489.
12. S. Weinberg, Phys. Rev. 26D(1982)287.
13. L.J. Hall and I. Hinchliffe, Phys. Lett. 112B(1982)285.

14. T. Inami, C.S. Lim and N. Sakai, in preparation.
15. R. Barbieri et al., Z. f. Phys. 13(1982)267; CERN TH. 3309.
16. L. Alvarez-Gaumé et al., HUTP-81/A063;
M. Dine and W. Fischler, IAS prep.;
C.R. Nappi and B.A. Ovrut, IAS prep.;
L. Ibáñez and G.G. Ross, Phys. Lett. 110B(1982) 215;
J. Ellis et al., Phys. Lett. 113B(1982)283.
17. S. Dimopoulos and S. Raby, LA-UR-82-1282.
18. M. Dine and W. Fischler, IAS prep.
19. E. Witten, Phys. Lett. 105B(1981)267.
20. T. Banks and V. Kaplunovsky, TAUP 1028-82; T. Banks, talk at this conference.
21. J. Polchinski and L. Susskind, SLAC-Pub-2924.
22. S. Weinberg, Phys. Rev. Lett. 43(1979)1566;
F. Wilczek and A. Zee, ibid. 1571.
23. N. Sakai and T. Yanagida, Nucl. Phys. B197(1982)533.
24. S. Dimopoulos, S. Raby and F. Wilczek, UMHE 81-64.
25. J. Ellis, D.V. Nanopoulos and S. Rudaz, Nucl. Phys. B202(1982)43.
26. N. Sakai, KEK-TH 53.
27. J. Polchinski, SLAC-Pub-2931-T.
28. J.P. Derendinger and C.A. Savoy, UGVA-DPT 1982/07-357.

ASPECTS OF GRAND UNIFIED MODELS WITH SOFTLY BROKEN SUPERSYMMETRY

K.Inoue, A.Kakuto*, H.Komatsu** and S.Takeshita

Department of Physics, Kyushu University 33, Fukuoka 812
*The Second Department of Engineering, Kinki University, Iizuka 820
**Institute for Nuclear Study, University of Tokyo, Tanashi, Tokyo 188

Supersymmetric theories have an outstanding property of no-renormalization for F-terms. This may give an important key to the solution of the gauge hierarchy problem in the grand unified theories (GUTs).

If nature chooses a supersymmetric theory, the supersymmetry must be broken at low energies spontaneously or explicitly. In spontaneously broken supersymmetric theories, however, there is a severe constraint on masses of component fields in a given supermultiplet, and in order to construct a realistic model, we must introduce disgusting complexity in the model[1].

On the other hand in the explicit breaking scheme, it is possible to construct a realistic model with minimal set of supermultiplets[2]. The minimal supersymmetric SU(3)×SU(2)×U(1) model contains the following supermultiplets:

					SU(3)	SU(2)	Y/2
vector multiplets							
	V_1 :	V_1^μ,	λ_1,	D_1	($\underline{1}$,	$\underline{1}$,	0)
	V_2 :	V_2^μ,	λ_2,	D_2	($\underline{1}$,	$\underline{3}$,	0)
	V_3 :	V_3^μ,	λ_3,	D_3	($\underline{8}$,	$\underline{1}$,	0)
left-handed Higgs multiplets							
	H_1 :	A_1,	ψ_1,	F_1	($\underline{1}$,	$\underline{2}$,	-1/2)
	H_2 :	A_2,	ψ_2,	F_2	($\underline{1}$,	$\underline{2}$,	1/2)
left-handed matter multiplets							
	ℓ_r :	$A(\ell_r)$,	$\psi(\ell_r)$,	$F(\ell_r)$	($\underline{1}$,	$\underline{2}$,	-1/2)
	$\bar{e}_r$:	$A(\bar{e}_r)$,	$\psi(\bar{e}_r)$,	$F(\bar{e}_r)$	($\underline{1}$,	$\underline{1}$,	1)
	q_r :	$A(q_r)$,	$\psi(q_r)$,	$F(q_r)$	($\underline{3}$,	$\underline{2}$,	1/6)
	$\bar{u}_r$:	$A(\bar{u}_r)$,	$\psi(\bar{u}_r)$,	$F(\bar{u}_r)$	($\underline{3}^*$,	$\underline{1}$,	-2/3)
	$\bar{d}_r$:	$A(\bar{d}_r)$,	$\psi(\bar{d}_r)$,	$F(\bar{d}_r)$	($\underline{3}^*$,	$\underline{1}$,	1/3)

where r(=1,2,3) are generation indices.

In this scheme, gauge fermion masses (M_1, M_2 and M_3 for λ_1, λ_2 and λ_3 respectively) and all mass terms of the scalar components of matter multiplets ($m^2(\phi)$ for

$A(\phi)$ with $\phi = \ell_r, \bar{e}_r, q_r, \bar{u}_r, \bar{d}_r$) and those of Higgs multiplets (m_1^2 for A_1, m_2^2 for A_2 and m_3^2 for A_1-A_2 mixing terms) are freely adjustable parameters of the theory.

If the explicit breaking scheme is the case in nature, the low energy phenomena give considerable constraints on the values of these breaking parameters as we will see below.

All $m^2(\phi)$ for $\phi = \ell_r, \bar{e}_r, q_r, \bar{u}_r, \bar{d}_r$ must be large ($\sim O((10^2\ \mathrm{GeV})^2)$) and positive. If any one of them (except ℓ_r) are negative, $U(1)^{EM}$ or color SU(3) conservation breaks down.

In order to obtain desirable symmetry breaking $SU(2)\times U(1) \to U(1)^{EM}$, Higgs scalars A_1 and A_2 must acquire non-vanishing vacuum expectation values through the minimization of the Higgs potential

$$V = \frac{g^2}{2}(A_1^\dagger \frac{\tau^a}{2} A_1 + A_2^\dagger \frac{\tau^a}{2} A_2)^2 + \frac{g'^2}{8}(A_1^\dagger A_1 - A_2^\dagger A_2)^2$$
$$+ m_1^2 A_1^\dagger A_1 + m_2^2 A_2^\dagger A_2 - m_3^2 (A_1 A_2 + A_1^* A_2^*) .$$

The existence of the Higgs vacuum requires the following conditions[3]:

$$m_1^2 + m_2^2 > 2|m_3^2| , \quad m_3^4 > m_1^2 m_2^2 ,$$

that is, m_3^2 must lie between the algebraic and geometrical average of m_1^2 and m_2^2. Since we expect that our $SU(3)\times SU(2)\times U(1)$ model is embedded in some grand unified group G_{GUT}, we reject the possible existence of U(1)-D term in the lagrangian. Therefore if there are no soft-breaking effects, which implies $m_1^2 = m_2^2$ and $m_3^2 = 0$, $SU(2)\times U(1)$ does not break down.

The existence of the superpartners of leptons, quarks and gauge bosons induces the flavor changing neutral interactions such as $s + \bar{d} \to d + \bar{s}$ and $\mu \to e\gamma$. In order to suppress such effects, the following stringent conditions must be satisfied to validate the super-GIM mechanism[4]:

$$|m^2(q_1) - m^2(q_2)| \;/\; m^2(q_1) < O(10^{-3}),$$
$$|m^2(\ell_1) - m^2(\ell_2)| \;/\; m^2(\ell_1) < O(10^{-3}).$$

It is implausible to expect all these conditions are satisfied by accident. It may be desirable to find some kind of systematic treatment of breaking parameters which guarantees all the requirements. It may well be likely that all soft-breaking terms come from the single origin.

Here we examine the exciting possibility that at the unification energy scale ($\mu \cong M_X$), the theory is almost supersymmetric and the soft-breaking terms exist only in the G_{GUT} invariant mass terms of gauge fermions λ's[5]. At lower energies, all the other soft-breaking terms are generated through radiative corrections[6].

In order to clarify whether this "minimal" soft-breaking scheme really gives the required soft-breaking parameters at low energy ($\mu \cong M_W$), we must examine the renormalization group analysis.

The supersymmetric part of the lagrangian consists of the usual "kinetic" terms and the F-component of the super potential[7)]

$$W = f(\ell H_1 \bar{e}) + h(qH_1\bar{d}) + \tilde{h}(qH_2\bar{u}) + m(H_1H_2).$$

The soft-breaking terms are given by[8)]

$$\mathcal{L}_{break} = \{-M_1\lambda_1\lambda_1 - M_2\lambda_2\lambda_2 - M_3\lambda_3\lambda_3 + h.c.\} - \sum_{\phi=\ell,\bar{e},q,\bar{u},\bar{d}} m^2(\phi)A(\phi)^\dagger A(\phi)$$
$$- \Delta_1^2 A_1^\dagger A_1 - \Delta_2^2 A_2^\dagger A_2 + m\Delta_3(A_1A_2 + A_1^* A_2^*)$$
$$+ \{fm_f A(\ell)A_1A(\bar{e}) + hm_h A(q)A_1A(\bar{d}) + \tilde{h}m_{\tilde{h}}A(q)A_2A(\bar{u}) + h.c.\}.$$

The Higgs scalar mass terms are given as $m_1^2 = m^2 + \Delta_1^2$, $m_2^2 = m^2 + \Delta_2^2$, $m_3^2 = m\Delta_3$.

By examining the renormalization group equations for this minimal model, we see that all the supersymmetry breaking parameters are generated through radiative corrections starting from a boundary condition[5)]

$$m \sim O(10^2 \text{ GeV})$$
$$M_1 = M_2 = M_3 \equiv M \sim O(10^2 \text{ GeV})$$
$$m(\phi) = \Delta_1 = \Delta_2 = \Delta_3 = m_f = m_h = m_{\tilde{h}} = 0 \qquad \text{at } \mu = M_X.$$

The gauge fermion loop contributions give large positive masses of order 10^2 GeV to the scalar partners of leptons and quarks through the following renormalization group equation:

$$(4\pi)^2 \mu\frac{\partial}{\partial\mu} m^2(\phi) = -8 \sum_{i=SU(3),SU(2),U(1)} g_i^2 C_2(R)_i M_i^2 + [\text{Yukawa coupling}].$$

Since the Yukawa coupling contributions are negligibly small for the scalars of the first and second generations, those with the same SU(3)×SU(2)×U(1) quantum numbers are almost degenerate. Therefore in our scheme the super-GIM mechanism works well to suppress the dangerous flavor changing neutral interactions. Combining with the renormalization group equations for gauge fermion masses

$$(4\pi)^2\mu\frac{\partial}{\partial\mu}M_1 = 22g'^2M_1, \quad (4\pi)^2\mu\frac{\partial}{\partial\mu}M_2 = 2g^2M_2, \quad (4\pi)^2\mu\frac{\partial}{\partial\mu}M_3 = -6g_c^2M_3,$$

we get the following mass relations at $\mu = M_W$:

$$M_2/M_1 = 2.01, \qquad M_3/M_1 = 7.19,$$
$$m(\ell_r)/M_1 = 1.78, \qquad m(\bar{e}_r)/M_1 = 0.95,$$
$$m(q_r)/M_1 = 6.60, \qquad m(\bar{u}_r)/M_1 = 6.40,$$
$$m(\bar{d}_r)/M_1 = 6.38, \qquad (r = 1,2)$$

where $M_X = 2.37\times 10^{16}$ GeV and $\alpha_G^{-1} = 4\pi/g^2(\mu = M_X) = 24.1$ are used.

In order to obtain desired Higgs vacuum, which is characterized by the conditions $m_1^2 + m_2^2 > |m_3^2|$ and $m_3^4 > m_1^2 m_2^2$ at $\mu \doteqdot M_W$, parameters m, M and Yukawa couplings at $\mu = M_X$ must be in the appropriate domain. If all Yukawa couplings are negligibly small, Higgs scalars A_1 and A_2 receive the same renormalization effects and the relation $m_1^2 = m_2^2$ follows. Therefore the breakdown of SU(2)×U(1) does not occur. It is indispensable in our minimal scheme that at least one of the Yukawa couplings is large enough to generate sizable mass difference between m_1^2 and m_2^2. The only candidate which can be freely large is the top quark Yukawa coupling $\tilde{h}$. Therefore the occurrence of the spontaneous breakdown $SU(2)\times U(1) \to U(1)^{EM}$ requires the existence of the lower bound of $\tilde{h}$. The detailed calculation shows $\tilde{h}(\mu = M_X) \gtrsim 0.095$. This lower bound turns out to be that of top quark mass,

$$m_t \simeq \langle A_2\rangle \tilde{h}(\mu = M_W) \gtrsim 60 \text{ GeV}.$$

Matter scalars acquire additional mass terms through vacuum expectation values $\langle A_1\rangle$ and $\langle A_2\rangle$. Especially their couplings to the D-components of the gauge multiplets give the following mass terms,

$$M_W^2 \cos 2\theta \; A^\dagger [I_3 - Y(\tan^2\theta_W)/2] A$$

where $\theta = \cot^{-1}(\langle A_1\rangle / \langle A_2\rangle)$. Their contributions are not always positive. The most dangerous is that for the lightest scalars $A(\bar{e}_r)$ (r=1,2). Their physical masses are given by

$$m^2(\bar{e}_r) - M_W^2 \cos 2\theta \; \tan^2\theta_W .$$

Since $m^2(\bar{e}_r) = (0.95)^2 M_1^2$, the positivity of this physical mass requires the lower bound of M_1. The detailed computation gives

$$M_1 \gtrsim 30 \text{ GeV}.$$

In conclusion, our "minimal" soft-breaking scheme in the minimal supersymmetric SU(3)×SU(2)×U(1) model embedded in the standard GUT group works well. The spontaneous breakdown $SU(2)\times U(1) \to U(1)^{EM}$ occurs through radiative corrections. The super-GIM mechanism works well to naturally suppress the dangerous flavor changing neutral interactions. In order for our scheme to work, following constraints must be satisfied:

$$m_t \gtrsim 60 \text{ GeV},$$

$$M_1 = M_2/2.01 = M_3/7.19 \gtrsim 30 \text{ GeV}.$$

References

1) P.Fayet, Phys. Lett. 69B(1977), 489.
P.Fayet, Phys. Lett. 70B(1977), 461.
S.Weinberg, Harvard preprint HUTP-81/A047.
2) N.Sakai, Z. Phys. C11(1981), 153.
S.Dimopoulos and H.Georgi, Nucl. Phys. B193(1981), 150.
3) K.Inoue, A.Kakuto, H.Komatsu and S.Takeshita, Prog. Theor. Phys. 67(1982), 1889.
4) J.Ellis and D.V.Nanopoulos, Phys. Lett. 110B(1982), 44.
5) K.Inoue, A.Kakuto, H.Komatsu and S.Takeshita, preprint KYUSHU-82-HE-5, to be published in Prog. Theor. Phys. 68(1982), No.3.
6) Similar approach is proposed by L.Ibâñez and G.G.Ross, Phys. Lett. 110B(1982),215.
7) For a review, see for example P.Fayet and S.Ferrara, Phys. Reports 32(1977), 249.
8) L.Girardello and M.T.Grisaru, Nucl. Phys. B194(1982), 65.

Supersymmetric Dipole Mechanism and Vacuum Energy

Nobuyoshi Ohta*)

Department of Physics, Faculty of Science

University of Tokyo, Tokyo 113, Japan

Much attention has been paid recently to supersymmetric theories in the hope of resolving the gauge hierarchy problem in grand unified theories. Quadratic divergences in elementary scalar theories make it necessary to adjust a parameter to 38 decimal precision to achieve the large hierarchy.[1] The remarkable non-renormalization theorem in supersymmetric theories may provide a natural resolution of this problem.[2]

One of major questions in this approach is how to break supersymmetry taking the following points into account:

i) Supersymmetry must be broken in realistic models,

ii) if supersymmetry is broken spontaneously at the tree level, the unrealistic mass formula $\sum_J (-1)^{2J}(2J+1)m_J^2 = 0$ always follows,

iii) Supersymmetry remains unbroken by perturbation if it is so at the tree level.

It then seems more realistic to accept the following explicit breakings which are soft in the sense that no quadratic divergences are generated.[3]

$$\begin{aligned}
&\text{a)}\quad L_a = \mu_1^2(A^2 + B^2),\\
&\text{b)}\quad L_b = \mu_2^2(A^2 - B^2),\\
&\text{c)}\quad L_c = \mu_3\bar{\lambda}\lambda,\\
&\text{d)}\quad L_d = \mu_4(A^3 - 3AB^2),\\
&\text{e)}\quad L_e = \mu_5^3 A,
\end{aligned} \tag{1}$$

---

*) Fellow of Japan Society for the Promotion of Science.

where A and B are the scalar and pseudoscalar fields of a chiral multiplet, respectively, while λ is the spinor field of a vector multiplet.

We shall show, however, that all these soft breakings can also be derived by the mechanism of spontaneous breaking and discuss the vacuum energy in this scheme,[4] which was originally proposed by Slavnov.[5]

Let L_0 be a supersymmetric Lagrangian of chiral multiplets and/or vector multiplets, which are called the matter fields. we also introduce left handed chiral superfields S_- and $\tilde{S}_-$, and consider the additional supersymmetric Lagrangian

$$\Delta L = \frac{1}{8}(\bar{D}D)^2(S_-^+\tilde{S}_-) - \frac{1}{2}(\bar{D}D)(\xi S_-^+\tilde{\Phi}_-^+ + \eta\tilde{S}_-) + \text{h.c.}, \tag{2}$$

where ξ and η are real constants and $\tilde{\Phi}_-$ is a left-handed chiral superfield made of the matter fields. Expressed in terms of component fields $S_- = (a_-, \alpha_-, f_-)$, $\tilde{S}_- = (\tilde{a}_-, \tilde{\alpha}_-, \tilde{f}_-)$ and $\tilde{\Phi}_- = (\tilde{A}_-, \tilde{\psi}_-, \tilde{F}_-)$, eq.(2) becomes

$$\begin{aligned}\Delta L = {} & \partial_\mu a_-^+\partial^\mu\tilde{a}_- + \bar{\alpha}_- i\partial\!\!\!/\tilde{\alpha}_- + f_-^+\tilde{f}_- \\ & + \xi\{\tilde{A}_-^+ f_-^+ - \bar{\alpha}_-\tilde{\psi}_-^c + \tilde{F}_-^+ a_-^+\} + \eta\tilde{f}_- + \text{h.c.},\end{aligned} \tag{3}$$

where $\tilde{\psi}_-^c = C\bar{\tilde{\psi}}_-^T$. The first line can be diagonalized by orthogonal transformations

$$\begin{pmatrix}\phi_1 & \psi_1 & f_1 \\ \phi_2 & \psi_{2-} & f_2\end{pmatrix} = \frac{1}{\sqrt{2}}\begin{pmatrix}1 & 1 \\ -1 & 1\end{pmatrix}\begin{pmatrix}a_- & \alpha_- & f_- \\ \tilde{a}_- & \tilde{\alpha}_- & \tilde{f}_-\end{pmatrix} \tag{4}$$

resulting in

$$\begin{aligned}\Delta L = {} & \partial_\mu\phi_1^+\cdot\partial^\mu\phi_1 - \partial_\mu\phi_2^+\cdot\partial^\mu\phi_2 + \bar{\psi}_{1-}i\partial\!\!\!/\psi_{1-} - \bar{\psi}_{2-}i\partial\!\!\!/\psi_{2-} + f_1^+f_1 - f_2^+f_2 \\ & + \frac{\xi}{\sqrt{2}}[\tilde{A}_-^+(f_1^+ - f_2^+) - (\bar{\psi}_{1-} - \bar{\psi}_{2-})\tilde{\psi}_-^c + \tilde{F}_-^+(\phi_1^+ - \phi_2^+) + \text{h.c.}] \\ & + \frac{\eta}{\sqrt{2}}(f_1 + f_2 + f_1^+ + f_2^+).\end{aligned} \tag{5}$$

From eq. (5) we derive the field equations

$$\Box\phi_1 = \Box\phi_2 = -\frac{\xi}{\sqrt{2}}\tilde{F}_-^+, \tag{6a}$$

$$i\partial\!\!\!/\psi_{1-} = i\partial\!\!\!/\psi_{2-} = \frac{\xi}{\sqrt{2}}\tilde{\psi}_-^c, \tag{6b}$$

$$f_1 = -\frac{\eta}{\sqrt{2}} - \frac{\xi}{\sqrt{2}}\tilde{A}_-^+, \quad f_2 = \frac{\eta}{\sqrt{2}} - \frac{\xi}{\sqrt{2}}\tilde{A}_-^+. \tag{6c}$$

Clearly, the fields ϕ_1 and ψ_{1-} are fields of positive metric while ϕ_2 and ψ_{2-} are ghosts of negative metric, forming a supersymmetric set of dipole fields. We show that these dipole fields are unobservable yet break supersymmetry spontaneously. For this reason we call this scheme "supersymmetric dipole mechanism."

It is easy to see that $\phi_1 - \phi_2$ and $\psi_{1-} - \psi_{2-}$ are free fields:

$$\Box(\phi_1 - \phi_2) = i\partial\!\!\!/(\psi_{1-} - \psi_{2-}) = 0, \tag{7}$$

which allow us to impose the subsidiary conditions

$$(\phi_1 - \phi_2)^{(+)}|\text{phys}\rangle = (\psi_{1-} - \psi_{2-})^{(+)}|\text{phys}\rangle = 0. \tag{8}$$

The combinations $\phi_1 - \phi_2$ and $\psi_{1-} - \psi_{2-}$ turn out to be zero-norm fields, while $\phi_1 + \phi_2$ and $\psi_{1-} + \psi_{2-}$ are removed from physical states by eq. (8), and the net result is that they produce the breaking term spontaneously

$$-\xi\eta(\tilde{A}_- + \tilde{A}_-^+), \tag{9}$$

which is obtained by substituting (6c) into (5). Note that, although supersymmetry is broken spontaneously, there appear no physical Goldstino nor vacuum energy. This situation arises because one of the Goldstinos is of negative metric. We can also show that the mass splittings between bosons and fermions are generated without contradicting supercurrent conservation.

Eq. (8) gives the desired breaking terms L_e, L_b, L_d and L_c in

(1) by setting $\tilde{\Phi}_- = \Phi_-$, Φ_-^2, ϕ_-^3 and $\bar{\Psi}_{++}\Psi_{--}$, respectively, where Ψ_{++} and Ψ_{--} are the spinor superfields of the gauge field strengths. The term L_a can be obtained essentially in the same way by replacing the chiral multiplets S_- and $\tilde{S}_-$ by two vector multiplets V and $\tilde{V}$.[4)]

Now we wish to point out another peculiar feature of this model by choosing the Wess-Zumino model for the matter part and Φ_- for $\tilde{\Phi}_-$.

$$L_0 = \frac{1}{2}(\partial_\mu A)^2 + \frac{1}{2}(\partial_\mu B)^2 + \frac{1}{2}\bar{\psi}i\not{\partial}\psi + \frac{1}{2}F^2 + \frac{1}{2}G^2$$

$$+ m(AF - BG - \frac{1}{2}\bar{\psi}\psi) + g(A^2F - B^2F - 2ABG - A\bar{\psi}\psi - B\bar{\psi}i\gamma_5\psi). \quad (9)$$

The additional Lagrangian gives a linear A term and thus the scalar field A acquires a nonvanishing vacuum expectation value a. The vacuum energy is

$$V(a) = \frac{1}{2}(m + ga)^2a^2 + \sqrt{2}\xi\eta\ a, \quad (10)$$

and a is determined by the minimization condition of this potential. Obviously V(a) may vanish or become negative, contrary to naive expectation. This is precisely due to the presence of negative-metric fields. To see this, we have only to consider the linear part of the supercurrent

$$J_\mu = i\gamma_\mu[\frac{1}{\sqrt{2}}(m + ga)a\psi + (\frac{\eta}{\sqrt{2}} + \frac{\xi a}{2})\psi_1 + (\frac{\eta}{\sqrt{2}} - \frac{\xi a}{2})\psi_2]. \quad (11)$$

By using the canonical equal-time anticommutators of ψ, ψ_1 and ψ_2, we get

$$\frac{1}{4}\langle 0|\sum_\alpha\{S_\alpha, S_\alpha^+\}|0\rangle = \int d^3x[\{\frac{1}{\sqrt{2}}(m + ga)a\}^2 + (\frac{\eta}{\sqrt{2}} + \frac{\xi a}{2})^2 - (\frac{\eta}{\sqrt{2}} - \frac{\xi a}{2})^2], \quad (12)$$

where the negative sign in the last term is due to the negative metric of ψ_2. Eq. (12) agrees with (10), implying that the naive argument for positive vacuum energy must be modified in the presence of negative-metric fields.

In conclusion, we have shown that all the explicit soft breakings of supersymmetry may be regarded as spontaneous, providing with a theoretical basis for their softness. This model is a first nontrivial example with supersymmetry broken spontaneously but with not necessarily positive vacuum energy owing to an unobservable Goldstino of negative metric. Finally, from our results, we conjecture that the absence of quadratice divergences may lead to supersymmetry either unbroken or broken spontaneously. Some results in this direction have been obtained recently.[6)]

The author wishes to thank Professor Y. Fujii for useful advices and careful reading of the manuscript.

References

1) L. Susskind, Phys. Rev. D20 (1979) 2619.

2) S. Dimopoulos and H. Georgi, Nucl. Phys. B193 (1981) 150;
N. Sakai, Z. für Physik, C11 (1981) 153.

3) L. Girardello and M. T. Grisaru, Nucl. Phys. B194 (1982) 65;
J. Illiopoulos and B. Zumino, Nucl. Phys. B76 (1974) 310;
K. Harada and N. Sakai, Prog. Theor. Phys. 67 (1982) 1877.

4) N. Ohta, Phys. Letters 112B (1982) 215;
N. Ohta and Y. Fujii, Nucl. Phys. B202 (1982) 477 and references therein.

5) A. A. Slavnov, Nucl. Phys. B124 (1977) 301.

6) T. Inami, H. Nishino and S. Watamura, Univ. of Tokyo preprint (1982).

NOSONOMY OF AN INVERTED HIERARCHY MODEL

T. Banks
Department of Physics and Astronomy
Tel Aviv University
Ramat Aviv, Tel Aviv 69978
Israel

This is a report of an unsuccessful attempt to build a realistic Grand Unified model based on Witten's[1] inverted hierarchy mechanism. The work was a collaboration with Vadim Kaplunovsky[2]. Witten's phenomenon is a common property of supersymmetric gauge theories in which supersymmetry (SUSY) is spontaneously broken by the vacuum expectation value of an F term.

Details of the mechanism:

1) $\frac{\partial W}{\partial \phi_i} \neq 0$ for all ϕ_i and $W = C_{ijk}\phi_i\phi_j\phi_k + C_{ij}\phi_i\phi_j + C_i\phi_i$

   imply that the potential $V = \left|\frac{\partial W}{\partial \phi_i}\right|^2$ has a line of degenerate minima extending to ∞. Call this flat direction X.

2) The one loop radiative corrections to the potential V(X) for large X have the form

   $M^4 C \ln X/M$ (M is the Lagrangian mass scale).

   In a gauge theory C can be negative. Renormalization group analysis[2] shows that in many cases V(X) develops a minimum at $X = 0(Me^{1/g^2})$ with g a small coupling and M the SUSY breaking scale.

3) Since $\frac{\partial^2 W}{\partial X^2} = 0$ the partner of X is the Goldstone fermion.

4) When X>>M some particles get mass of 0(X). These are precisely those particles which couple to X (to order $\frac{M}{X}$). The large part of the mass matrix is

$$\langle X\rangle C_{xij}\phi_i\phi_j$$

So the couplings of particles of mass $\leq$M to the Goldstone fermion are 0(M/X). This implies that supersymmetry breaking in the spectum has the pattern

Supersymmetric mass	breaking
0(X)	0(M)
0(M)	$0(M^2/X)$
$\leq 0(M^2/X)$	$0(M^2/X)$

so only particles of mass $\leq 0(M^2/X)$ can be identified with the real world.

The model:

Fields A, Y in 24 of SU_5

$\underline{X}, Z$ in 1 of SU_5

$H, \bar{H}$ in $5, \bar{5}$ of SU_5

+ quarks-lepton superfields

$$W = \lambda \mathrm{tr} A^2 Y + \lambda'(\mathrm{tr}(A^2) - M^2) X + \beta \bar{H} A H + \gamma Z \bar{H} H + \text{quark lepton couplings to } H + \bar{H}$$

When X, Y get large expectation values all chiral superfields except $A_{1,1}$, $Y_{8,1}$, $Y_{1,3}$, X, Z, H, $\bar{H}$+quarks and leptons get masses of $0(X)$ (indices refer to $SU_3 \times SU_2$ transformation properties). Also

$$SU_5 \to SU_3 \times SU_2 \times U_1$$

Coset gauge boson masses are $0(X)$. $A_{1,1}$ gets mass of $0(M)$. The other masses are $0(M^2/X)$ or 0.

Diseases of the Model

I. Renormalization

With 3 generations of quarks and leptons color is not asymptotically free above 10 TeV and becomes strong before the GUT scale is reached. (2 loop terms $\gtrsim$ 1 loop)

Most realistic model has 2 generations but

$$\alpha_{e.m} \sim \frac{1}{137}$$

$$\Lambda_{QCD} \sim 100 \text{ MeV}$$

$$\Rightarrow M_{GUT} = 10^{24} \text{ GeV}$$

Best we can do with $M_{GUT} \leq 10^{18}$ GeV is

$$M_{GUT} = 10^{18} \text{ GeV}$$

$$\Lambda_{QCD} \sim 100 \text{ MeV}$$

$$\alpha_{e.m.} \sim \frac{1}{113}$$

choosing $\alpha_{e.m} \sim \frac{1}{137}$ forces $\Lambda_{QCD} \sim 10$ keV

The problem is caused by the light $M \sim 10$ TeV color octet superfield $Y_{8,1}$ which is essential to the inverse hierarchy mechanism. We have not been able to construct a model without this field.

II. Gravitational Problems

1) Gravitino mass density of the Universe

Pagels, Primack[3], Weinberg[4] bounds on the SUSY breaking scale

$$M < 2\times 10^6 \text{ GeV} \quad \text{or} \quad M > 10" \text{ GeV}$$

(actually the lower bound on M may be higher depending on the details of R symmetry breaking. We make the <u>optimistic</u> assumption that our model's R symmetry can be arranged so that this mimimal bound is satisfied).

In an inverse hierarchy model

$$\alpha \frac{M^2}{X} \sim 100 \text{ GeV and } 10^{14} \text{ GeV} \leqslant X \leqslant 10^{18} \text{ GeV}$$

The first estimate is due to the fact that $SU_2 \times U_1$ breaking in these models is a 2 loop radiative correction[2]. So

$$10^9 \text{ GeV} \leqslant M \leqslant 10^{11} \text{ GeV}$$

The model is barely compatible with the minimal bound. Factors of π^2 could be important. Note compatibility ⇒ $X \gtrsim 10^{18}$ GeV.

2) Gravitational corrections to the effective potential classical coupling to supergravity implies[5]

$$V = \left|\frac{\partial W}{\partial \phi}\right|^2 \to \left[\left|\frac{\partial W}{\partial \phi} + \frac{\phi}{M_p^2} W\right|^2 - \frac{1}{M_p^2}|W|^2\right] e^{\frac{\phi^+\phi}{M_p^2}}$$

In Witten's original scenario $M_p = O(M)$ in the fundamental gravitational Lagrangian and the physical Planck mass arises from a non-minimal gravitational coupling $\sqrt{g}\, X\, R$ to the X field. But in this case the above gravitational correction completely swamps the radiative corrections which create the inverse hierarchy effect. So we must give up Witten's attractive explanation of the ratio between the supersymmetry breaking scale M and the Planck scale (this problem was pointed out to me by Ed Witten).

III. The Revenge of Hierarchy

$SU_2 \times U_1$ breaking must be a radiative effect since tree level scales M, X are both >> 100 GeV. To obtain a nontrivial effective potential for H, $\overline{H}$ we must couple them to SUSY breaking. There are two (?) possibilities

$\delta W = \beta \overline{H}_i Y^i_j H^j$ ⇒ wrong SU_5 breaking (Witten)[1]

$\delta W = \beta \overline{H}_i A^i_j H^j$

$\langle A \rangle \propto 2_{2_{2_{-3_{-3}}}}$

gives mass of O(M) to all components of H, $\overline{H}$. Solution by fine tuning

$\delta' W = \gamma \overline{H} H$

tune γ (accuract $10^{-10} \sim 10^{-11}$) to cancel doublet mass. Brilliant solution of fine tuning problem - the sliding scalar:

$$\delta' W = \gamma Z\bar{H}H$$

(proposed independently by Witten, T.B. and Kaplunovsky, Dine + Fischler, Dimopoulos + Raby)[6,2]

If the radiative effective potential favors parallel expectation values of the doublet components of H, $\bar{H}$ (and it does for some values of parameters)[2,6] z will adjust itself to exactly cancel the doublet mass.

Unfortunately there are radiative corrections to the effective potential for Z which are much bigger than this term, so $\langle Z \rangle$ will probably not sit at the right point. We still have a fine tuning problem of $0(10^{-10})$. This problem is more serious than the others but it may be curable by a change in the treatment of H, $\bar{H}$. The other problems seem to be quite general features of inverse hierarchy models at least for an SU_5 gauge group. We have not been able to find models based on SO_{10}, Thus we conclude that Witten's inverted hierarchy does not work.

References

1. E. Witten, Phys. Lett. 105B, 271 (1981).
2. T. Banks and V. Kaplunovsky, Tel Aviv preprints TAUP 1028-82 TAUP 1011-82 and manuscript in preparation.
3. H. Pagels and J. Primack, Phys. Rev. Lett. 48, 223 (1982).
4. S. Weinberg, Phys. Rev. Lett. 48, 1303 (1982).
5. J. Bagger and E. Witten, Princeton preprint (1982).
6. M. Dine and W. Fischler, IAS preprint (1982).
   S. Dimopoulos and S. Raby, Los Alamos preprint (1982).
   E. Witten, (private communication).

GRAND UNIFIED THEORIES WITH SYMMETRY AND LOCAL SUPERSYMMETRY

R. Arnowitt, Pran Nath, and A.H. Chamseddine
Department of Physics
Northeastern University
Boston, Massachusetts 02215, USA

Abstract

The properties of local supersymmetric (LS) grand unified models based on the gauge group (N=1 supergravity)xSU(5) are described. The gravitational interactions of supergravity cause a dynamical symmetry breaking of both SU(2)xU(1) and supergravity at the mass $m_s = \kappa m^2 \sim 300$ GeV where m is the scale of supersymmetry breaking. These results are maintained at the one loop level. The models described are realistic in that they are in accord with low energy phenomenology.

1. Introduction

Over the past year, supersymmetry has played an increasingly important role in model building in grand unified theory. The intrinsic appeal of a symmetry which combines fermions and bosons into irreducible multiplets, along with the fact that realistic supersymmetric (SUSY) GUT models now exist(1) has greatly stimulated interest in such theories. Thus SUSY models can correctly account for the b quark to τ lepton mass ratio m_b/m_τ, the electro-weak mixing parameter $\sin^2\Theta_W$, and yield proton lifetimes $\tau_p \sim 10^{31}$ years. Perhaps most remarkable is the fact that supersymmetry offers a possible way of maintaining a mass hierarchy without the necessity of fine tuning the parameters of the theory in each order of perturbation theory(2). Of course, supersymmetry must be broken in the real world, and if the mass hierarchy is to be maintained in the presence of this breaking, the scale of supersymmetry breaking m_s must obey $m_s \sim 1$ TeV, so that Higgs masses remain small. It has been pointed out, however, that cosmological considerations require that the actual scale of supersymmetry breaking occur at a much higher mass(3), i.e., at $m \sim 10^{11}$ GeV, and a number of models have recently been proposed(4) where the low energy phenomena are "protected" against seeing directly the effects of the supersymmetry breaking phenomena at the intermediate mass scale m.

The above SUSY models are all based on global supersymmetry. For most of these theories, the GUT mass M is quite large, i.e., $M \sim 10^{16}$ GeV and hence $\kappa M \sim 10^{-2}$ (where the Newtonian constant is $G = \kappa^2/8\pi$). Thus gravitational effects may no longer be negligible. We present in this report a description of some of the properties of a new class of GUT models based on local supersymmetry(5,6,7) (LS) constructed by coupling N=1 at supergravity(8) to a set of scalar left multiplets, and a gauge vector multiplet. The gauge group of the theory is thus (N=1 supergravity)xG. We will choose G to be SU(5) here (though other possibilities can be considered). Gravity is thus included into the grand unification, and gravitational affects are indeed found to be important. Some of the material presented here overlaps with other work in this area. Thus the idea that local supersymmetry should be incorporated in GUT models has also been suggested by Weinberg[9] and by Ovrut and Wess[10]. The coupling of supergravity to an arbitrary number of chiral multiplets and a gauge multiplet has independently been worked out by Cremmer et al.[11]

2. Survey of Results

Before describing the details of LS GUT models, we list here briefly some of the results obtained, and how LS GUT differs from the global SUSY GUT: (1) If supersymmetry is not spontaneously broken in SUSY models, the different vacua arising from the breaking of G [e.g., SU(5), SU(4)xU(1) and SU(3)xSU(2)xU(1)] remain degenerate. IN LS GUT, however, the gravitational interactions produce a splitting of these vacua[5,9,10] of size $\sim\kappa^2 M^6$ (energy/vol). Thus one of the problems of SUSY models is automatically overcome (without introducing any ad hoc terms in the superpotential to break supersymmetry). (2) If one introduces a super-Higgs potential[12,13] with mass scale m, into the superpotential,

$$g_{S-H} \equiv m^2(Z+b) \tag{1}$$

(where Z is a scalar superfield and b is chosen to adjust the cosmological constant to zero) then one can construct LS models where there is simultaneous spontaneous breaking of both supersymmetry and SU(2)xU(1) at the same mass scale m_s where[5]

$$m_s = \kappa m^2 \tag{2}$$

Since experimentally, electroweak breaking occurs at $m_s \sim 300$ GeV one has $m \sim 10^{10}$-10^{11} GeV in agreement with the cosmological bounds[3] on m. Eq. (2) is a remarkable result in that it is the first known example of gravitation effecting low energy phenomena. It can occur because supergravity allows "semi-gravitational" phenomena. (Recall that $\kappa \sim G^{1/2}$.) The LS model thus relates supersymmetry breaking to electro-weak breaking, the low energy regime being dynamically "protected" from the relatively high super-Higgs mass scale m by the factor κm in Eq. (2). It is interesting to note that Eq. (2) represents a form of "geometric hierarchy" (arising here at the tree level) in that $m \sim \sqrt{m_s m_p}$ where m_p is the Planck mass. (3) As discussed in (1) above LS GUT automatically produces $O(\kappa^2)$ splitting of the different gauge vacua even before supersymmetry breaking. If one arranges the physically interesting vacuum [e.g., SU(3)xSU(2)xU(1)] to have zero cosmological constant, then unfortunately it will lie highest. While Weinberg[9] has pointed out that this vacuum is most likely stable against finite bubble formation leading to decay into the lower vacua (since they are in anti-deSitter spaces) it is of interest to ask whether one may circumvent this problem more directly. When the super-Higgs potential is included, this is indeed the case, and one may construct

LS models where the SU(3)xSU(2)xU(1) vacuum is the absolute minimum[6]. In these models the $O(\kappa^2)$ corrections to the splitting cancels, and the degeneracy is lifted in the κ^4 order by the super-Higgs effect. One finds that the vacuum splitting is now $\sim m_s^4 = \kappa^4 m^8$ (instead of $\sim \kappa^2 M^6$). Models of this type may have interesting cosmological consequences since the smallness of m_s implies that the Universe will not choose the physical vacuum until relatively late in its development. (4) A key feature of the above tree results is that those fields whose vacuum expectation values (VEV) vanish in the global limit, $\kappa \to 0$ (e.g., the electro-weak Higgs fields) grow VEV's of size $\sim m_s = \kappa m^2$ (rather than $\sim m$ or κM^2). It is clearly important that these features be maintained when loop corrections are included for the theory to be physically viable. We have examined the one loop terms[7] and have established the following:

(i) As is well-known in <u>global</u> supersymmetry[14] if supersymmetry is not spontaneously broken at the tree level, it does not break perturbatively at the loop level. A similar theorem holds at least to one loop in LS GUT models: if the super-Higgs potential is set to zero (so no tree spontaneous symmetry breaking of supersymmetry occurs) and if the cosmological constant is set to zero, there is no perturbative dynamical symmetry breaking at one loop to <u>arbitrary</u> order in κ. The result follows from a remarkable set of cancelations of κ-dependent terms (over and above the global cancellations) in the loop expression and may in fact hold also at higher order.

(ii) When the super-Higgs potential is present, the one loop contributions produce non-vanishing contributions to the effective potential of size $\sim\lambda^4 m_s^4 = \lambda^4\kappa^4 m^8$ where λ is a characteristic dimensionless coupling constant. The tree effective potential is of size $\sim\lambda^2 m_s^4$ and so the loop corrections do not destabilize the tree results (provided the coupling constants are small enough so that the loop expansion is valid). The significant phenomena here is the cancellation of terms of size $\kappa M^3 m^2$ and $\kappa^2 M^2 m^4$ which would have swamped the tree results and hence negated the previous results. Thus, even in the presence of gravitational effects, the loop corrections shield the low energy regime from the large mass m, and produce supersymmetry breaking effects of size $m_s = \kappa m^2$.

3. Supergravity Couplings

Standard grand unified models are based on a set of left-handed spinors χ_L^a, $a = 1 \ldots N$ describing quarks and leptons, gauge vector mesons V_μ^α in the adjoint representation of the group G (α is the group index), and a set of scalar Higgs mesons. For global SUSY models, each of these fields must become members of a supersymmetry multiplet. The two basic multiplets for constructing supersymmetric theories are the scalar left handed chiral (F-type) multiplet,

$$\sum = (z, \chi_L, h) \tag{3}$$

and the vector (D-type) multiplet with components

$$(C, \xi, H, K, V_\mu, \lambda, D) \tag{4}$$

In Eq. (3), z = A+iB is a complex scalar field and h = F+iG are auxilliary (constraint) fields. In Eq. (4), ξ and λ are Majorana spinors, C,H,K,D are scalar fields with D also an auxilliary field. In constructing models, both the matter spinors $\chi_L{}^a$ and the Higgs mesons are placed in left handed scalar multiplets. Thus, the quarks and leptons gain supersymmetric scalar partners ("squarks" and "sleptons") and the Higgs mesons gain spinor partners ("higgsinos"). The gauge vector mesons are placed in a vector multiplet. In the Wess-Zumino gauge(15) only the last three components of Eq. (4) are non-zero:

$$V \equiv (V_\mu{}^\alpha, \lambda^\alpha, D^\alpha) \tag{5}$$

Thus, the gauge mesons gain spinor partners ("gauginos"). A convenient tensor calculus of multiplets exists whose rules follow naturally using superspace methods(16).

We now discuss the coupling of these multiplets to supergravity. The coupling of a single chiral multiplet and of the vector multiplet to supergravity had previously been obtained(17,18). For GUT models, it is necessary to generalize this to an arbitrary number of chiral multiplets with simultaneously a gauge vector multiplet being present(5,11). We briefly summarize the analysis and some of the results. The basic rules for coupling F and D type multiplets of Eqs. (3) and (4) to supergravity maintaining supergravity gauge invariance are given in Refs. (17,18). To construct the most generally allowed Lagrangian then, one forms the most general F and D type multiplets [of Eqs. (3) and (4)] out of the matter and Higgs chiral multiplets Σ^a and the gauge multiplet V^α introduced in Eq. (5) which are also invariant under the gauge group G. One then couples these to supergravity according to rules of Refs. (18,19). A convenient way of carrying out the calculation is to first couple the gauge multiplet V to the chiral multiplets Σ^a (according to how the Σ^a transform under G) and then couple the resultant structure to supergravity(5).

Using the rules for multiplet multiplication, computing the F and D contributions to the Lagrangian, eliminating all auxilliary variables, and making various simplifying point transformations, one obtains after much labor the final form for the Lagrangian(5,11). It is, of course, too long to record here! We give now the tree effective potential of the theory which arises from the non-derivative couplings of the scalar fields z_a:

$$V(z, z^+) = \frac{1}{2} E[G_a G_a{}^+ - \frac{3}{2} \kappa^2 g g^+] + \frac{g_e{}^2}{32} [z_a{}^+ (T^\alpha z)_a]^2 \tag{6}$$

where $(T^\alpha)_{ab}$ are the group generators,

$$G_a = g(z),_a + \frac{\kappa^2}{2} z_a{}^+ g(z); \quad E = \exp \frac{1}{2} \kappa^2 z_a{}^+ z_a; \text{ and } g_{,a} \equiv \partial g/\partial z_a \tag{7}$$

Here $g(z_a)$ is the superpotential whose specification describes the specific model. (We have also chosen for simplicity interactions that normalize the spin zero kinetic energy to unity.) We will give below the fermion mass matrices of the Lagrangian, after the question of spontaneous breaking of supersymmetry is discussed.

4. Spontaneous Symmetry Breaking

Eq. (6), of course, correctly reduces to the global supersymmetry effective potential in the limit $\kappa \to 0$, and exhibits the gravitational clothing produced by supergravity. In all spontaneous symmetry breaking solutions that we will discuss, the second (T^α) term of Eq. (6) vanishes, and so we will neglect it in the following. The extremum condition, $V_{,a} = 0$, yields on the real manifold the relation

$$T_{ab}G_b = 0 \tag{8}$$

where

$$T_{ab} = g_{,ab} + \frac{\kappa^2}{2}(z_a g_{,b} + z_b g_{,a}) + \frac{\kappa^4}{4} z_a z_b g - \kappa^2 \delta_{ab} g \tag{9}$$

If $\det T_{ab} \neq 0$, then Eq. (8) requires that $G_a = 0$. These conditions imply that supersymmetry is not broken (since the supersymmetry transformation reads[17] $\delta\chi_L{}^a = -\frac{1}{2}\sqrt{E}\, G_a{}^+ \varepsilon_L + \ldots$ and so $\langle\delta\chi_L{}^a\rangle$ vanishes at the minimum). The gauge group G, however, would in general be broken. The requirement $G_a = 0$ leads to a set of solutions $z_a = z_a{}^{(i)}$ at which the effective potential becomes

$$V_{min}(z^{(i)}) = -\frac{3}{4}\kappa^2 E(z^{(i)})|g(z^{(i)})|^2 \leq 0 \tag{10}$$

Thus the gravitational interactions remove the degeneracy in $O(\kappa^2)$ of the different gauge vacua found in SUSY models even before supersymmetry is broken[5,9,10]. (The spacing of the vacua are generally of size $\kappa^2 M^6$, where M is the GUT mass, for most models.) However, it may turn out that G_a cannot vanish for some channels, implying that T_{ab} has at least one zero eigenvalue at the minimum. Under this circumstance, supersymmetry will have broken spontaneously. This, of course, is the interesting case, and what happens depends in part on the model chosen.

As an interesting case we consider a model based on the global SU(5) SUSY GUT of Sakai and Georgi and Dimopoulos[1] where $g = g_1 + g_2$ with

$$g_1 = \lambda_1(\frac{1}{3}T_r \Sigma^3 + \frac{M}{2}T_r\Sigma^2) + \lambda_2 H'_x(\Sigma^x{}_y + 3M\delta^x{}_y)H^y + \lambda_3 U H'_x H^x$$

$$+ \varepsilon_{uvwxy}H^u M^{vw} m_1 M^{xy} + H'_x M^{xy} M'_y \qquad g_2 = m^2(Z + B) \tag{11}$$

where $\Sigma^x{}_y$ is in the adjoint 24 representation, H'_x and H^x are $\overline{5}$ and 5 Higgs multiplets, U is a singlet and M^{xy} and M'_y are the matter 10 and $\overline{5}$ representations. g_2 is the super-Higgs potential[12,13] of Eq. (1). If g_2 is set to zero, one finds to lowest order three global solutions[1], i.e., H^x, $H_x{}'$, U, M^{xy} and M'_y vanish and

$$\text{(i) } \Sigma^x{}_y{}^{(0)} = 0; \text{ (ii) } \Sigma^x{}_y{}^{(0)} = \frac{M}{3}[\delta^x{}_y - 5\delta^x{}_5\delta^5{}_y];$$

$$\text{(iii) } \Sigma^x{}_y{}^{(0)} = M[2\delta^x{}_y - 5(\delta^x{}_5\delta^y{}_5 + \delta^x{}_4\delta^y{}_4)] \tag{12}$$

These solutions preserve SU(5), SU(4)xU(1) and SU(3)xSU(2)xU(1) respectively as well as supersymmetry. If we set $g_1 = 0$ then g_2 breaks supersymmetry and Eq. (8) yields

$$Z^{(0)} = a(\sqrt{2} - \sqrt{6})/\kappa \ ; \quad B^{(0)} = -a(2\sqrt{2} - \sqrt{6})/\kappa; \ a = \pm 1 \tag{13}$$

where $B^{(0)}$ has been chosen to set the cosmological constant to zero. The general solution of Eq. (8) for the full $g = g_1 + g_2$ can be obtained as a perturbation around Eqs. (12) and (13). We find(5) that the solution generated out of case (iii) of Eq. (12) simultaneously breaks SU(3)xSU(2)xU(1) down to the physical vacuum $SU(3)^c xU(1)$ and supersymmetry at the same mass scale $m_s \equiv \kappa m^2$ by growing VEV's for U, H_x' and H^x:

$$U = -a\, x\, m_s/(\sqrt{2}\lambda_3); \quad H^x = H'_x = y\, m_s/(\sqrt{2}\lambda_3)\delta^x{}_5 \tag{14}$$

where x and y obey the algebraic equations

$$x^2 + x(3 - \sqrt{3} - 6\lambda) + y^2 + (1 - 3\lambda)^2 = 0 \tag{15}$$

$$y^2(3 - \sqrt{3} - 6\lambda) + 2xy^2 + x = 0; \ \lambda \equiv \lambda_2/\lambda_1$$

Eq. (14) implies

$$m_s = \xi(246 \text{ GeV}); \quad \xi \equiv \lambda_3/y \sim 1 \tag{16}$$

and hence the super-Higgs mass scale is

$$m = \sqrt{\xi}\,(2.4 \times 10^{10} \text{ GeV}) \tag{17}$$

which is consistent with the cosmological estimates(3). The gravitino mass is

$$m_{3/2} = \frac{m_s}{\sqrt{2}} \exp(2 - \sqrt{3}) = \xi(228 \text{ GeV}) \tag{18}$$

It should be stressed that relations such as Eqs. (16)-(18) are obtainable only because the theory correlates the supersymmetry breaking with the SU(2)xU(1) breaking. Otherwise one would not be able to theoretically estimate $m_{3/2}$.

The above theory is "realistic" in the sense that the breaking of SU(2)xU(1) occuring allows one to recover all the usual low energy phenomenology. Flavor changing neutral currents are suppressed and there are no light scalar bosons (since they gain a common mass $\sim m_s$). The gluino and photino are massless at the tree level but presumably grow mass $\sim 10 \sim 100$ GeV at the loop level.

5. Ground State In LS GUT

As was pointed out in Sec. IV, even if supersymmetry is not broken, the degeneracy of the different gauge vacua is generally lifted in $O(\kappa^2)$. The different solutions will in general have large cosmological constants, however. One can, by adding a constant to the superpotential g(z) cancel the cosmological constant in any one solution (i.e., choose the constant so that $g(z^{(1)})$ vanishes). Eq. (10), however, shows that all other solutions will lie lower in anti-deSitter spaces(5,9,10). Weinberg has pointed out that most likely the Minkowski solution is stable against decay into anti-deSitter spaces by finite size bubble formation(9). However, it is important to note that the above result that the Minkowski solution lie highest is actually model dependent and not a general property of LS GUT.

To see this, we again write $g(z) = g_1(z) + g_2(z)$ where g_2 is the super-Higgs potential. Then one may evade the above result if we choose $g_1(z)$ such that $g_1(z^{(i)})$ all vanish at least in the global limit $\kappa \to 0$, re-establishing the degeneracy to $O(\kappa^2)$. Further, the existance of the super-Higgs potential implies that supersymmetry has broken and so that one linear combination of the G_a, call it G_o, is non-zero. Thus Eq. (10) is modified to read

$$V_{min} = \frac{1}{2} E(z^{(i)})[G_o^2(z^{(i)}) - \frac{3}{2}\kappa^2|g(z^{(i)})|^2] \sim O(\kappa^4 m^8) \tag{19}$$

It is clear that the r.h.s. of Eq. (19) is no longer negative definite, and which state lies lowest depends on the model. It is indeed possible to construct a $g_1(z)$ for which the SU(3)xSU(2)xU(1) is the absolute minimum (with SU(4)xU(1) lying higher). In models of this type, the different vacua are separated by $O(m_s) \sim 1$TeV. Thus the cosmology of such models may have interesting new features.

6. Loop Corrections

When loop corrections to the effective potential are included, Eq. (8) is modified to read

$$E\, T_{ab}G_b + L_a = 0 \tag{20}$$

where at the one loop level L_a has the form

$$L_a = \sum_J (-1)^{2J}(2J + 1)\mathrm{Tr}M_J^2 M_J^2{}_{,a}\, \ell n M_J^2/\mu^2 \tag{21}$$

In Eq. (21) M^2_J is the mass matrix for the particles of spin J. One may obtain the spin 0 mass matrix by differentiating Eq. (6). Writing $z_a = A_a + iB_a$, and evaluating all VEV's on the real manifold we find

$$(M^2)_{ab}^{A} = \tilde{g}_{,ac}\, \tilde{g}_{,bc} + \tilde{g}_{,abc}\tilde{g}_{,c} - \frac{3}{2}\kappa^2(\tilde{g}_{,ab}\, \tilde{g} + \tilde{g}_{,a}\, \tilde{g}_{,b}) \tag{22}$$

$$(M^2)_{ab}^{B} = \tilde{g}_{,ac}\, \tilde{g}_{,bc} - \tilde{g}_{,abc}\, \tilde{g}_{,c} - \frac{1}{2}\kappa^2(\tilde{g}_{,ab}\, \tilde{g} - \tilde{g}_{,a}\, \tilde{g}_{,b})$$
$$+ \kappa^2\delta_{ab}(\tilde{g}_{,c}\, \tilde{g}_{,c} - \frac{\kappa^2}{2}\tilde{g}^2) \tag{23}$$

where we have introduced the notation

$$\tilde{g} = \exp(\frac{\kappa^2}{4} z_a^2)g(z) \tag{24}$$

The fermion mass matrices are defined by

$$-L_{mass}^{F} = \frac{1}{2}\bar{\chi}^a\, m_{ab}\chi^b + g_e\, \bar{\chi}^a \mu_{a\alpha}\lambda^\alpha \tag{25}$$

The formulation of Sec. (3) yields the values

$$m_{ab} = \tilde{g}_{,ab} - \frac{\kappa^2}{2}\delta_{ab}\, \tilde{g} - \frac{2}{3}\tilde{g}_{,a}\, \tilde{g}_{,b}/\tilde{g}; \quad \mu_{\alpha a} = (T^\alpha)_{ab}z^b \tag{26}$$

We note the mass matrices contain κ dependence both implicitly in the function $\tilde{g}$ and explicitly. Since

$$\tilde{g}_{,b} = (\exp \frac{\kappa^2}{4} z_a{}^+ z_a) G_b \tag{27}$$

the vanishing of $\tilde{g}_{,a}$ is equivalent to no supersymmetry breaking. For this solution the vanishing of the cosmological constant then implies $\tilde{g} = 0$ at the minimum.

We first consider the situation where $g(z)$ does not contain a super-Higgs. One may then show(7) that if at the tree level there exist solutions with zero cosmological constant and no breaking of supersymmetry ($\tilde{g}_{,a} = 0 = \tilde{g}$) then perturbatively this situation is maintained at the one loop level to all orders in κ. This result follows from the fact that Eqs. (22), (23) and (26) imply that L_a has the form

$$L_a = T_{ab}{}^{(1)} \tilde{g}_{,b} + T_a{}^{(1)} \tilde{g} \tag{28}$$

where $T_{ab}{}^{(1)}$ and $T_a{}^{(1)}$ are regular in the limit $\tilde{g}_{,a}, \tilde{g} \to 0$. The special structure of Eq. (28) arises from the fact that there is a remarkable cancellation of κ dependent terms in L_a that do not have either one factor of $\tilde{g}_{,b}$ or $\tilde{g}$. Eq. (20) now clearly has the solution $\tilde{g}_{,a} = 0 = \tilde{g}$ maintaining the tree situation. All other possible solutions with zero cosmological constant would have to be non-perturbative (i.e., with loop corrections cancelling the tree pieces).

We next include in the super-Higgs potential. We saw in Sec. (4) that supersymmetry then spontaneously breaks at the tree level with the Higgs mesons growing vacuum expectation values $\sim m_s = \kappa m^2$ and all $\tilde{g}_{,a} \sim m_s{}^2 = \kappa^2 m^4$ (except in the super-Higgs channel itself). There is again a cancellation of all the large terms of order $\kappa M m^3$ and $\kappa^2 M\, m^4$ in L_a yielding a remaining part $L_a \sim \lambda^4 \kappa^3 m^6 = \lambda^4 m_s{}^3$ where λ is one of the dimensionless coupling constants(7). This is to be compared with the tree terms where $T_{ab} G_b \sim \lambda^2 m_s{}^3$. Thus the loop corrections again "protect" the low energy phenomena from both the GUT mass $M \sim 10^{16}$ GeV and the intermediate mass scale $m \sim 10^{11}$ GeV and preserve the general structure of the tree results!

Finally, one might ask if it is possible to grow the super-Higgs term dynamically and deduce the size of the mass m rather than inserting it by hand. The above discussion shows that this can only be achieved non-perturbatively and it is in fact possible to construct models that do this(7). In this case one finds

$$m \sim \kappa^2 M^3 [\ell n M^2/\mu^2]^{1/2} \sim 10^{11} \text{ GeV} \tag{29}$$

Theories of this kind contain only two mass scales: the Planck mass κ^{-1} and the GUT mass M, both m and m_s being deduced from these.

References

1. E. Witten, Nucl. Phys. B188 (1981) 513; N. Sakai, Z. für Phys. C11 (1981) 153; S. Dimopoulos and H. Georgi, Nucl. Phys. B193 (1981) 150; M. Dine and W. Fischler, Phys. Lett. 110B (1982) 227; D.V. Nanopoulos and K. Tamvakis, CERN Preprint TH. 3247 (1982).
2. A recent comparison of SUSY GUT models and standard GUT models is given in D.V. Nanopoulos, Second Europhysics Study Conference on Unification of Fundamental Interactions, Erice (1981) (CERN TH. 3249).
3. S. Weinberg, Phys. Rev. Lett. 48 (1982) 1303.
4. S. Dimopoulos and S. Raby, Los Alamos Preprint LA-UR-82-1282 (1982); J. Polchinski and L. Susskind, SLAC Preprint 2924 (1982); M. Dine and W. Fischler, Institute for Advanced Study Preprint and Ref. (1); R. Barbieri, S. Ferrara and D.V. Nanopoulos, CERN Preprint TH. 3309 (1982).
5. A.H. Chamssedine, R. Arnowitt and Pran Nath, Northeastern University Preprint NUB#2559 (1982). Detailed derivations are given in a longer version "Formulation of Locally Supersymmetric Unified Theories" (in preparation).
6. Pran Nath, R. Arnowitt and A.H. Chamseddine, Northeastern University Preprint NUB#2563 (1982).
7. R. Arnowitt, A.H. Chamseddine and Pran Nath, Northeastern University Preprint NUB#2569 (1982).
8. D.Z. Freedman, P. van Nieuwenhuizen and S. Ferrara, Phys. Rev. D13 (1976) 3214; S. Deser and B. Zumino, Phys. Lett 62B (1976) 335.
9. S. Weinberg, Phys. Rev. Lett. 48 (1982) 1176.
10. B. Ovrut and J. Wess, Institute for Advanced Study Preprint in preparation.
11. E. Cremmer, S. Ferrara, L. Girardello, and A. Van Proeyen, CERN Preprints TH. 3312 (1982) and TH. 3348 (1982).
12. J. Polony, Budapest Preprint KFK1-1977-93 (1977) (unpublished).
13. E. Cremmer, B. Julia, J. Scherk, S. Ferrara, L. Girardello and P. van Nieuwenhuizen, Nucl. Phys. B147 (1978) 175.
14. S. Weinberg, Phys. Lett. 62B (1976) 111; W. Lang, Nucl. Phys. B114 (1976) 123.
15. J. Wess and B. Zumino, Nucl. Phys. B70 (1974) 39.
16. A. survey of the superspace approach to global supersymmetry is given in A. Salam and J. Strathdee, Fortschritte der Physik 26 (1978) 57.
17. E. Cremmer, B. Julia, J. Scherk, S. Ferrara, L. Girardello and P. van Nieuwenhuizen, Nucl. Phys. B147 (1979) 105.
18. S. Ferrara and P. van Nieuwenhuizen, Phys. Lett. 76B (1978) 404; K.S. Stelle and P.C. West, Phys. Lett. 77B (1978) 376.
19. K.S. Stelle and P.C. West, Phys. Lett. 74B (1978) 330; S. Ferrara and P. van Nieuwenhuizen, Phys. Lett. 74B (1978) 333.
20. S. Ferrara, J. Scherk and P. van Nieuwenhuizen, Phys. Rev. Letters 37 (1976) 1035.

THE AUXILIARY FIELD / ULTRAVIOLET FINITENESS CONNECTION

P.K. Townsend

Laboratoire de Physique Théorique de l'Ecole Normale Supérieure
24, rue Lhomond, 75231 PARIS CEDEX 05, FRANCE

ABSTRACT :

The status of the auxiliary field program for supersymmetric field theories is reviewed, with emphasis on the implications for improved ultraviolet behaviour, and conversely on the implications of ultraviolet behaviour for auxiliary fields.

1. INTRODUCTION

For many years one of the central problems of supersymmetric field theories has been that of auxiliary fields. Generally, supersymmetry transformations that leave invariant a given action will close to form the usual supersymmetry algebra only when the field equations are used. This is called "on-shell supersymmetry". If field equations are not needed to close the algebra then the supersymmetry is "off-shell". For this to happen we must have equal numbers of boson and fermion field components as well as equal numbers of boson and fermion propagating modes (i.e. states). Since fermions always have more components than they propagate modes, a balance in the number of boson and fermion states will usually mean an imbalance in the number of boson and fermion field components. Hence the need for boson "auxiliary" i.e. non-propagating, fields to balance the number of boson and fermion field components off-shell while not disturbing the balance of states. Of course equality of the numbers of bosons and fermions is only a necessary condition for off-shell supersymmetry, not a sufficient one. In general, a solution to an auxiliary field problem will require fermion as well as boson auxiliary fields.

There are several reasons why we should want auxiliary fields. The most obvious one is that if the algebra of transformations includes field equations then these transformations are tied to the particular action that yields these equations. The introduction of auxiliary fields frees the transformation rule of reference to a particular action. This allows, inter alia, the addition of invariant actions to produce new invariant actions. Without auxiliary fields this involves a laborious procedure in which terms are added order by order to the action and transformation rules to obtain a new invariant action with new transformation rules.

A second, and more important, motivation for auxiliary fields is that they allow us to consider superfield perturbation theory in which component field Feynman graph calculations can be done together, and with more ease, as supergraph calculations.

This formalism allows us to deduce immediately, just from the form of the Feynman rules, many "non-renormalization theorems" which have remarkable consequences for the ultraviolet behaviour of the N-extended supersymmetric theories.

Many of the authors that have contributed to our understanding of auxiliary fields have advanced as one of their motivations a better understanding of ultraviolet behaviour while, at the same time, most of the real progress in this field has been technical. Given some recent advances the time now seems ripe for a survey of what we can expect auxiliary fields to say about the quantum theory, and vice versa.

2. AUXILIARY FIELDS

Consider the simplest example of an on-shell supersymmetric theory, the Wess-Zumino model[1],

$$\mathcal{L} = -\tfrac{1}{2}(\partial_\mu A)^2 - \tfrac{1}{2}(\partial_\mu B)^2 - \tfrac{1}{2}\bar{\lambda}\not{\partial}\lambda \tag{2.1}$$

with scalar A, pseudoscalar B and spinor λ_α . One easily establishes that the algebra of supersymmetry transformation rules that leaves $I = \int \mathcal{L}\, d^4x$ invariant, closes to give the usual supersymmetry algebra only if the λ field equation, $\not{\partial}\lambda = 0$, is used. To extend (2.1) to an off-shell supersymmetric model we require auxiliary fields, and a count of components shows that two auxiliary scalar bosons would be the simplest solution. To find these auxiliary fields we can investigate the various irreducible representations of supersymmetry on fields for $N = 1$, which can be found by, e.g. superfield methods. One easily finds a representation with maximum spin 1/2 with the fields (A, B, λ , F, G), corresponding to an $N = 1$ chiral superfield). F and G are obviously the wanted dimension two auxiliary fields and they will occur in the action as $F^2 + G^2$. The new, F and G dependent, transformation rules will close off-shell, i.e. without the use of field equations.

What makes this example so simple is the fact that the highest propagated spin, i.e. 1/2, is also the highest spin of the off-shell multiplet of fields. In general, for <u>massless</u> theories the maximum (propagating) spin, s, is bounded from below by

$$s\ (\text{propagating}) \geqslant \frac{N}{4} \tag{2.2}$$

where N is the number of supersymmetries. For <u>massive</u> on-shell multiplets, or off-shell multiplets of fields, the bound is

$$s\ (\text{off-shell}) \geqslant \frac{N}{2} \tag{2.3}$$

which is more stringent. If s (propagating) $\geqslant$ N/2 we can always find an off-shell multiplet of fields with maximum spin s, and this will give a set of auxiliary fields. For those models for which $N/4 \leqslant s(\text{propagating}) < N/2$ the auxiliary field problem is much harder, and probably without a solution for most cases. At any rate, it is clear that in such cases the off-shell multiplet of fields must contain spins higher than the maximum propagated spin. This can happen in two ways.

(i) We may have unconventional field representations with spin "discontinuities" in the massless limit. The simplest case is the gauge antisymmetric tensor $A_{\mu\nu}$ which is spin 1 off-shell but which propagates a massless spin-zero mode (if the kinetic term is the usual $\partial_\rho A_{\mu\nu} \partial^{[\rho} A^{\mu\nu]}$ one).

(ii) We may have high spin (i.e. higher than the physical propagated spin), non-gauge auxiliary fields.

For those cases for which s(propagating) $\geqslant$ N/2 there is a systematic way to construct the relevant off-shell representations by means of "supercurrents"[2]. The best known example is the N = 1 spin 2 conformal supercurrent multiplet, containing the (traceless) energy momentum tensor, $T_{\mu\nu}$, the supersymmetry current S_μ and the axial current j_μ^5. For the model of (2.1), for example, these currents are bilinears in the fields A, B, λ and they are conserved as a consequence of the A, B, λ field equations. The currents themselves ($T_{\mu\nu}$, S_μ, j_μ^5) are not subject to field equations, however, and therefore form an off-shell multiplet. The corresponding "contragredient" multiplet of fields $\{h_{\mu\nu}, \psi_\mu, A_\mu\}$ is also an off-shell representation with gauge transformations determined by the conservation conditions on the currents via the Noether coupling $I(\text{int}) = \int h \cdot J$. That is, δh is whatever leaves I(int) invariant. The multiplet $\{h_{\mu\nu}, \psi_\mu, A_\mu\}$ is that of N = 1 conformal supergravity.

Given a maximum desired spin, s, e.g. 2 in the above case, the supercurrent construction yields an off-shell multiplet with this spin provided there is a "matter" multiplet of maximum spin s/2 that can be used to form it. Since this maximum spin is bounded by (2.2) we again arrive at the bound s $\geqslant$ N/2 for the off-shell multiplet of currents, or fields. Thus we can say that the really difficult auxiliary field problems start when there is no lower spin "matter" to which the given off-shell model could couple. Precisely, systematic methods, such as supercurrents, yield auxiliary fields for :

(i) matter theories (spin $\leqslant$ 1/2) for N = 1 only

(ii) gauge theories (spin $\leqslant$ 1) for N $\leqslant$ 2

(iii) conformal supergravity (spin $\leqslant$ 2) for N $\leqslant$ 4

The restriction to conformal supergravity in (iii) is because the supercurrent method yields an <u>irreducible</u> multiplet which is what is needed for conformal supergravity. The Poincaré supergravity theories are based on reducible (partially locally reducible) multiplets because such theories couple to a tracefull energy momentum tensor; $T_\mu{}^\mu \neq 0$. To find the Poincaré supergravity theories we must add additional multiplets. For N = 1 and N = 2 these additional multiplets have spin $<$ 2 and the passage from conformal to Poincaré supergravity is relatively straightforward. For N $\geqslant$ 3 there are no lower spin multiplets. Any multiplet added to the spin 2 conformal multiplet would itself have at least spin 2. This is a problem because these higher spin fields will generally be gauge fields and so could not appear easily as auxiliary fields and even if they could they would probably not allow consistent interactions. Because of this we can add to the list of off-shell theories constructible by systematic methods, only

(iv) Poincaré supergravity (spin $\leqslant 2$) for $N \leqslant 2$.

There have been recently two additional successful constructions of off-shell supersymmetric theories that fall outside the scope of the above mentioned methods. They both are auxiliary field solutions with high spin (non gauge) auxiliaries. The first is a set of auxiliary fields for linearized 10-dimensional supergravity[3] which implies, by dimensional reduction, a solution of the auxiliary field problem for linearized N = 4 Poincaré supergravity coupled to 6 N = 4 super-Maxwell multiplets. The maximum spin of the off-shell multiplet is 4. The construction makes essential use of the 10-dimensional supercurrent obtained as a bilinear of on-shell 10-dimensional super-Yang-Mills fields[4], but the final set of fields is more than those of the supercurrent alone. The second[5] is a set of auxiliary fields for the N=2 matter theory with spins 1/2 (sometimes called the hypermultiplet). The problem with a previous off-shell version of this model[6] is that it required one of the spin zero states to be represented by a gauge antisymmetric tensor, or equivalently a conserved vector. In interaction with Y-M fields this becomes a covariantly conserved vector, which is a constraint that cannot be solved in a useful way. The new solution has only conventional field assignments and can be consistently coupled to N = 2 super Y-M theory. The maximum spin of the off-shell multiplet is 1.

Another approach to the auxiliary field problem is to allow off-shell central charges[7]. Here one allows field equations to remain in the supersymmetry algebra if they can be interpreted as new central charge transformations. For massless theories these transformations must vanish on-shell, but need not off-shell. This approach is less restrictive, allowing for example a set of auxiliary fields for the N = 4 Maxwell (i.e. non-interacting) theory, and general methods have been developed for finding such auxiliary field sets[7,8]. But for superspace perturbation theory such auxiliary field solutions are not useful because they cannot be used to develop an unconstrained superfield formalism. The importance of this point is what concerns us next.

3. ULTRAVIOLET FINITENESS

A set of auxiliary fields, without an off-shell central charge, is the first and critical step towards an unconstrained superfield formulation. One proceeds first to a constrained superfield formulation according to one of several well established techniques. For example, each field, or field strength in the case of a gauge field, can be considered as the $\theta = 0$ component of a superfield. The transformation rules can then be used to determine the higher θ-components of these superfields. Of course only the lowest dimension field strength superfields constructed in this way will be independent. The higher dimension field strengths will occur in the θ-expansion of the lower dimension field strength superfields. These superfields are subject to various constraints, or rather identities if one arrives at them in the above fashion, analogous to the constraint $\partial_\mu {}^*F^{\mu\nu} = 0$ for the Maxwell field strength.

If these constraints are solved, e.g. $F_{\mu\nu} = \partial_\mu A_\nu - \partial_\nu A_\mu$ in the above illustration, one obtains an unconstrained superfield formulation in terms of a "prepotential" (the analogue of A_μ). Thus, in the simplest well-known case of the N = 1 (1, 1/2) multiplet the lowest dimension covariant field is the spinor λ_α which becomes the $\theta = 0$ component of the field strength superfield $W_\alpha(x, \theta)$. These constraints on W_α are solved by $W_\alpha = \bar{D}^2 D_\alpha V$, thereby introducing the unconstrained prepotential V.

These constraints can easily be solved for the free theory but with more difficulty for the interacting theory, particularly for $N>1$. For N = 1 the solutions to both the Yang-Mills and the supergravity constraints are known, but for N = 2 neither is yet known in supersymmetric form. This is probably not a serious problem in principle. One can obtain a solution to the constraints of the interacting theory as a perturbation series about the known solution to the free theory constraints. By solving these constraints either in closed form or perturbatively we arrive finally at an unconstrained superfield formulation of the interacting theory. This is the starting point for the development of superfield perturbation theory.

There is one exception to the need for unconstrained superfields to do superfield perturbation theory, and that is the N = 1 chiral field ϕ , satisfying $\bar{D}_{\dot\alpha}\phi = 0$ (two component spinor notation). The solution to this constraint, for the free field, is $\phi = \bar{D}^2 X$, introducing a complex scalar prepotential X . The chirality constraint is sufficiently simple that with a few modifications to the usual methods of extracting Feynman rules one can deal directly with ϕ without having to use X . Indeed the introduction of the prepotential X illustrates a problem that can be avoided for N = 1 but is endemic to $N>1$ theories[9]. The prepotential X has a gauge transformation $X \rightarrow X + \bar{D}_{\dot\alpha}\bar{\Lambda}^{\dot\alpha}$, but the parameter $\bar{\Lambda}^{\dot\alpha}$ is only determined up to a transformation $\bar{\Lambda}^{\dot\alpha} \rightarrow \bar{\Lambda}^{\dot\alpha} + \bar{D}_{\dot\beta}\,\xi^{(\dot\beta\dot\alpha)}$ because this transformation leaves X invariant. Hence, the ghosts associated with the gauge transformation of X, themselves have a gauge transformation with parameter $\bar{\Lambda}^{\dot\beta\dot\alpha} = \bar{\Lambda}^{\dot\alpha\dot\beta}$. But the secondary ghosts associated with this gauge transformation will again have an even larger invariance, with a symmetric tri-spinor parameter, and so on ad infinitum. This is analogous to the series of ghosts associated with gauge antisymmetric tensors. In that case, however, the series involves an increasing or decreasing number of antisymmetrized vector indices and so terminates.

An infinite series of ghosts, apparently unavoidable for $N>1$, presents a problem because the effective action from which Feynman rules are read off cannot be written down in closed form. However, one can arrange for all but a finite number of ghosts to decouple, or, in a background field approach, for all but a finite number to couple only to the background field[10]. In this case the infinite series of ghosts can only contribute at one loop. This makes one loop a special case that must be dealt with separately in the background field method. It can then be shown, with the exception of one loop and on the assumption that an N-extended unconstrained superfield formalism exists, that all counterterms in N-extended superfield form must be such that

(i) they are full superspace integrals $I = \int d^Dx\, d^{4N}\theta\, \mathcal{L}$ (counterterm)

(ii) $\mathcal{L}$(counterterm) is a local product of superfields

(iii) Only "covariant" superfields occur in $\mathcal{L}$ and in such a way that $\mathcal{L}$ is background gauge invariant.

The first two criteria[11] follow directly from the Feynman rules and the usual properties of counterterms. The third criterion[10] is more subtle and depends on the deployment of the background field method. The precise formulation of this method is what determines the meaning of "covariant". It means pre-gauge invariant, which means that the superspace dimension 1/2 potential A_α^i for super-Yang-Mills theories, or the vielbein $E_M{}^A$ and super-connections $\Omega_{MA}{}^B$ for supergravity, may appear in gauge covariant quantities, but not the unconstrained prepotentials.

If we choose the Y-M action in D dimensions to be $g^{-2}\int d^Dx\,(F_{\mu\nu})^2$ such that $\dim(A_\mu) = 1$ then $\dim(g^2) = 4 - D$ (in units of mass). The counterterm at ℓ loops has the form

$$(g^2)^{\ell-1} \int d^Dx\, d^{4N}\theta\, \mathcal{L} \tag{3.1}$$

and the same result holds for both super-matter (spins $\leq 1/2$) and conformal supergravities if the coupling constant is again chosen to appear as a g^{-2} factor multiplying a g-independent $\mathcal{L}$. For Poincaré supergravity at ℓ loops we have a counterterm of the form

$$(\kappa^2)^{\ell-1} \int d^Dx\, d^{4N}\theta\, \mathcal{L} \tag{3.2}$$

where κ is the gravitational constant of dimension $(1 - D/2)$. The minimum dimension for $\mathcal{L}$ in (3.1) is 2 for super Y-M theories, corresponding to the choice $\mathcal{L} = A_\alpha^i \lambda^\alpha_i$; 2 for super-matter, corresponding to $\mathcal{L}$ = (physical scalar)^2 ; zero for conformal supergravity, corresponding to $\mathcal{L} = E = \mathrm{Ber}(E_M{}^A)$. Similarly, the minimum dimension for $\mathcal{L}$ in (3.2) is zero.

Since counterterms have dimension zero, (after account is taken of the integration measure and factors of coupling constants), the minimal dimensions of $\mathcal{L}$ imply the absence of allowed counterterms at ℓ loops in D dimensions if

$$D < 4 + \frac{(2N-2)}{\ell} \quad \text{for super-Y-M and super-matter} \tag{3.3a}$$

$$D < 2 + \frac{(2N-2)}{\ell} \quad \text{for Poincaré supergravity} \tag{3.3b}$$

$$D < 4 + \frac{(2N-4)}{\ell} \quad \text{for conformal supergravity} \tag{3.3c}$$

N is the number of supersymmetries and ℓ the loop order. This is meant to hold beyond one loop. These results would imply[10] that both N = 2 and N = 4 super Y-M theories in D = 4 are finite beyond one loop, which appears to be the case up to three loops[12,13]. The N = 4 theory happens to be also one loop finite, while the N = 2 theory is not.

For Poincaré supergravity (3.3b) would imply[10] finiteness through 6 loops in $D = 4$. For conformal supergravity (3.3c) would imply[14] finiteness of the $N \geq 3$ theories in $D = 4$ beyond one loop (and it appears that the $N = 4$ theory is finite also at one loop[15]. In fact, we can do somewhat better for conformal supergravity because the counterterm available for $N = 2$ is $\int d^4x \, d^8\theta \, E$ which vanishes[16], so the $N = 2$ conformal supergravity should also be finite beyond one loop.

All these restrictions on counterterms assume that an N-extended unconstrained superfield formalism exists. This critical assumption is very probably true for the conformal supergravity theories because their auxiliary fields are all known[17], but appears to be false[18,19], apart from a few exceptions, for the $N>2$ super Y-M and (Poincaré) supergravity theories. In this case, we may make a weaker assumption :

(A) There exists an unconstrained M-extended superfield formalism (and hence M-extended auxiliary fields) for N-extended super Y-M or Poincaré supergravity, with $M \leq N$ to be determined.

With this assumption the formulae (3.3) are replaced, in the relevant cases, by

$$D < 4 + \frac{(2M - 2)}{\ell} \qquad \text{for N-extended super Y-M} \tag{3.4a}$$

$$D < 2 + \frac{(2M - 2)}{\ell} \qquad \text{for N-extended Poincaré supergravity} \tag{3.4b}$$

But what is the correct value of M ? For super-matter and conformal supergravity we have, or expect, $M = N$, and the same is true for $N = 2$ Y-M theory. Given the restrictions on auxiliary fields found in refs.(18, 19) and given the recent progress in the auxiliary field search of refs.(3, 5) it would appear that we already have auxiliary fields for almost all those cases for which they can be found, and that M is restricted by

$$M \leq N/2 \quad \text{for } N = 4 \text{ Y-M, } N = 4 \text{ pure Poincaré supergravity, } N = 8 \text{ Poincaré supergravity} \tag{3.5}$$

(By "pure" supergravity I mean with no matter multiplets). I say "almost" because in order to have an $M = 4$ formulation of $N = 8$ Poincaré supergravity one needs an off-shell version of the $N = 4$ "spin 3/2 multiplet" which is not yet known. But let us be optimistic and assume that the bound of (3.5) is saturated. Then we arrive at the conclusion that, for $\ell > 1$ there is no allowed counterterm in D dimensions at ℓ loops if

$$D < 4 + \frac{2}{\ell} \qquad N = 4 \text{ Y-M theory} \tag{3.6a}$$

$$D < 2 + \frac{6}{\ell} \qquad N = 8 \text{ Poincaré supergravity} \tag{3.6b}$$

We observe that (3.6a) is still sufficient to ensure finiteness[5]. It is interesting to compare (3.6) with similar results obtained from the superstring approach[20] by extrapolation from one loop :

$$D < 4 + \frac{4}{\ell} \qquad N = 4 \text{ Y-M theory} \tag{3.7a}$$

$$D < 2 + \frac{6}{\ell} \qquad N = 8 \text{ Poincaré supergravity} \tag{3.7b}$$

The results (3.6b) and (3.7b) agree. They also agree with the result of yet another approach[21] and imply that an ultraviolet divergence can be expected to appear at three loops. The relevant three loop counterterm is also known[22], so a calculation for supergravity at three loops remains the critical test of whether we can expect further "miraculous" cancellations in these theories.

There is a discrepancy between (3.6a) and (3.7a). This is not necessarily a contradiction because (3.6a) is valid only beyond one loop while (3.7a) is known to be valid only <u>at</u> one loop. If (3.7a) is valid beyond one loop also then it would appear that divergence cancellations are slightly better than one has a right to expect from superspace perturbation theory. I remark that if M = 3 were possible for the N = 4 Y-M theory, then (3.4a) would yield a formula in agreement with (3.7a), but this possibility is ruled out according to ref. 19.

In all of the above discussion some kind of supersymmetric regularization has been assumed. But is this reasonable ? The scheme used successfully in practice is dimensional regularization by dimensional reduction but its internal consistency is questionable[23]. For general arguments (if not explicit calculations) it is preferable to avoid it. There is an alternative scheme ; higher derivative regularization. For gauge theories this fails at one loop but Slavnov has shown[24] how a two stage regularization method can be devised in which one loop is regulated separately and supersymmetrically. When this method is applicable its consistency is not in question, but the trouble is that it is not applicable <u>unless</u> we have auxiliary fields. The reason is that the higher derivative terms used to regulate the theory propagate additional massive ghost modes (whose mass goes to infinity as the regulator is switched off), and these modes will be in supermultiplets only if auxiliary fields were initially present. This may seem to crush hopes of applying this method to N = 4 Y-M and N = 8 supergravity theories, but fortunately auxiliary fields for M = N/2 is just sufficient to allow the method to work. The point is that in this case the ghost modes of the regulated theory will be in N-supersymmetric multiplets with a central charge equal to the mass, and so the N-extended supersymmetry will <u>not</u> be broken by the regulator. It has been checked that the ghost modes for the N = 4 theory with M = 2 do indeed form N = 4 multiplets with a central charge[5].

REFERENCES

1. J. Wess and B. Zumino, Nucl. Phys. B49, 52 (1974)
2. S. Ferrara and B. Zumino, Nucl. Phys. B87, 207 (1975) ; M.F. Sohnius, Phys. Lett. B81, 8 (1979) ; E. Bergshoeff, M. de Roo, and B. de Wit, Nucl. Phys. B182, 173 (1981) ; P.S. Howe, K.S. Stelle and P.K. Townsend, Nucl. Phys. B192, 332 (1981)
3. P.S. Howe, H. Nicolai and A. van Proyen, Phys. Lett.B 112, 446 (1982)

4. E. Bergshoeff, M. de Roo, B. de Wit and P; van Nieuwenhuizen, Nucl. Phys. B195, 97 (1982) ; E. Bergshoeff and M. de Roo, Phys. Lett.112, 53 (1982)
5. P.S. Howe, K.S. Stelle and P.K. Townsend, CERN preprint TH 3271 (1982) to appear in Nucl. Phys.B
6. P. Breitenlohner and M.F. Sohnius, Nucl. Phys. B165, 483 (1980) ; M.F. Sohnius K.S. Stelle and P.C. West in "Supergravity and Superspace" eds. S.W. Hawking and M. Rocek, C.U.P. (1981)
7. M.F. Sohnius, K.S. Stelle and P.C. West, Phys. Lett. B92, 123 (1980) ; W. Siegel, Nucl. Phys. B173, 51 (1980)
8. M.F. Sohnius, K.S. Stelle and P.C. West, Nucl. Phys. B173, 127 (1980) ; J.G. Taylor in "Quantum Structure of Spacetime", eds. M.J. Duff and C.J. Isham, C.U.P. to appear
9. W. Siegel and S.J. Gates, Nucl. Phys. B189, 295 (1981)
10. M.T. Grisaru and W. Siegel, Nucl. Phys. B201, 292 (1982)
11. M.T. Grisaru, W. Siegel and M. Rocek, Nucl. Phys. B159, 429 (1979)
12. O.V. Tarasov and A.A. Vladimirov, Phys. Lett. B96, 94 (1980) ; M.T. Grisaru, M. Rocek and W. Siegel, Phys. Rev. Lett. 45, 1063 (1980) ; W. Caswell and D. Zanon Phys. Lett. B100, 152 (1980)
13. L.V. Adveev and O.V. Tarasov, Phys. Lett. B112, 356 (1982)
14. K.S. Stelle and P.K. Townsend, Phys. Lett. B113, 25 (1982)
15. E.S. Fradkin and A.A. Tseytlin, Nucl. Phys. B203, 157 (1982)
16. E. Sokatchev, Phys. Lett. B100, 466 (1981)
17. E. Bergshoeff, M. de Roo and B. de Wit, Nucl. Phys. B182, 173 (1981)
18. W. Siegel and M. Rocek, Phys. Lett. B105, 278 (1981)
19. V.O. Rivelles and J.G. Taylor, "Off-shell no-go theorems for higher dimensional supersymmetries and supergravities", Kings College preprint (1982)
20. M. Green, J. Schwarz and L. Brink, Nucl. Phys. B198, 474 (1982); for a review see J. Schwarz, preprint CALT 68-911 (1982)
21. T. Curtright, Phys. Rev. Lett. 48, 1704 (1982)
22. R. Kallosh, Phys. Lett. B99, 122 (1981) ; P.S. Howe, K.S. Stelle and P.K. Townsend Nucl. Phys. B191, 445 (1981)
23. See W. Siegel, P.K. Townsend and P. van Nieuwenhuizen in "Superspace Supergravity" eds. S.W. Hawking and M. Rocek (C.U.P.)
24. A.A. Slavnov in "Superspace and Supergravity" eds. S. Hawking and M. Rocek (C.U.P.)

Consistency of Coupling in Supergravity with Propagating Lorentz Connexion

Hitoshi Nishino

Institute of Physics
University of Tokyo, Komaba 3-8-1
Meguro-ku, Tokyo 153, Japan

§1. Introduction

Gauge principle is nowadays the most elegant and successful scheme in grand unifications of three kinds of interactions other than the gravitational interaction. The recent hierarchy problem of grand unification theories motivates us to consider their supersymmetric extensions (super-unifications). Once supersymmetry is introduced into the description of nature, there is no theory other than supergravity [1], which is consistent with the gravitational interaction.

On the other hand, there has been another approach to the gravitational interaction, i.e. Poincaré gauge theory [2,3]. This theory is a kind of gauge theory of Poincaré group (P_m, J_{rs}), where the generators P_m and J_{rs} are treated as equally as possible. In particular, the Lorentz connexion gauge field $\omega_\mu{}^{rs}$ is treated as an independent propagating field.

With these developments in mind, we have tried in our previous papers [4∿6] to unify these two approaches, i.e. supergravity theory and Poincaré gauge theory. In other words, we attempted to present "a super Poincaré gauge theory", or "a supergravity theory with propagating Lorentz connexion" [4∿6].

This kind of theory is relevant to the problem of hidden symmetry in extended supergravity theories. Consider for example the recent result of N=8 extended supergravity showing the appearance of the hidden local SU(8) symmetry from the original local Lorentz group SO(1,10) in the 11-dimensional space-time before the dimensional reduction [7]. The study of the propagation mechanism of SO(1,10) Lorentz connexion may be hence as important as that of SU(8) gauge fields.

Our supergravity theory with propagating Lorentz connexion is based on the nonminimal and reducible multiplet of Breitenlohner (126 Breitenlohner potentials) [8]. It consists of two multiplets: the supergravity multiplet (SG) and the Lorentz connexion multiplet (LC). In our previous papers [4∿6], we have presented the Lagrangians for these multiplets: $\mathcal{L}_{SGa}$ and $\mathcal{L}_{LCa}$. The former is a supersymmetric extension of the "massless case" of Poincaré gauge theory [3] and the latter contains the kinetic terms of LC.

The simplest example of $\mathcal{L}_{LCa}$ was already proposed by Breitenlohner, which contains the YM-type kinetic term of $\omega_\mu{}^{rs}$, i.e. $-(1/4)eR_{\mu\nu}{}^{rs}(\omega)\times R^{\mu\nu}{}_{rs}(\omega)$ [8]. We named this Lagrangian $\mathcal{L}_{LCa1}$ in our papers [4∿6].

As was noticed in our paper [4], theories with the Lagrangian of the "massless case" in Poincaré gauge theory may generally have the problem of inconsistent couplings of $\omega_\mu{}^{rs}$ to spinor fields. In our theory, the coupling of $\omega_\mu{}^{rs}$ to its spinor partner $\lambda_\alpha{}^{tu}$ is shown to be consistent to all orders in the total Lagrangian $\mathcal{L} = \mathcal{L}_{SGa} + \mathcal{L}_{LCa1}$ in spite of the "masslessness" of the Lagrangian $\mathcal{L}_{SGa}$. We regard this as an important result of supersymmetry of our theory.

The Lagrangian $\mathcal{L}_{LCa1}$, however, has the problem of negative energy ghosts reflecting noncompactness of the Lorentz group. To circumvent this difficulty we have presented two new Lagrangians named $\mathcal{L}_{LCa2}$ and $\mathcal{L}_{LCa3}$, which are written in local forms in superspace as [4]

$$\mathcal{L}_{LCa3}(z) = \frac{1}{16} V(z)(R(z))^2, \tag{1.1}$$

$$\mathcal{L}_{LCa3}(z) = - V(z)(\bar{R}(z))^2, \tag{1.2}$$

with $\bar{R}(z) \equiv (1/24)\varepsilon_{mnrs}V^{m\mu}(z)V^{n\nu}(z)R_{\mu\nu}{}^{rs}(z)$ in the notation of Ref. [9]. By examining the free Lagrangians $\mathcal{L}^{(0)}_{LCa2}(x)$ and $\mathcal{L}^{(0)}_{LCa3}(x)$ in component fields, we have shown the absence of negative energy ghosts. The physical helicity states of these Lagrangians are $(\partial_\mu v^\mu(0^+), \gamma_\rho\partial_\sigma\lambda^{\rho\sigma}(\pm 1/2), \tilde{D}^{\rho\sigma}(0^-))$ and $(\partial_\mu a^\mu(0^-), \gamma_\rho\partial_\sigma\lambda^{\rho\sigma\prime}(\pm 1/2), \tilde{D}^{\rho\sigma\prime}(0^+))$, respectively, where $v^\mu(a^\mu)$ is the vector (axial vector) component of $\omega_\mu{}^{\rho\sigma}$, and $\lambda^{\rho\sigma\prime}(\tilde{D}^{\rho\sigma\prime})$ is the dual component of $\lambda^{\rho\sigma}(\tilde{D}^{\rho\sigma})$ [4,6]. The consistency check of $\omega_\mu{}^{rs}$-$\lambda_\alpha{}^{tu}$ couplings, however, is very difficult, since interaction terms in $\mathcal{L}_{LCa}$ and $\mathcal{L}_{LCa3}$ are far complicated than those in $\mathcal{L}_{LCa1}$ [4,6].

In order to obtain at least trilinear couplings in $\mathcal{L}_{LCa2}$, we have set up a local tensor calculus for the Breitenlohner potentials. I consists of a multiplication rule of two scalar multiplets, a D-type Lag rangian and a correspondence rule between a scalar curvature superfield R(z) and a scalar multiplet. These rules are established up to the firs correction terms (required by local supersymmetry) to the global rules

[5]. By using these results, we derived the interaction terms necessary for consistency check of $\omega_\mu{}^{rs}$-$\lambda_\alpha{}^{tu}$ trilinear couplings in $\mathcal{L}_{LCa2}$ [5].

The consistency check of trilinear couplings of $\omega_\mu{}^{rs}$ to its spinor partner $\lambda_\alpha{}^{tu}$ is to show that there is a nontrivial solution satisfying the simultaneous equations: $T^{[\mu\nu]}(\lambda) = 0$ and the free field equation of $\lambda_\alpha{}^{rs}$ [4,6]. Intuitively the former originates from the absence of the antisymmetric part of the Einstein tensor in the "massless case" in Poincaré gauge theory. In the next section we perform this consistency check by using the explicit form of $T^{[\mu\nu]}(\lambda)$ from trilinear terms in $\mathcal{L}_{LCa2}$.

§2. Consistency of Couplings

Our check of coupling consistency of $\omega_\mu{}^{rs}$ to $\lambda_\alpha{}^{tu}$ is to show the existence of nontrivial and propagating solutions of the simultaneous equations: $T^{[\mu\nu]}(\lambda) = 0$ and the free field equation $\not{\partial}\gamma_\rho\partial_\sigma\lambda^{\rho\sigma} = 0$ [6]. The form of $T^{[\mu\nu]}(\lambda)$ is obtained from trilinear $e_\mu{}^m$-$\lambda_\alpha{}^{rs}$-$\lambda_\beta{}^{tu}$ terms in $\mathcal{L}_{LCa2}$

In order to simplify the caluculation, we decompose $\lambda_\alpha{}^{\rho\sigma}$ into the following three components:

$$\lambda_\alpha{}^{\rho\sigma} = \phi_\alpha{}^{\rho\sigma} + \frac{1}{2}(\gamma^\rho\phi^\sigma - \gamma^\sigma\phi^\rho)_\alpha + \frac{i}{12}(\sigma^{\rho\sigma}\phi)_\alpha, \tag{2.1}$$

$$\phi^\rho = \gamma_\sigma\lambda^{\sigma\rho} - \frac{1}{4}\gamma^\rho\phi, \qquad \phi \equiv -i\sigma_{\rho\sigma}\lambda^{\rho\sigma}, \qquad (\gamma_\rho\phi^\rho \equiv 0, \quad \gamma_\rho\phi^{\rho\sigma} \equiv 0). \tag{2.2}$$

The field equation of $\lambda_\alpha{}^{\rho\sigma}$ then takes the form

$$\not{\partial}(\gamma_\rho\partial_\sigma\lambda^{\rho\sigma}) = \not{\partial}(\partial_\rho\phi^\rho + \tfrac{1}{4}\not{\partial}\phi) = 0, \tag{2.3}$$

which is invariant under the following gauge transformation with the paramater $\varepsilon^{\rho\sigma}$:

$$\delta\lambda^{\rho\sigma} = \delta\phi^{\rho\sigma} = \varepsilon^{\rho\sigma} \quad (\gamma_\rho\varepsilon^{\rho\sigma} \equiv 0). \tag{2.4}$$

This implies that the $\phi^{\rho\sigma}$ component of $\lambda^{\rho\sigma}$ is unphysical and can be always gauged away. We then obtain

$$T^{[\mu\nu]}(\lambda) = [\ \tfrac{1}{16}\bar{\zeta}\gamma_5\sigma^{\mu\nu}\not{\partial}(\partial_\rho\phi^\rho - \tfrac{1}{4}\not{\partial}\phi)$$
$$+(\text{terms containing } \phi^{\rho\sigma} \text{ or vanishing by the use of the free field equation } \not{\partial}\gamma_\rho\partial_\sigma\lambda^{\rho\sigma} = 0)] - (\mu\leftrightarrow\nu). \tag{2.5}$$

The simultaneous equations for our consistency check are now

$$\bar{\zeta}\gamma_5\sigma^{\mu\nu}\not\partial(\partial_\rho\phi^\rho - \tfrac{1}{4}\not\partial\not\phi) = 0 \tag{2.6}$$

$$\not\partial(\partial_\rho\phi^\rho + \tfrac{1}{4}\not\partial\not\phi) = 0 \tag{2.7}$$

We can find nontrivial solutions of (2.6) and (2.7) satisfying

$$\not\partial\partial_\rho\phi^\rho = 0 \tag{2.8}$$

$$\not\partial\not\phi = 0 \tag{2.9}$$

Since these equations give nontrivial propagating solutions, we conclude that the trilinear $\omega_\mu{}^{rs}$-$\lambda_\alpha{}^{tu}$ couplings in our theory are indeed consistent couplings [6].

§3. Physical Helicity States and Global Supersymmetry

We show here that the $\partial_\mu\phi^\mu$ component (but not ϕ) of $\lambda_\alpha{}^{\rho\sigma}$ describes the physical helicity states $J=\pm 1/2$ by studying global supersymmetric co variances.

Under global supersymmetry, the left hand side of $\not\partial\partial_\mu\phi^\mu = 0$ (2.8) i transformed into

$$\delta(\not\partial\partial_\rho\phi^\rho) = \tfrac{1}{2}\varepsilon\Box\partial_\rho v^\rho + \tfrac{1}{2}(\gamma_5\sigma_{\rho\sigma}\varepsilon)\partial^{[\rho}\partial_\tau\tilde{D}^{\sigma]\tau}$$
$$-\tfrac{i}{2}(\sigma^{\rho\sigma}\varepsilon)\Box\partial_\rho v_\sigma - \tfrac{i}{4}\varepsilon^{\rho\tau\upsilon\omega}(\sigma_{\rho\sigma}\varepsilon)\partial^\sigma\partial_\tau\tilde{D}_{\upsilon\omega} \tag{3.1}$$

The first line contains the left sides of the free field equations of v^μ and $\tilde{D}^{\rho\sigma}$, while the second line contains the terms removed by appropriate gauge conditions [6]. In a similar way, the left sides of the free fiel equations of v^μ and $\tilde{D}^{\rho\sigma}$ are transformed into that of $\lambda_\alpha{}^{\rho\sigma}$ [6]. These facts imply that $\partial_\mu\phi^\mu$ describes the physical helicity states $J=\pm 1/2$, tha are transformed into $\partial_\mu v^\mu(0^+)$ and $\tilde{D}^{\rho\sigma}(0^-)$ under global supersymmetry.

§4. Concluding Remarks

We summarize here our main conclusions:

(1) There is no negative energy ghost in $\mathcal{L}_{LCa2}$ and $\mathcal{L}_{LCa3}$.

(2) Although $\mathcal{L}_{SGa}$ is a supersymmetric extension of the "massless case" Lagrangian of Poincaré gauge theory, the $\omega_\mu{}^{rs}$-$\lambda_\alpha{}^{tu}$ couplings are consistent up to the trilinear interaction order.

As far as we know, no other theory of supergravity with propagating Lorentz connexion possesses both of these two properties (1) and (2).

Future studies include

(1) Checking coupling consistency of other interactions and to higher orders.

(2) Quantizing fields and discussing renormalizability.

(3) Making our theory more realistic.

References

[1] D.Z. Freedman, P. van Nieuwenhuizen and S. Ferrara, Phys. Rev. D13 (1976) 3214;
S. Deser and B. Zumino, Phys. Lett. 62B (1976) 335;
For a review, P. van Nieuwenhuizen, Phys. Rep. 68 (1981) 189.

[2] H. Weyl, Phys. Rev. 77 (1950) 699;
R. Utiyama, Phys. Rev. D191 (1956) 1597;
D.W. Sciama, in "Recent Development in General Relativity", Pergamon, New York, 1962, p415;
T.W.B. Kibble, J. Math. Phys. 2 (1961) 212;
K. Hayashi and A. Bregman, Ann. Phys. 75 (1973) 562.

[3] K. Hayashi and T. Shirafuji, Prog. Theor. Phys. 66 (1981) 866, 883, 1435, 2222, ibid. 65 (1981) 525, ibid. 66 (1981) 318.

[4] H. Nishino, Prog. Theor. Phys. 66 (1981) 287.

[5] H. Nishino, Tokyo Univ. preprint, UT-Komaba 82-5, to be published in Sept. 1982 Issue of Prog. Theor. Phys..

[6] H. Nishino, Tokyo Univ. preprint, UT-Komaba 82-6.

[7] E. Cremmer and B. Julia, Phys. Lett. 80B (1978) 48;
Nucl. Phys. B159 (1979) 141.

[8] P. Breitenlohner, Phys. Lett. 67B (1977) 49;
Nucl. Phys. B124 (1977) 500.

[9] L. Brink, M. Gell-Mann, P. Ramond, and J.H. Schwarz, Phys. Lett. 74B (1978) 336;
Nucl. Phys. B145 (1978) 93.

COSMOLOGICAL PHASE TRANSITION IN MICROCANONICAL GRAVITY

G. Horwitz
Racah Institute of Physics
The Hebrew University
Jerusalem, Israel

Phase transitions in cosmology whether first or second order, spontaneous or dynamic were first studied by methods which ignore or exclude gravitational effects (Linde, 1974). The basis for the exclusion was the assumption that the effects are local, and that since the relevant times are much later than Planck times quantum gravity is unimportant and thus gravity enters only as a background metric which can be removed by studying results in a local inertial frame. Gradually various gravitational effects have been added: the effect of curvature in addition or instead of the Higgs mass for symmetry breaking (Grib, Mostepanenko and Frolov, 1977), topological effects (Avis and Isham, 1978), gravitational effects in growth rate of bubbles (Coleman and De Luccia, 1980).

In the present work we introduce a completely new element: reasonably realistic models in which there exists a nonquantum gravitational contribution to the entropy based on a microcanonical (MCE) rather than grandcanonical treatment of the statistical mechanics. The Hawking thermal states of de Sitter cosmology (Gibbons and Hawking, 1977) is an example of such gravitational entropy, but the latter is not a relevant model for phase transitions. Horwitz and Weil (1982) have developed a self-consistent generalization of such thermal states for a closed FRW universe with a positive cosmological constant Λ which has both scalar "radiation" and gravitational entropy. The dynamics is that of the usual classical models except that Λ and the invariant temperature are not independent parameters. In the present work we apply this method to a hyperbolic universe with $\Lambda < 0$ and a conformally coupled scalar field with a ϕ^4 self interaction. Such systems have been found to have a ground state with broken symmetry due to negative curvature even without mass.

The basis of our approach is the use of a Gibbsian definition for a thermodynamic equilibrium (TDE) MCE, generalized to apply to appropriate cosmological mocels and extended to include gravity. The standard Gibbs entropy is based on equal weight sums over states in a shell of (nearly) fixed constants of motion (COM) including energy, angular momentum, etc. For cosmological models, the energy which is not conserved can be replaced by the so-called dilatation operator D. Projecting the stress energy tensor $T_{\mu\nu}$ on the conformal time-like Killing vector ξ^{μ}_{c} characterizing FRW universes, a conserved quantity is obtained giving the COM D. This also requires a vanishing trace, hence invariance to Weyl transformations of the action. Thus instead of the standard definition for static systems, we have (units: $c = \hbar = G = 1$)

$$\exp S/k = \mathrm{Tr}\,\delta(D - D_0) \tag{1}$$

where

$$D = \int_{\Sigma_0} d\Sigma_\mu \sqrt{-g}\, T^{\mu}_{\nu}\, \xi^{\nu}_{c} \tag{2}$$

with Σ_0 a space-like hypersurface, $T_{\mu\nu}$ the functional derivative of the action with respect to the metric, the conformal time-like Killing vector satisfying $\xi_{c\mu;\nu} + \xi_{c\nu;\mu} = f\, g_{\mu\nu}$ where in appropriate coordinates f depends only on time. D is then a COM provided the action $I(\phi,\psi,g_{\mu\nu})$ is invariant to Weyl transformations by an arbitrary local scale function $\Omega(x^{\mu}):\phi \to \phi/\Omega$, $\psi \to \psi/\Omega$, $g_{\mu\nu} \to \Omega^2\, g_{\mu\nu}$. For the purposes of our model, we have included two scalar bosons, in addition to the metric.

For a FRW universe, with gravity appearing only as a background metric, and considering two alternative coordinate systems: one involving cosmic time $ds^2 = dt^2 - a^2\, d\sigma^2$ and the other conformal time: $ds^2 = a^2(d\eta^2 - d\sigma^2)$, where $d\sigma^2$ is a given time-independent element depending only on a parameter k which is +1 for closed, -1 for open and 0 for spatially flat spaces. Applying our entropy definition (1) we readily find the Tolman-Ehrenfest result for entropy:

$$S = 8\pi/3\, a_s\, T^3(\eta) = 8\pi/3\, a_s\, T^3(t) a^3(t) \tag{3}$$

their condition for TDE being

$$T(t)a(t) = \text{const.} = T(\eta)\ . \tag{4}$$

We wish to stress the interpretation of this situation as a state of global and not only local (in time) equilibrium. This is a maximum entropy, collisions preserving the state which is homogeneous and isotropic, but temperatures and various densities as functions of proper (cosmic) time vary with the expansion; the fixed entropy is stretched over growing proper volume elements in a self similar fashion. The truly new element of our analysis which goes beyond the above rederivation and reinterpretation of known results involves the inclusion of gravitational contributions with the use of a MCE and not based on quantum gravity.

Thus in the above mentioned work of Horwitz and Weil (1982) for a k = +1 and $\Lambda > 0$ FRW universe with conformally coupled massless bosons or in the presently discussed k = -1, $\Lambda < 0$ FRW universe with massless conformally coupled bosons having a ϕ^4 self interaction, the microcanonical TDE state found includes gravitational contributions. In both cases, these are self-consistent analogues of the TDE states of a BH in a radiation cavity (Hawking, 1976; Gibbons and Perry, 1978). Thermal fluctuations of the quantum radiation field couple to the classical fields of gravity through the microcanonical constraint, leaving an overall gravitational contribution to the entropy unrelated to quantum gravity or Planck times. In order to define such a TDE state which includes gravity one requires a Weyl invariant gravitational theory defined with an extra scalar field ψ in which the gravitational action

$$I_G' = (8\pi)^{-1}\int d^4x\sqrt{-g}\,[\tfrac{1}{2}R - \Lambda] + \text{surf. terms} \tag{5}$$

is replaced by

$$I_G = -3(8\pi)^{-1}\int d^4x(-g)^{\frac{1}{2}}[g^{\mu\nu}\psi_{,\mu}\psi_{,\nu} - \tfrac{1}{6}R\psi^2 + \tfrac{1}{3}\Lambda\psi^4]\ . \tag{6}$$

This leads to an expression for an action including the scalar, massive bosons conformally coupled with self interaction. The action is Weyl invariant even with the mass, but we shall treat here only $m^2 = 0$.

$$I = \tfrac{1}{2}\int d^4x(-g)^{\frac{1}{2}}\Big\{[g^{\mu\nu}\phi_{,\mu}\phi_{,\nu} - \tfrac{1}{6}R\,\phi^2 - m^2\,\phi^2\,\psi^2 - \lambda\,\phi^4] - 3(4\pi)^{-1}[g^{\mu\nu}\psi_{,\mu}\psi_{,\nu} - \tfrac{1}{6}R\,\psi^2 + \tfrac{\Lambda}{3}\psi^4]\Big\} \qquad (7)$$

where the line element for $k = -1$ FRW is

$$ds^2 = a^2(\eta)[d\eta^2 - dx^2 - \sinh^2 x(d\theta^2 + \sin^2\theta\, d\Phi^2)]\,. \qquad (8)$$

The standard choice of conformal gauge $\Omega = \psi^{-1}$ reduces I to the standard form (5) with the mass term $-m^2\phi^2$. Instead, after limiting our discussion to the case where $g_{\mu\nu}$ retains the form (8) with arbitrary $a(\eta)$, we now choose our conformal gauge with $\Omega = a(\eta)$, which then reduces the background metric to a static one - that in (8) with the $a(\eta)$ deleted. In the quantized version of this theory, this means we shall neglect spin-2 excitations (gravitons), although the method can readily be extended to include them. (It is a peculiar situation that presently to carry out classical SM for gravity one must introduce quantum gravity.) We subsequently interpret the transformed ψ to be the dynamic factor of the cosmic scale function, in our quasi-classical theory. Notice, in such a static background metric, there are no conformal trace anomalies although mass generation can be retained with a $m^2\phi^2\psi^2$ term. This choice also leads to a static energy operator, and the result that the Hamiltonian defined by (2) is equivalent to the canonical definition.

We now outline our calculation using (1), (2) and (7), choosing $D = 0$ corresponding to Einstein's equations when ψ is identified as the scale function, and taking a Fourier-Laplace transform of the delta function (1) becomes

$$\exp S/k = \mathrm{Tr}\int d\beta/2\pi i \exp - \beta D \qquad (9)$$

$$= \int[d\phi]\int[d\psi]\int d\beta/2\pi i \exp - I_E \qquad (10)$$

where the Euclideanized action I_E

$$I_E = \tfrac{1}{2}V\int_0^\beta du[\dot{\phi}^2 + \phi_{,i}\phi^{,i} - \phi^2 + \lambda\,\phi^4] - 3/8\pi\, V\int_0^\beta du[\dot{\psi}^2 - \psi^2 + \psi^4/R_0^2] \qquad (11)$$

with the boundary conditions $\phi(0) = \phi(\beta) = \phi_0$; $\psi(0) = \psi(\beta) = \psi_0$ and V is the (infinite) spatial 3-surface of the hyperbolic universe, and we have assumed that ψ depends only on the (imaginary) time. Since we seek only extremal values of ψ, which remain homogeneous and isotropic of course, significant fluctuations depend on x^i, but we ignore them for the present. We have written our cosmological constant $\Lambda = -3/R_0$ and $R/6 = -1$ for the static hyperbolic universe of scale length unity. This kind of model is known to lead to a ground state with broken symmetry even without a Higgs mass. The extremal solution of the ϕ and ψ satisfy the equations

$$\ddot{\phi}_{cl} + \phi_{cl} - 2\lambda\,\phi_{cl}^3 = 0 \qquad (12)$$

$$\ddot{\psi}_{cl} + \psi_{cl} - 2/R_0^2\,\psi^3 = 0 \qquad (13)$$

respectively. The functional integrals for fixed boundary values of ϕ_0 and ψ_0 are expanded about these classical solutions. The quadratic quantum fluctuations of the ψ's are dropped as we seek only classical gravity contributions. The quadratic expansion of the ϕ's is the one-loop term for the scalar conformal bosons, and we will take here the crudest high temperature approximation, that its contribution to the entropy is A/β^3 where $A = \pi^2/45$. Properly one should calculate it self-consistently, but we are here only looking for the qualitative modification of standard treatment by our gravitational inclusive MCE formalism. There remain the integrals over ψ_0, ϕ_0 and $\delta\beta$: where we also require a quadratic expansion around the extremum of β. The first integrals of (12) and (13) yield

$$\tfrac{1}{2}[\dot{\phi}^2 + \phi^2 - \lambda\phi^4] = \tfrac{1}{2}[\dot{\phi}^2 + V_1] = \omega \tag{14}$$

$$\tfrac{1}{2}[\dot{\psi}^2 + \psi^2 - \psi^4/R^2] = \tfrac{1}{2}[\dot{\psi}^2 + V_2] = \varepsilon \ . \tag{15}$$

The self-consistent solution if obtained by solving

$$3A/\beta^4 = \omega - 3/4\pi\varepsilon = \bar{\omega}/\lambda - 3R_0^2/8\pi\bar{\varepsilon} \tag{16}$$

$$\beta = 4\int^{y_-} dy[\bar{\omega} - y^2 + y^4]^{-\frac{1}{2}} = 4K(t^2)/y_+ \tag{17}$$

$$\beta = 2\int_0^{x_-} dx[\bar{\varepsilon} - x^2 + x^4]^{-\frac{1}{2}} = 2K(s^2)/x_+ \tag{18}$$

where the K are complete elliptic integrals of the first kind and $\bar{\omega} = \dot{y}^2 + y^2 - y^4$ with $y = \lambda^{\frac{1}{2}}\phi$; and $\bar{\varepsilon} = \dot{x}^2 + x^2 - x^4$ with $x = \psi/R_0$. Furthermore, $x_\pm^2 = \frac{1}{2}[1 \pm (1 - 4\bar{\varepsilon})^{\frac{1}{2}}]$ and $y_\pm^2 = \frac{1}{2}[1 \pm (1 - 4\bar{\omega})^{\frac{1}{2}}]$.

We then find two kinds of solutions which are shown to be local entropy maxima for appropriate values of ,R . One preserves the symmetry for the interacting bosons (Fig. 1a and Fig. 1b).

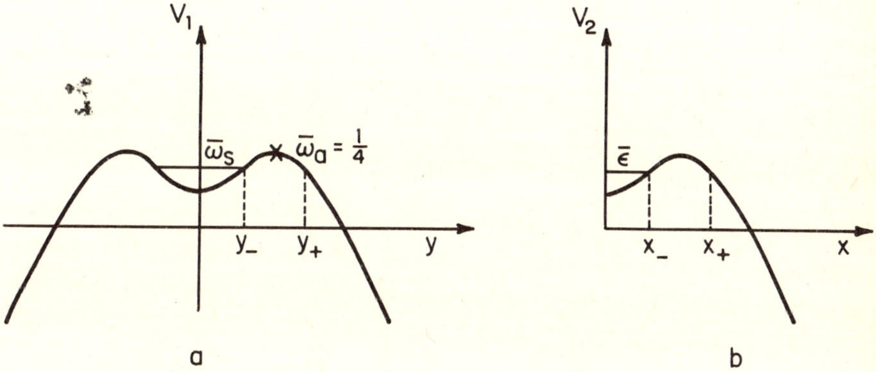

Figure 1

$\bar{\omega}_s$ represents a solution which preserves symmetry and $\bar{\omega}_a = 1/4$ the broken symmetry solution. The $\bar{\varepsilon}$ solution is similar for both cases.

ψ is positive since it represents the scale function. The other is the broken symmetry solution shown in Fig. 1a. The entropy for the two solutions are respectively, (E, elliptic integral of second kind)

$$S_{sym}/V = -3R_0^2/8\pi[\tfrac{1}{3}x_+][K(s^2)] + \tfrac{4}{3}y_+/3[K(t^2) - E(t^2)] \tag{19}$$

$$S_{asym}/V = \beta/8\lambda - [R_0^2/8\pi]\ x_+[K(t^2) - E(t^2)] \ . \tag{20}$$

There is no second-order phase transition possible since the temperature β^{-1} is a constant, for given (λ, R_0). Both symmetric and broken symmetry solutions can be locally stable in certain cases for equal values of λ and R_0. The symmetric solution is generally of lower entropy than the broken symmetry solution, so that it would appear that a first-order phase transition is possible, although the details of the thermodynamic analysis have yet to be carried out.

The cosmological solution is found by analytic continuation to real time

$$a(t) = 2^{-\frac{1}{2}} R_0 (1 - \gamma \cos^2 2t/R_0)^{\frac{1}{2}} \tag{21}$$

with

$$\gamma \equiv \sqrt{1 - 4\varepsilon} \tag{22}$$

range

$$x_- \leq a/R_0 \leq x_+$$

period

$$0 \leq t \leq \pi/2\, R_0 \quad .$$

Thus we have evaluated the MCE entropy of this model and found its dynamical solution, finding a possible first-order phase transition. This method is capable of wide generalization and the study of its consequences is being pursued.

References

Avis S. J. and Isham C. J., 1978, Proc. R. Soc. London A363, 581.

Coleman S. and De Luccia F., 1980, Phys. Rev. D21, 21, 3305.

Gibbons G. W. and Hawking S.W., 1977, Phys. Rev. D15, 2738, 2752.

Gibbons G. W. and Perry M.J., 1978, Proc. R. Soc. London A358, 467.

Grib A. A., Mostepanenko V. M. and Frolov, V. M., 1977, Theor. and Math. Phys. 33, 869.

Hawking S. W., 1976, Phys. Rev. D13, 191.

Horwitz G. and Weil D., 1982, Phys. Rev. Lett. 48, 219.

Linde A., 1974, Pism'a Zh. Eksp. Fiz. 19, 320; 1975, JETP Lett. 19, 183.

DIMENSIONAL REDUCTION

Peter G. O. Freund
The Enrico Fermi Institute and the Department of Physics
The University of Chicago, Chicago, Illinois 60637

Higher dimensional unified theories (i.e., generalized Kaluza-Klein theories)[1] have been considered for quite some time now. What is a simple gravity (or supergravity[2]) theory in a N+4-dimensional space-time becomes a theory involving gravity, and lower spin Bose (and Fermi) fields in four dimensions, upon the compactification of N space-like dimensions. Besides the number N of these extra dimensions, 4-dimensional Physics depends on the nature of the compact (and very small) N-dimensional manifold M_N. Depending on M_N we can envisage the low energy gauge group $G_{LE} = SU(3)_{color} X (SU(2) X U(1)_{electroweak}$ as the isometry group G_{KK} of M_N. Then the simplest possibility[3] is N=7 with $M_N = CP_2 X S^2 X S^1$. In this case there is no grand unification of the usual kind[4] whatsoever, $G_{KK} = G_{LE}$. Grand unification can be included by choosing a manifold M_N with isometry group G_{KK} larger than G_{LE}, e.g., $G_{KK} = SU(5)$, or by acquiring in the process of dimensional reduction gauge symmetries beyond those corresponding to G_{KK} (this happens when reducing 11-dimensional supergravity[5] to 4-dimensions when an additional Cremmer-Julia gauged SU(8) symmetry is nonlinearly realized by the scalar fields).[6] If the 4-dimensional theory possesses more gauged symmetry than G_{LE}, the question is at what scale this symmetry is restored. Standard renormalization group reasoning claims[7] this to occur at a length scale of $\ell_{GUT} \sim 10^{-15}\ GeV^{-1}$ or $\ell_{SUSY} \sim 10^{-16}\ GeV^{-1}$ in the minimal supersymmetric case. These scales being very far from present-day "physical" scales, the question can be asked as to whether at such scales 4-dimensional Physics is still applicable.

To answer this question we have to estimate the size of the small dimensions.[1] Requiring the Yang-Mills piece of the reduced 4-dimensional lagrangian to have the correct normalization relative to the Einstein piece and to have the proper minimal coupling say to charged scalar fields determines the size of the small dimension

$$\ell = 4\pi\, G^{1/2} (g^2/4\pi)^{-1/2} \tag{1}$$

where G is Newton's gravitational constant and g the Yang-Mills coupling constant. Equation (1) yields $\ell \sim 10^{-17} GeV^{-1}$, very close to ℓ_{SUSY}. It thus appears that in generalized Kaluza-Klein theories by the time the grand unification scale is reached (4+N)-rather than 4-dimensional Physics applies.[8] In particular, dimensionality is increased even <u>before</u> quantum gravity effects become large. This has prompted me[9] and also Ramond[10] to consider cosmologies in which the "effective" space-dimensionality is time dependent. Earlier work on such cosmologies is due to Chodos and Detweiler.[11]

Specifically we start from 11-dimensional supergravity[5] in which case supersymmetry requires an antisymmetric tensor Bose-matter field $A_{\mu\nu\rho}$ which is known[12] to produce preferential compactification of either 7 or 4 space-like dimensions. The first case is of course interesting and we generalize the Ansatz of reference 12) to include a cosmological time-dependence. The splitting of space-time into a 4-dimensional physical space-time and a 7-dimensional manifold M_7 is then automatic. A solution in which the cosmological scale of ordinary (3-dimensional) space increases linearly with cosmological time t, whereas the scale of M_7 increases only as $t^{1/7}$ is found. The 4-dimensional gravitational constant turns out to decrease at 1/t as proposed by Dirac.[13]

An alternative solution has M^7 a sphere of small time-independent radius, and 4-dimensional space time, an anti-de-Sitter universe with cosmological factor $R(t)=R(0)\cos\alpha t$ where α is determined by the 11-dimensional gravitational constant and by the, here time-independent, scale of the antisymmetric tensor field.

These solutions do not involve the Fermi-matter fields and as such should not be valid in the matter dominated era. They also break down at too early times where quantum gravity matters, but they can reasonably be expected to be relevant in some time interval around the dimensional transition where the 11-dimensional manifold splits into $M_4 \times M_7$. Other cosmological solutions along similar lines have been discussed in references 9)-11). New scales are introduced in these solutions: the time t_s at which all d-dimensions have comparable sizes, the actual size $\ell(t_s)$ of the dimensions at time t_s and the strength of gravity at that time. Depending on the details of the evolution, these scales can considerably exceed the present day Planck (10^{-19} GeV^{-1}) and Kaluza-Klein (10^{-17} GeV^{-1}) scales. The cosmology of the very early universe is thus seriously affected.

The preferential compactification of 7 space dimensions has been achieved here at the classical level as in reference 12). The alternative could be entertained that the choice of 4 large dimensions occurs at the quantum level, by somehow summing over all possible dimensionalities and for dynamical reasons ending up with four large physical dimensions in some approximation. Such a possibility was considered by others as well.[14]

This work was supported in part by the U. S. National Science Foundation.

References

1) Th. Kaluza, Sitzungsber. Preuss. Akad. Wiss. Phys. Math. Kl. 966 (1921);
O. Klein, Z. Phys. 37, 895 (1926);
B. de Witt, Dynamical Theories of Groups and Fields,
(Gordon and Breach, New York, 1965) p. 139;
R. Kerner, Ann. Inst. H. Poincaré 9, 143 (1968);
A. Trautman, Rep. math. Phys. 1, 29 (1970);
Y.-M. Cho and P.G.O. Freund, Phys. Rev. D12, 1711 (1975).

2) D.Z. Freedman, P. van Nieuwenhuizen and S. Ferrara, Phys. Rev. D13, 1314 (1976);
S. Deser and B. Zumino, Phys. Lett. 62B, 335 (1976).

3) E. Witten, Nucl. Phys. B186, 412 (1981).

4) J.C. Pati and A. Salam, Phys. Rev. D8 (1973) 1240;
H. Georgi and S.L. Glashow, Phys. Rev. Lett. 32 (1974) 438.

5) E. Cremmer, B. Julia and J. Scherk, Phys. Lett. 76B, 409 (1978).

6) E. Cremmer and B. Julia, Nucl. Phys. B159 (1979) 141;

7) H. Georgi, H. Quinn and S. Weinberg, Phys. Rev. Lett. 33 (1974) 451.
S. Dimopoulos, S. Raby and F. Wilczek, Phys. Rev. D24 (1981) 1681;
S. Weinberg, Harvard preprint HUTP-81/A047 (1981);
N. Sakai and T. Yanagida, Nucl. Phys. B197 (1982) 533;
M.B. Einhorn and D.R.T. Jones, Nucl. Phys. B196 (1982) 475;
J. Ellis, D.V. Nanopoulos and S. Rudaz, CERN preprint TH 3199 (1981).

8) P.G.O. Freund, preprint EFI 82/23.

9) P.G.O. Freund, preprint EFI 82/24.

10) P. Ramond, Univ. of Florida preprint UFTP 82-21.

11) A. Chodos and S. Detweiler, Phys. Rev. D21, 2167 (1980).

12) P.G.O. Freund and M.A. Rubin, Phys. Lett. 97B, 233 (1980).

13) P.A.M. Dirac, Proc. Roy. Soc. (London) A165, 199 (1935).

14) R. Geroch, private communication.
S. Shenker, private communication.

KALUZA-KLEIN TYPE THEORY

H. Sugawara
KEK, Tsukuba, Japan

I would like to discuss two problems related to the recently proposed unified theory of Kaluza-Klein[1] type by T. Kaneko and myself[2]. One is Mach's principle and the other is a classical solution to Einstein's equation which has a $[\delta(X)]^2$ singularity.

Let us start with Mach's principle. Mach stated this principle rather vaguely implying that the inertial mass may be determined by the distant matter or by the entire matter in the universe. I would like to show in this talk that this situation can be realized in a very simple model. I am aware of the existence of the long history of discussions on Mach's principle, mostly within the framework of Einstein theory[3]. I ask this learned audience to forget about these discussions for the time being and listen to may naive approach.

Suppose that our universe is composed of only two particles. How can we write down an action which incorporates Mach's principle in this simplified case? The conventional classical action which describes two particle system is;

$$L = \int L(t)dt = \int [\frac{1}{2}m_1(\vec{X}_1')^2 + \frac{1}{2}m_2(\vec{X}_2')^2 - V(\vec{X}_1, \vec{X}_2)]dt \qquad (1)$$

Here the masses m_1 and m_2 are free parameters and there is no way to relate these parameters to the motion of the particles. Let us consider, therefore, the following action which might look weird to most of you.

$$L = \int L_1(q_1', q_1, q_2)dt \times \int L_2(q_2', q_1, q_2)dt \quad , \qquad (2)$$

where the form of L_1 and L_2 will be fixed later. Here we restrict ourselves to one space dimension for simplicity. We also assume that there exists a natural mass unit which we take to be 1. We get the following set of Euler's equations from equation (2):

$$\frac{\partial L_1}{\partial q_1}\bar{L}_2 + \frac{\partial L_2}{\partial q_1}\bar{L}_1 - \frac{\partial}{\partial t}(\frac{\partial L_1}{\partial q_1'})\bar{L}_2 = 0 \quad , \qquad (3a)$$

and

$$\frac{\partial L_2}{\partial q_2}\bar{L}_1 + \frac{\partial L_1}{\partial q_2}\bar{L}_2 - \frac{\partial}{\partial t}(\frac{\partial L_2}{\partial q_2'})\bar{L}_1 = 0 \quad , \qquad (3b)$$

where

$$\bar{L}_1 = \int L_1(q_1', q_1, q_2)dt \tag{4a}$$

and

$$\bar{L}_2 = \int L_2(q_2', q_1, q_2)dt \quad . \tag{4b}$$

Let us now write down an explicit form of L_1 and L_2:

$$L_1 = \frac{1}{2} q_1'^2 - V(q_1-q_2) \quad , \tag{5a}$$

and

$$L_2 = \frac{1}{2} q_2'^2 - V(q_2-q_1) \quad . \tag{5b}$$

Then equations (3a) and (3b) reduce to the following forms respectively:

$$\bar{L}_2 q_1'' = - \frac{\partial V_1}{\partial q_1} \quad , \tag{6a}$$

and

$$\bar{L}_1 q_2'' = - \frac{\partial V_2}{\partial q_2} \quad , \tag{6b}$$

where

$$V_1 = V_2 = (\bar{L}_1+\bar{L}_2)V \quad . \tag{7}$$

The meaning of equations (6a) and (6b) is clear: The mass of particle 1 is equal to $\bar{L}_2$ and the mass of particle 2 is equal to $\bar{L}_1$ implying that the mass of a particle is determined by the motion of the other particle which is interacting with it. By adding equation (6a) to equation (6b) and taking equation (7) into account we obtain

$$\bar{L}_2 \dot{q}_1 + \bar{L}_1 \dot{q}_2 = \text{constant} \quad , \tag{8}$$

which is nothing but momentum conservation. Equations (6a) and (6b) reduce to the following equation in terms of the relative coordinate $q = q_1 - q_2$:

$$mq'' = - \frac{\partial V(q)}{\partial q} \quad , \tag{9}$$

where

$$m = \bar{L}_1\bar{L}_2/(\bar{L}_1+\bar{L}_2)^2 \quad , \tag{10}$$

and

$$V(q) = V_1 = V_2 \quad . \tag{11}$$

We obtain from equation (9) the following energy integral:

$$\frac{1}{2} mq'^2 + V(q) = E \quad . \tag{12}$$

From equation (12) and the definition of L_1 we get

$$L_1 = \frac{1}{2}\left(\frac{\bar{L}_1}{\bar{L}_1+\bar{L}_2}\dot{q}\right)^2 - V(q)$$

$$= \left(\frac{\bar{L}_1}{\bar{L}_2}\right)E - \left(1 + \frac{\bar{L}_1}{\bar{L}_2}\right)V(q) \quad . \tag{13}$$

we have, therefore,

$$\bar{L}_1 = \frac{\sqrt{k}}{1+k}\int [kE - (1+k)V]\,\frac{da}{\sqrt{2(E-V)}} \quad , \tag{14a}$$

and

$$\bar{L}_2 = \frac{\sqrt{1/k}}{1+k}\int [E - (1+k)V]\,\frac{dq}{\sqrt{2(E-V)}} \quad , \tag{14b}$$

with

$$k = \bar{L}_1/\bar{L}_2 \quad . \tag{15}$$

We get from equations (14a), (14b) and (15);

$$k = 1 \quad . \tag{16}$$

Let us now consider the special case when

$$V(q) = \frac{1}{|q| - a} \quad . \tag{17}$$

Two particles start separating from each other after the 'big bang' at $t = 0$ ($q(t=0) = 0$) and reach the largest separation distance at $t = \infty$ ($q(t=\infty) = a$). From equation (14a) and (14b) we obtain

$$m_1 = m_2 = \bar{L}_1 = \bar{L}_2 = \frac{1}{2\sqrt{2}}\int_o^a \left[\frac{2\sqrt{a}}{\sqrt{q(a-q)}} - \frac{1}{\sqrt{a}}\sqrt{\frac{a-q}{q}}\right]dq \quad . \tag{18}$$

We, therefore, conclude that the mass of a particle is directly related to the size of the universe. To show that the form given in equation (2) is not too peculiar I will prove that the ordinary Maxwell's theory with an unspecified value of the electric charge can be written in this form. We take scalar electrodynamics as an example:

$$L = \left[-\frac{1}{4}\int d^4X\; F_{\mu\nu}F^{\mu\nu}\right] \times \int [(D_\mu\phi)^\dagger(D^\mu\phi) - V(|\phi|)]d^4X \quad , \tag{19}$$

$$\equiv L_A \cdot L_\phi \quad ,$$

with $D_\mu = \partial_\mu - iA_\mu$ where A_μ is the usual vector potential multiplied by the charge e. ϕ is a complex scalar field.

The equation of motion we get for A_μ is:

$$\partial_\mu F^{\mu\nu} + \frac{L_A}{L_\phi} J^\nu(x) = 0 \quad , \tag{20}$$

where

$$J^\nu(x) = i(\phi^\dagger D^\nu\phi - (D^\nu\phi)^\dagger\phi) \quad . \tag{21}$$

The equation for scalar field is conventional. Multiplying $A_\nu(X)$ to equation (20) and integrating over the whole 4-dimensional space, we obtain

$$2L_\phi = -\int J^\nu A_\nu d^4X \quad . \tag{22}$$

The equation (20), therefore, can be rewritten as

$$\partial_\mu F^{\mu\nu} - \frac{2L_A}{\int J^\nu A_\nu d^4x} J^\nu = 0 \quad . \tag{23}$$

This equation shows that we have

$$e^2 = \frac{-2L_A}{\int J^\nu A_\nu d^4x} \quad . \tag{24}$$

We can easily check that the value of e^2 is completely arbitrary just like the boundary values to equation (23). We conclude that the action (19) is equivalent to scalar electrodynamics at least at the classical level with an arbitrary value for the fine structure constant. We do not discuss the quantum version in this talk. The action we considered in our paper[(2)] is the same kind of action as in equation (2) or in equation (19). It is the product of Kaluza's action and the action for the minimal 4-dimensional surface in the n-dimensional Riemannian space. For the detail see paper (2).

Let me now turn to the next topic which is the solution to Einstein's equation with a $[\delta(x)]^2$ singularity. Although our task is to obtain a solution to the Kaluza-Klein equation[(2)] I will show in this talk that the usual 4-dimensional Einstein equation with a single particle as the source has a solution with a $[\delta(x)]^2$ singularity.

Einstein's equation in this case reads

$$R^{\mu\nu} - \frac{1}{2} g^{\mu\nu}R + 8\pi G T^{\mu\nu} = 0 \quad , \tag{25}$$

with

$$T^{\mu\nu} = m\int d\tau \delta^4(x-x(\tau)) \frac{\frac{dx^\mu}{d\tau}\frac{dx^\nu}{d\tau}}{[-g_{\mu\nu}\frac{dx^\mu}{d\tau}\frac{dx^\nu}{d\tau}]^{\frac{1}{2}}} \frac{1}{\sqrt{g}} \quad , \tag{26}$$

where $g = \det(g_{\mu\nu})$. For $X^0 = f(\tau)$ and $X^i = 0$ we have

$$T^{oo} = m\delta^3(x) \frac{1}{[-g_{00}]^{\frac{1}{2}}} \frac{1}{\sqrt{g}} \quad , \tag{27}$$

and

$T^{ij} = T^{io} = 0$. We now make the ansatz

$$g_{\mu\nu} = \begin{pmatrix} -a(t) & \\ & b(t)\begin{pmatrix} \delta(x_1)^2 & & \\ & \delta(x_2)^2 & \\ & & \delta(x_3)^2 \end{pmatrix} \end{pmatrix} \quad , \tag{28}$$

to solve equation (25). We then have

$$T^{\circ\circ} = m/\{a(t)b(t)^{\frac{3}{2}}\} \ , \tag{29}$$

and

$$\sqrt{g} = [a(t)b^3(t)]^{\frac{1}{2}}\delta(x_1)\delta(x_2)\delta(x_3) \ . \tag{30}$$

After some calculations we can rewrite equation (25) in the following form with the ansatz (28):

$$-\frac{d}{dt}(\frac{b'}{2a}) + 2(\frac{b'}{2a})(\frac{b'}{2b}) - \frac{b'}{2a}(\frac{a'}{2a} + \frac{3b'}{2b}) + \frac{b}{2}[\lambda + \frac{8\pi Gm}{\sqrt{b^3}}] = 0, \tag{31}$$

and

$$3\frac{d}{dt}(\frac{b'}{2b}) + 3(\frac{b'}{2b})^2 - \frac{a'}{2a}(\frac{3b'}{2b}) - \frac{a}{2}[\lambda - \frac{8\pi Gm}{\sqrt{b^3}}] = 0, \tag{32}$$

where we have added the cosmological term λ for the sake of completeness. We can easily see that $a = a_0t^4$ and $b = b_0t^4$ give a solution to equations (31) and (32) when $\lambda = 0$ if the following condition is satisfied:

$$4\pi Gma_0/b_0^{\frac{3}{2}} = 6 \quad . \tag{33}$$

For this solution we have

$$\int T^{\circ\circ}\sqrt{g}\ d^3x = \frac{m}{\sqrt{a_0}}\frac{1}{t^2} \quad .$$

The physical meaning of this solution in the above 4-dimensional case is not very clear but it is rather straight forward in the Kaluza-Klein case as has been extensively discussed by T. Kaneko and myself in reference (2).

(references)

1. Th. Kaluza, Sitzungsber. Preuss. Akad. Wiss. Berlin, Math.-Phys. Kl, 966(1921)
2. T. Kaneko & H. Sugawara, to be published in Prog. Theor. Phys.
3. See, for example, paper by N. Rosen in "To Fulfill a Vision', Jerusalem Einstein Centennial Symposium, Edited by Y. Ne'eman and published by Addison-Wesley, Inc., (1981)

PREGEOMETRY

K. Akama
Department of Physics, Saitama Medical College
Moroyama, Saitama, 350-04, Japan

All the existing experimental evidences, though not so many, clearly support the general relativity of Einstein as a theory of gravitation. So far, extended investigations have been made, based on the premise of general relativity. Even the theories of induced gravity,[1] or pregeometry, where the Einstein action is derived from a more fundamental stage, are not free of this premise. However, if the principle of general relativity is true, it should be a manifestation of some underlying dynamics, just like the Kepler's law for the Newtonian gravity, or like the law of definite proportion in chemical reactions for atoms, etc. So we would like to ask here why the physical laws are generally relative, instead we premise it. The purpose of this talk is to propose a model to give a possible answer to the question. By general relativity, we mean the general covariance of physical laws in the curved spacetime. Our solution, in short, is that it is because our four-spacetime is a four-dimensional vortex-like object in a higher-dimensional flat spacetime, where only the special relativity is assumed. To be specific, we adopt the dynamics of the Nielsen-Olesen vortex[2] in a six-dimensional flat spacetime, and show that general relativity actually holds in the four-spacetime. Furthermore we will show that the Einstein equation in the four-spacetime is effectively induced through vacuum fluctuations, just as in Sakharov's pregeometry.[1]

We start with the Higgs Lagrangian in a six-dimensional flat spacetime

$$\mathcal{L} = -\frac{1}{4} F_{MN} F^{MN} + D_M \phi^\dagger D^M \phi + a|\phi|^2 - b|\phi|^4 + c \tag{1}$$

where $F_{MN} = \partial_M A_N - \partial_N A_M$ and $D_M \phi = \partial_M + ieA_M$.

This has the 'vortex' solution[2]

$$A_M = \epsilon_{0123MN} A(r) X^N/r, \quad \phi = \varphi(r) e^{in\theta}, \tag{2}$$

$(r^2 = (X^5)^2 + (X^6)^2)$, where $A(r)$ and $\varphi(r)$ are the solutions of the differential equations,

$$-\frac{1}{r}\frac{d}{dr}\left(r\frac{d}{dr}\varphi\right) + \left[\left(\frac{n}{r} + eA\right)^2 - a + 2b\varphi^2\right]\varphi = 0$$
$$-\frac{d}{dr}\left(\frac{1}{r}\frac{d}{dr} rA\right) + \varphi^2\left(e^2 A + \frac{en}{r}\right) = 0 \tag{3}$$

The 'vortex' is localized within the region of $O(\epsilon)$ $(\epsilon = 1/\sqrt{a})$ in two of the space dimensions (X^5, X^6), leaving a four-dimensional subspacetime $(X^0 - X^3)$ inside it. For large a , the curved 'vortices' with curvature $R \ll a$

become approximate solutions,[3] which we denote by A_M^0 and ϕ^0. Let the center of the 'vortex' be $X^M = Y^M(\xi^\mu)$ $(\mu = 0-3)$, and take the curvilinear coordinate x^M such that, near the 'vortex',

$$X^M = Y^M(x^\mu) + n_m^M x^m, \quad (M = 0-3, 5, 6,\ \mu = 0-3,\ m = 5, 6) \tag{4}$$

where X^M is the Cartesian coordinate, and n_m^M are the normal vectors of the 'vortex'. (Hereafter Greek suffices stand for 0-3, small Latin, 5, 6, and capital, 0-3, 5, 6). Then, the solution is

$$A_M^0 = \epsilon_{0123MN} A(r) x^N / r, \quad \phi^0 = \varphi(r) e^{in\theta}. \ (r^2 = x^m x^m). \tag{5}$$

The S-matrix element between the states Ψ_i and Ψ_f is given by

$$S_{fi} = \int \prod_{X^M} dA_M\, d\phi\, d\phi^\dagger \exp\left[i\int \mathcal{L}\, d^6X\right] \Psi_f^* \Psi_i \prod_{X^M} \delta(\partial_M A^M) \tag{6}$$

We assume that the path-integration is dominated by the field configurations of the approximate solutions (5) and small quantum fluctuations around it. To estimate it, we first extract the collective coordinate by inserting

$$1 = \int \prod_{X_{/\!/}} dY^M(\xi^\mu)\, \delta(Y^M(\xi^\mu) - C^M(\xi^\mu)) \tag{7}$$

where $C^M(\xi^\mu)$ is the center of the distribution of $|\tilde{\phi}|^2$ $(\tilde{\phi} = \phi - \sqrt{a/2b})$ in the normal plane $N(\xi^M)$ of the 'vortex' at $x^\mu = \xi^\mu$,

$$C^M(\xi^M) = \int_{N(\xi^\mu)} X^M |\tilde{\phi}|^2 d^2X_\perp \Big/ \int_{N(\xi^\mu)} |\tilde{\phi}|^2 d^2X_\perp \tag{8}$$

By $\prod_{X_{/\!/}}$, we mean the product over the four parameters ξ^μ with the invariant measure. Then, we transform them into the representation in the curvilinear coordinate x^M, and we change the path-integration variables A_M and ϕ to their quantum fluctuations $B_{\bar{N}} = A_{\bar{N}} - A_{\bar{N}}^0$ and $\sigma = \phi - \phi^0$, retaining the terms up to quadratic in them.

$$\begin{aligned} S_{fi} = \int \prod_{X_{/\!/}} dY^M \prod_{x^M} dB_{\bar{N}}\, d\sigma\, d\sigma^\dagger\, \delta(\sqrt{-g}\, \nabla_{\bar{N}} B^{\bar{N}}) \prod_{X_{/\!/}} \delta(\tilde{C}^M) \\ \exp\left[i\int (\mathcal{L}_0 + \mathcal{L}_2)\sqrt{-g}\, d^6x\right] \Psi_f^* \Psi_i , \end{aligned} \tag{9}$$

with

$$\mathcal{L}_0 = \mathcal{L}(\phi = \phi^0, A_M = A_M^0) \tag{10}$$

$$\begin{aligned} \mathcal{L}_2 = & -\tfrac{1}{2} g^{LM} \nabla_L B_{\bar{N}} \nabla_M B^{\bar{N}} + B_{\bar{N}} B^{\bar{N}} e^2 |\phi^0|^2 \\ & + g^{LM} (D_L^0 \sigma)^\dagger (D_M^0 \sigma) - 4ie V^{\bar{N}M} B_{\bar{N}}\, \mathrm{Im}(\sigma^\dagger D_M^0 \phi^0) \\ & + a|\sigma|^2 - b[4|\phi^0 \sigma|^2 + 2\,\mathrm{Re}(\sigma^\dagger \phi^0)^2] , \end{aligned} \tag{11}$$

$$\tilde{C}^m = \int x^m |\tilde{\Phi}|^2 dx^5 dx^6 \Big/ \int |\tilde{\Phi}|^2 dx^5 dx^6 \tag{12}$$

$$= \frac{1}{J_0}\int x^m [|\sigma|^2 + 2\,\mathrm{Re}(\tilde{\Phi}^0 \sigma^\dagger)\{1 - \frac{2}{J_0}\int \mathrm{Re}(\tilde{\Phi}^0 \sigma^\dagger) dx^5 dx^6\}]\, dx^5 dx^6; \tag{13}$$

and $\tilde{C}^\mu = 0$, where the barred suffices stand for the local Lorentz frame indices, $V^{\bar{N}M}$, the vielbein, g^{LM}, the metric tensor, ∇_M, the covariant differentiation, $D^0_M = \nabla_M + ieA^0_M$, and $J_0 = \int |\Phi^0|^2 dx^5 dx^6$. The Lagrangian $\mathcal{L}_2$ indicates that, outside the 'vortex', any low energy fields are suppressed because of the high barrier of $|\phi^0|^2$. Inside the 'vortex', $g_{m\mu} = O(R/a) \ll 1$, $g_{mn} = -\delta_{mn} + O(R/a)$ and $B_{\bar{M}}$ reduces to the four-vector $B_{\bar{\mu}}$ and the two scalers $B_{\bar{m}}$. Thus, the spacetime looks like four-dimensional and curved to observers with large scale. It is easily checked that the action is invariant under the general coordinate transformation of the curved four-spacetime, i.e. the physical laws are generally relative!

Now we see that the Einstein action is induced through vacuum polarizations. The effective action S^{eff} for it is given by

$$S^{eff} = -i\ln\int \prod_{x^M} dB_{\bar{N}}\, d\sigma\, d\sigma^\dagger\, \delta(\sqrt{-g}\,\nabla_{\bar{N}} B^{\bar{N}}) \prod_{X_{/\!/}} \delta(\tilde{C}^M) \exp[i\int\sqrt{-g}\,\mathcal{L}_2 d^6x]. \tag{14}$$

Exponentiating the argument of the δ-functions by $\delta(x) = \int dk\, e^{ikx}$, we get

$$S^{eff} = -i\ln\int \prod_{\xi^\mu} dw_m \prod_{x^M} dB_{\bar{M}}\, d\sigma\, d\sigma^\dagger\, dv\, \exp[i\int(\Xi\Phi + \Phi^\dagger\Delta\Phi)d^6x] \tag{15}$$

with $\Phi^\dagger = (B^{\bar{M}}, \sigma^\dagger, \sigma)$ (16)

$$\Xi = \sqrt{-g}\,(\nabla_{\bar{M}} v,\ w_m x^m \tilde{\Phi}^{0\dagger}/J_0,\ w_m x^m \tilde{\Phi}^0/J_0) \tag{17}$$

$$\Delta = \sqrt{-g}\begin{pmatrix} \eta_{\bar{M}\bar{N}}(\frac{1}{2}\nabla_L\nabla^L + e^2|\phi^0|^2) & ieD^0_{\bar{M}}\phi^{0\dagger} & -ieD^0_{\bar{M}}\phi^0 \\ -ieD^0_{\bar{N}}\phi^0 & \frac{1}{2}D^0_L D^{0L} + \frac{a}{2} - 2b|\phi^0|^2 + \delta^m_{11} w_m & -b(\phi^0)^2 + \delta^m_{12} w_m \\ ieD^0_{\bar{N}}\phi^{0\dagger} & -b(\phi^{0\dagger})^2 + \delta^m_{21} w_m & \frac{1}{2}D^0_L D^{0L} + \frac{a}{2} - 2b|\phi^0|^2 + \delta^m_{22} w_m \end{pmatrix} \tag{18}$$

where δ^m is the nonlocal operator in 5-6 plane,

$$\delta^m(x,x') = \frac{1}{2J_0} x^m \delta(x-x') + \frac{1}{2J_0^2}(x^m + x'^m)\begin{bmatrix}\tilde{\Phi}^0(x) \\ \tilde{\Phi}^0(x)^\dagger\end{bmatrix}[\tilde{\Phi}^0(x')^\dagger\ \ \tilde{\Phi}^0(x')] \tag{19}$$

Performing the path-integration in $B_{\bar{N}}$, σ, $\sigma^\dagger$, and v, we get (with $\Xi_0 = \Xi|_{v=0}$)

$$S^{eff} = \frac{1}{2} i\,\mathrm{Tr}\ln\Delta + \frac{1}{2} i\,\mathrm{Tr}\ln[\partial_M\sqrt{-g}(\Delta^{-1})^{MN}\sqrt{-g}\,\partial_N] - \frac{1}{4}\int \Xi_0^\dagger \Delta^{-1}\Xi_0\, d^6x \tag{20}$$

S^{eff} in (20) is estimated perturbatively in $h^{MN} = g^{MN} - \eta^{MN}$ ($\eta^{MN} = \mathrm{diag}(1,-1,-1,-1,-1,-1)$) and w. The propagator is given by the inverse of $\Delta|_{h^{MN}=0,\, w=0} \equiv \Delta_0$. Δ_0 can be separated into two parts Δ_0^{sp} and Δ_0^{ex} which operates on four-space variables x^μ, and the extra space variables x^m, respectively. Furthermore, these Δ_0's are block-diagonalized into two parts Δ_0^V and Δ_0^S, which operate on the four-vector B^μ and the coupled scalars $(S^{(1)}, S^{(2)}, S^{(3)}, S^{(4)}) = (B^5, B^6, \sigma, \sigma^\dagger)$, respectively. They are given by

$$\Delta_0^{V,sp} = \frac{1}{2}\Box\ , \quad \Delta_0^{S,sp} = \frac{1}{2}\Box\ , \quad \Delta_0^{V,ex} = -\frac{1}{2}\partial_\ell\partial_\ell + e^2|\phi^0|^2$$

$$\Delta_0^{S,ex} = \begin{pmatrix} (-\frac{1}{2}\partial_\ell\partial_\ell + e|\phi^0|^2)\eta_{mn} & ieD_n^0\phi^{0\dagger} & -ieD_n^0\phi^0 \\ -ieD_m^0\phi^0 & -\frac{1}{2}D_\ell^0 D_\ell^0 + \frac{a}{2} - 2b|\phi^0|^2 & -b(\phi^0)^2 \\ ieD_m^0\phi^{0\dagger} & -b(\phi^{0\dagger})^2 & -\frac{1}{2}D_\ell^0 D_\ell^0 + \frac{a}{2} - 2b|\phi^0|^2 \end{pmatrix} \quad (21)$$

where $\Box = \eta^{\mu\nu}\partial_\mu\partial_\nu$. Then, the propagators for each class are given by

$$[\Delta_0^{V\,-1}]^{\mu\nu} = \eta^{\mu\nu}\sum_k (\Box + m_k^2)^{-1} V_k(x^m) V_k(x'^m)\ ,$$
$$[\Delta_0^{S\,-1}]^{(a)(b)} = \sum_k (\Box + m_k'^2)^{-1} S_k^{(a)}(x^m) S_k^{(b)}(x'^m)\ , \quad (22)$$

where V_k, $S_k^{(a)}$, m_k^2 and $m_k'^2$ are the solutions and the eigenvalues of the differential equations in the extraspace.

$$\Delta_0^{V\,ex} V_k = m_k^2 V_k \quad (23)$$
$$\Delta_0^{S,ex\,(a)(b)} S_k^{(b)} = m_k'^2 S_k^{(a)}$$

The argument of the logarithms in (20) is expanded as follows

$$\Delta = \Delta_0 (1 + \Delta_0^{-1}\Delta_{int})\ , \quad (24)$$

$$\partial_M\sqrt{-g}\,[\Delta^{-1}]^{MN}\sqrt{-g}\,\partial_N = 1 + \Delta_0'^{-1} + \partial_m(\Delta_0^{-1})^{mn}\partial_n + \Delta'_{int}\ , \quad (25)$$

where Δ_{int} and Δ'_{int} are the interaction parts including $h^{\mu\nu}$ and w, and

$$\Delta_0'^{-1} = \sum_k m_k^2 (\Box + m_k^2)^{-1} V_k(x^m) V_k(x'^m)\ . \quad (26)$$

We expand the logarithms in (20), and get series of one-loop diagrams with external $h^{\mu\nu}$ and w lines attached. These diagrams diverge quartically in the ultraviolet region. We introduce the momentum cutoff Λ much larger than $\sqrt{a}$, and calculate the divergent contributions. The diagrams with vertices which involve extra-space operators are less divergent.

After this, the same argument as in the pregeometry[1] leads to the Einstein action in the four-dimensional curved space. Namely, the divergent contributions are

$$S^{eff} = \int \sqrt{-g}\,[(N_0\alpha_0 + N_1\alpha_1 + \alpha_c)\Lambda^4 + (N_0\beta_0 + N_1\beta_1 + \beta_c)\Lambda^2 R]\,d^4x \tag{27}$$

plus less divergent terms, where N_0 and N_1 are the numbers of the scalar and the vector bound-states in (23), respectively, and $\alpha_0, \alpha_1, \beta_0, \beta_1$ are calculable constants of $O(1)$. The values are found in literatures,[1,4] though we should be careful, since they depend on the cutoff-method and even on the gauge.

α_c and β_c are the contributions from the continuum states in (23). Now, together with the contributions from $\mathcal{L}_0$, we finally get the Einstein action

$$S = \int \sqrt{-g}\,(\lambda + \frac{1}{16\pi G} R)\,d^4x \tag{28}$$

where

$$\lambda = \int \mathcal{L}_0\,dx^5 dx^6 + (N_0\alpha_0 + N_1\alpha_1 + \alpha_c)\Lambda^4, \tag{29}$$

$$\frac{1}{16\pi G} = (N_0\beta_0 + N_1\beta_1 + \beta_c)\Lambda^2.$$

In conclusion, in this model:

1) The principle of general relativity is induced, instead it is premised.
2) The Einstein equation is induced just as in Sakharov's pregeometry.
3) Two kinds of internal symmetries are induced, those of the transformation and the excitation in the extra-space. The former is somewhat like isospin, while the latter, generation. This suggests a new mechanism for unification of the interactions.
4) When the gravitational field is quantized, the ultraviolet divergences should be cut off at the inverse of the size of the 'vortex', which may be much smaller than the Planck mass. If this is the case, we can by-pass the problems of re-normalizability of gravity.
5) Particles with sufficiently high energy can penetrate into the extra dimensions.
6) At very high temperatures,[5] or high densities, the 'vortex' is spread out over the extra-space revealing the higher dimensional spacetime.

References

1. A.D. Sakharov, Dok. Akad. Nauk SSSR 177 (1967) 70;
Theor. Mate. Fiz. 23 (1975) 23;
K. Akama, Y. Chikashige, T. Matsuki and H. Terazawa, Prog. Theor. Phys. 60 (1978) 868;
K. Akama, Prog. Theor. Phys. 60 (1978) 1900;
For a review, S.L. Adler, Rev. Mod. Phys. 54 (1982) 729.
2. H.B. Nielsen and P. Olesen, Nucl. Phys. B61 (1973) 45.
3. D. Förster, Nucl. Phys. B81 (1974) 84;
J.L. Gervais and B. Sakita, Nucl. Phys. B91 (1975) 301.
4. K. Akama, Phys. Rev. D24 (1981) 3073.
5. K. Akama and H. Terazawa, Gen. Relat. Grav. to be published.

SCALE INVARIANT SCALAR-TENSOR THEORY AND THE ORIGIN OF GRAVITATIONAL CONSTANT AND PARTICLE MASSES

Yasunori Fujii
Institute of Physics
University of Tokyo, Komaba
Meguro-ku, Tokyo 153

Scalar-tensor theory seems to offer a natural alternative to the standard theory of gravitation, especially when one tries to unify the theory of gravitation and the theory of elementary particles. It also seems inevitable that any scalar-tensor theory results in a variable-G theory. This is, however, not always true. We present a simple viable scalar-tensor theory the result of which is not a variable-G theory in the usual sense. We suggest that the next simplest model will give a true variable-G theory.

We consider the fundamental Lagrangian[1)]

$$\mathcal{L}=\sqrt{-g}\left(\frac{1}{2}f^{-2}\phi^{2}R-\frac{1}{2}\epsilon g^{\mu\nu}\partial_{\mu}\phi\partial_{\nu}\phi+L_{M}+L_{I}\right), \qquad (1)$$

where ϕ is the scalar gravitational field; $\epsilon=\pm1$ depending on whether ϕ is a normal or ghost field, L_M and L_I being the matter and interaction Lagrangians, respectively. The coupling constant f^2 is dimensionless, expected to be of the order of unity. (We use the unit with $c=\hbar=1$.) For L_M we assume the Lagrangian of massless matter fields with dimensionless coupling constants, as in any of gauge theories. We also choose

$$L_{I}=-c_{0}\phi^{4}-g\bar{\psi}\psi\phi+\cdots, \qquad (2)$$

where ψ is a typical spinorial matter field. The Lagrangian (1) is characterized by the absence of dimensional constants, and is invariant under global scale transformation.

We assume a decomposition

$$\phi^{2}(x)=u(t)+\sigma(x), \qquad (3)$$

where $\sigma(x)$ is a usual spacetime-dependent field, while $u(t)$, called the cosmological background value (BGV), may depend only on the cosmic time t. The BGV $u(t)$ may change so slowly that it may be viewed as a constant in most physical phenomena except those that take place on a cosmological time scale. We then find from the first term of (1) that the effective time-dependent gravitational constant G(t) is given by

$$G(t) = (f^2/8\pi)\, u^{-1}(t). \tag{4}$$

The second term of (2) gives the effective mass

$$m(t) = g\, u^{1/2}(t). \tag{5}$$

From (4) and (5) we obtain the relation

$$G(t)\, m^2(t) = \text{const.} \tag{6}$$

This is a crucial consequence of this simplest model in which there is only one scalar field that plays a dual role ("single-scalar model").

We derive the field equations as given by

$$G_{\mu\nu} = f^2 \mathcal{T}_{\mu\nu} = f^2 \phi^{-2} \left[T_{\mu\nu} - f^2 (g_{\mu\nu}\Box - \nabla_\mu \nabla_\nu) \phi^2 \right], \tag{7a}$$

$$f^{-2}\phi R + \epsilon \Box \phi + \partial L_I / \partial \phi = 0, \tag{7b}$$

together with other matter field equations, where $T_{\mu\nu}$ is the symmetric energy-momentum tensor of the matter as well as ϕ. We must impose

$$\nabla_\mu \mathcal{T}^{\mu\nu} = 0, \tag{8}$$

in order to be consistent with LHS of (7a). The scalar field equation (7b) can be put into a simpler form

$$Z^{-1} \Box \phi^2 = 0, \tag{9}$$

with $Z^{-1} = \epsilon + 6f^{-2}$, where we have used the trace of (7a) and the matter field equations. On RHS of (9) we have the trace of the matter energy-momentum tensor which vanishes due to the scale invariance. The scalar field has now no direct matter source. No scalar long-range force occurs in the limit of weak gravitational field, thus leaving the experimental tests of general relativity unaffected. In this sense the theory is completely viable for any value of f^2.[2] This is in sharp contrast with Brans-Dicke theory;[3] in order to meet observational constraints, their coupling constant $\omega (= \epsilon f^2/4)$ is severely bounded almost to the extent that the theory does not make much sense. ($|\omega| \gg 60$.)

The factor Z^{-1} would vanish if one chooses a conformal coupling, $f^2 = 6$, $\epsilon = -1$, as in conformally invariant theories. Then, however, we have no field equation of ϕ^2. The scalar field no longer has a dynamical degree of freedom, and thus making the theory viable again. This is the way Dirac and other authors formulated their variable-G theories.[4] This approach is not satisfactory, however, because the theory has no built-in principle to determine the scalar field, and eventually G(t).[5] One has to appeal to some outside principle, like Dirac's Large Numbers Hypothesis. We insist that any physical quantity which develops with time must be dynamical. From this point of view we avoid the conformal coupling and assume $Z^{-1} \neq 0$. We hence obtain

$$\Box \phi^2 = 0. \tag{10}$$

We reiterate that requiring scale invariance and conformal noninvariance is an almost unique choice if we want a viable scalar-tensor theory maintaining a dynamical degree of freedom of the scalar field.[2]

We now assume the spatially flat Robertson-Walker metric with the pressureless matter. In accordance with the decomposition in (3), the BGV parts of (7a), (8) and (10) are calculated to be

$$3H^2 = f^2 \mathcal{T}_{00} = f^2 u^{-1}\left(\frac{1}{2}\epsilon \dot{v}^2 + c_0 v^4 - 3f^{-2} H \dot{u} + \rho\right), \tag{11a}$$

$$\dot{\mathcal{T}}_{00} - 3H u^{-1}\left(\epsilon \dot{v}^2 + f^{-2}\ddot{u} + \frac{1}{2} f^{-2} H \dot{u} + \rho\right) = 0, \tag{11b}$$

$$\frac{d}{dt}(a^3 \dot{u}) = 0, \tag{11c}$$

respectively, where $H = \dot{a}/a$ with $a(t)$ the scale factor of the universe, $u = v^2$, and ρ the density of the matter and $\sigma(x)$. We solve (11c) to obtain

$$a(t) = K \dot{u}^{-1/3}(t). \tag{12}$$

Substituting this into (11a) and (11b), and eliminating ρ, we obtain

$$\frac{\dddot{u} u}{\dot{u}} - \frac{3}{2}\frac{\ddot{u}^2 u}{\dot{u}^2} - \frac{1}{2}\ddot{u} - \frac{3}{16}\epsilon f^2 \frac{\dot{u}^2}{u} = -\frac{3}{2} c_0 f^2 u^2. \tag{13}$$

A solution u(t) of this third-order nonlinear differential equation will determine simultaneously G(t), m(t) and a(t) through (4), (5) and (12), respectively, being in conformity with Mach's principle.

In solving (13) we must give integration constants in the form of initial or boundary conditions. In this way we derive dimensional quantities in nature starting from the Lagrangian that has no dimensional constants. The same situation is typical in any spontaneous symmetry breaking.

Ignoring RHS of (13), for the moment, we solve (13) in a systematic way. Among the solutions, we find "asymptotically standard solutions" in which u(t) approaches a finite constant as $t \to \infty$. This implies that G(t) and m(t) also approach constant values and a(t) tends to the standard Einstein-de Sitter solution $t^{2/3}$. We may propose an interesting conjecture that the standard theory gives an accurate description of the present or recent universe just because we are already in the asymptotic region of t.

On the other hand, we may not be in the asymptotic epoch, or one of other solutions may be a true solution. Corresponding to Dirac's atomic gauge, we then apply a conformal transformation which brings $\phi(x)$ into a constant v_*, so that the particle mass $m_* = gv_*$ is also a constant. In this "microscopic unit system" in which the time is measured by using atomic clocks, the gravitational constant G_* is also a true constant due to (6). For this reason our theory is not a variable-G theory in the usual sense.

In spite of a truly constant G_* our theory is certainly different from the standard theory because the scalar field is still present, showing itself through a time-dependent cosmological term. We find that the Lagrangian after the conformal transformation is given by

$$\mathcal{L} = \sqrt{-g_*}\left(\frac{1}{16\pi G_*}R_* - \frac{1}{8\pi G_*}\Lambda + L_{*M} + L_{*I} - m_*\bar{\psi}_*\psi_*\right), \tag{14a}$$

where the stars indicate the quantities in the microscopic unit system, and the effective cosmological term Λ is found to be

$$\begin{aligned}\Lambda(t) &\sim |Z|^{-1/2} G_* g_*^{\mu\nu} \partial_\mu (v_* \ln\phi)\, \partial_\nu (v_* \ln\phi) \\ &\sim (\dot{v}/v)^2 \sim t^{-2}.\end{aligned} \tag{14b}$$

The relation (6) from which G_*= const follows may be avoided to give a true variable-G theory, if we include another scalar field ("two-scalar model"). As still another consequence of this next simplest model of a viable scalar-tensor theory, one of the scalar fields may acquire a non-zero mass μ, giving a finite-range Yukawa potential which is added to the Newtonian potential:[2)]

$$V(r) = \frac{GM}{r}\left(1 + \alpha e^{-\mu r}\right). \tag{15}$$

A plausibility argument gives $\mu^2 \sim (Gm^2)m^2$, where m is a typical particle mass. For a choice $m \sim 1$ GeV, we find the force-range $\mu^{-1} \sim 10^5$cm = 1 km. Experimental searches for any deviation from the purely Newtonian behaviors in this distance range are now under way.[6)]

References

1) For details, see Y. Fujii, UT-Komaba 82-8.

2) Y. Fujii, Phys. Rev. D9(1974), 874; GRG 13(1981), 1147.

3) C. Brans and R. H. Dicke, Phys. Rev. 124(1961), 925.

4) P. A. M. Dirac, Proc. Roy. Soc. A333(1973), 403.
F. Hoyle and J. V. Narlikar, Nature, 233(1971), 41.
V. Canuto et al., Phys. Rev. D16(1977), 1643.
H. Terazawa, Phys. Lett. 101B(1981), 43.

5) For a similar view, see also J. D. Bekenstein and A. Meisels, Phys. Rev. D22(1980), 1313.

6) See the papers cited in the second of refs.2.

CONCLUDING REMARKS

Y. Nambu

Enrico Fermi Institute
University of Chicago
5640 S. Ellis Ave
Chicago, IL 60637
U S A

I. I think it is quite fitting for Professor Utiyama to have organized this symposium on gauge theory and gravitation, for he is one of the pioneers who recognized in gauge theory a principle that could conceptually unify the various forces in nature, including gravity and electromagnetism.

The idea of the unification of forces of course dates back to earlier times. Following Einstein's theory of gravitation (1915), people like Weyl (1918) and Kaluza (1921) made an attempt to combine electromagnetic and gravitational forces in a geometrical principle, and supplied many of the key concepts that are being used by the physicists today. Einstein also devoted all his scientific efforts in his later years to the search for a correct unified theory. All these noble efforts, however, failed. Because of theoretical difficulties or lack of experimental support, the unification of forces remained the theorists' dream while the progress in particle physics uncovered more and more new particles and new phenomena. Thus, Nature seemed to be moving away farther and farther from the simple and elegant ideals of unification*.

This, I think, had been the state of affairs prevailing until the 1960's. Faced with the unexpected new structures of matter, physicists

---

*) There is an interesting article by Y. Fujii in Kagaku 7, 431 (1982). He relates that Professor Utiyama, who regarded himself a particle physicist, had to study gravity in secrecy. Actually, Utiyama developed a general gauge principle independently of Yang and Mills.

were kept busy trying to find a semblance of order in a chaos, building more or less ad hoc models in order to account for what they saw. What the physicists learned during this period is that Nature is much richer and more complex than had been thought previously. But they were also able to accumulate more knowledge, both experimental and theoretical, which eventually gave them confidence to try a renewed attack on the goals of unification.

One of the important theoretical concepts acquired in this period is that of non-Abelian gauge fields. When Yang and Mills discovered in 1954 the generalization of Maxwell's electromagnetic theory to a non-Abelian (SU(2)) variety, it was not seriously thought to be relevant to physics, because perfect non-Abelian internal symmetries or conserved quantum numbers did not exist. The only candidate for strictly conserved quantity other than energy-momentum and electric charge was baryon number, but there did not seem to be an associated baryon number gauge field, as was pointed out by Lee and Yang (1955). Indeed, our current belief is: no gauge field, no symmetry; hence baryon number should not be conserved in spite of the extremely high stability of matter. The gauge principle has become a pervasive dogma.

Professor Utiyama's contribution in 1956 was to recognize the value of Yang and Mills' ideas as a potential guiding principle in physics and to show that Einstein's gravity was also subject to a similar interpretation. He called the interactions that naturally follow from the gauge principle "the interactions of the first class", as opposed to "the interactions of the second class" like the Yukawa type interactions which do not follow from such a principle. Clearly the implication was that the latter was more arbitrary and less desirable, and should be gotten rid of if possible. The same attitude prevails now. It is true that we have replaced the Yukawa theory of strong interactions with the more fundamental gauge theory of color, but we have not succeeded in eliminating the second class interactions from the weak interactions. They are with us in the form of Higgs fields and couplings. Personally, I share with some of the theorists the view or expectation that the Higgs fields will eventually go the way of Yukawa's meson.

The persistence of the second class interactions, however, is because they are there. Even if they may not be the primary source of dynamics among truly elementary particles, they appear as secondary effects among composite systems. In principle they should be derivable from the former, but at the phenomenological level they are the ones responsible for the complexity of the real world. It is thus obvious that the gauge principle alone cannot explain everything. The world would look too simple and too symmetric if the gauge principle had to manifest itself in a straightforward manner.

II. This brings me to a brief discussion of some other theoretical ingredients out of which our current theoretical system of particle physics is made. Since this is not a detailed survey, I will pick only the ones directly relevant to the question raised above: How can the gauge principle be made to work in the real world? I would say there are two basic elements. One is renormalization, and the other is spontaneous symmetry breaking.

The renormalization theory removed the difficulties inherent in quantum field theory, and turned quantum electrodynamics into one of the most successful theories in physics. I suspect, however, that there are many physicists, especially of the older generation like myself, who regard renormalization theory as an imperfect and temporary measure. Although this may be so, I cannot help but be impressed by the extent of its successes. The discovery of the asymptotic freedom, or the antiscreening property in non-Abelian gauge theories is another milestone in this regard. It made the gauge principle really relevant to strong interactions. It demonstrated how quantized non-Abelian gauge fields behave very differently from the naive classical picture. I have no doubt about the basic correctness of the color gauge theory of strong interactions.

Turning now to the weak interactions, a different mechanism was necessary to make the gauge theory work in this realm, too. Such a mechanism has been found to be the spontaneous breakdown of symmetries, a phenomenon already familiar in condensed matter physics but not recognized as such until particle physicists started to use them. In any case, this gave us for the first time at least a theoretical possibility that Nature does not exhibit all the symmetries built into its fundamental laws. The Weinberg-Salam theory is a concrete realization of these ideas, and its validity has already been confirmed overwhelmingly, if not completely. However, Utiyama's second class interactions are still there, as I have mentioned before.

Weak interactions are the ones that cause the most trouble for us. They do not seem to rigorously observe any symmetry at all. Why do they look so irregular and arbitrary? Probably spontaneous breakdown alone is not enough to explain or derive everything. Specifically, I have in mind, for example, the problems of mass spectrum, CP violation, etc. In fact there exist already a few other mechanisms of symmetry breaking built into quantum field theory. The appearance of a renormalization mass scale and running coupling constants, the chiral anomaly, and the instanton and monopole effects belong to this category. The first of these gave physicists a real hope for a unification of forces, thereby starting the modern revival of unified theories. However, the other effects are yet to be fully exploited.

I am gradually turning my eyes toward the future. Most theorists are already busily working on the GUTS (grand unified theories). The SU(5) theory is their prototype which has a great deal of theoretical appeal and a few pieces of supporting evidence. Hopefully within the next few years, decisive events will take place which will confirm the basic tenets of the GUTS. One such event might be the proton decay, and another might be the detection of monopoles (or a confirmation of Cabrera's results). If either one should happen, physicists can peacefully sleep at night (at least for a day or two). But the GUTS are still leaving a lot of questions unanswered. Prominent among them are the problem of hierarchy and that of generation. The hierarchy problem is a product of the very successes of renormalization theory. Becauses of its logarithmic scale dependence, we have been able to extrapolate renormalization theory to enormous energies where true unification of forces is realized. At the same time, however, this created the problem of explaining why several vastly different mass scales exist. As for the existence of generations, most of the GUTS remain silent about it. We are not sure whether it is a manifestation of another broken symmetry or something else.

III. All these unanswered questions seem like minor ones in comparison to the grandeur and beauty of a unified theory. After all the GUTS have brought us to within a shouting distance from the Planck scale, the scale where the final unification of forces would take place. This connection of renormalization theory to gravity was anticipated by Landau in the 1950's but we now have its as a real possibility. We must be cautious, though, because such a rosy prospect is also fraught with dangers and pitfalls. Physicists should always keep one foot on the ground even when they are daydreaming.

At any rate, we are right now witnessing quite a bit of theoretical activity in GUTS and beyond. The basic principle underlying the gauge theories is a geometrical view of dynamics. The Maxwell-Yang-Mills type gauge theories embodying internal symmetries have been so interpreted in an abstract geometrical sense. It may be natural, however, to try to interpret these abstract geometries as something more concrete and akin to the real space-time, like in the attempts of Kaluza and Klein and their more recent followers. On the other hand, perhaps we need not try to carry the similarities too far. Only broad analogies may suffice.

Actually, the real problem we are facing at the moment is that the paradigms of the current particle physics, including among others the gauge principle, have worked, but not well enough to answer all important questions, nor badly enough to expose glaring contraditions. I am tempted to compare the situation with that of the early 1930's when particle physics was just being born. At that time the nature of nuclear forces was unknown and the validity of relativistic quantum theory uncertain. Some physicists like Heisenberg speculated that quantum mechanics would break down at nuclear scales. As it turned out, however, real progress was made by saving quantum theory through renormalization, but at the same time taking the radical step of postulating new particles. This strategy has worked so well that we are still following it today.

But now we are beginning to see the old problem again. Heisenberg's fundamental lenght is replaced by the Planck length. There is a slight difference of attitude, though, in that we are more preoccupied with the glorious outlook on this side of the limit than with the uncertainties on the other side. Will our strategy continue to carry us beyond the limit? Or will we have to squarely face up to the problem this time? Whatever the outcome, we certainly need new ideas. The supersymmetry and supergravity may very well play such rôles as those played by the renormalization theory before. The current frustrations we are having with regard to supersymmetry may be because we have not found the right way to use it.

To elaborate a little further, supersymmetry is subject to an abstract geometrical interpretation, and thus fits into the general spirit of unifying geometry and dynamics. It offers the possibility of unifying for the first time both fermions and bosons, or the conventional matter and the conventional forces. It also renders the self-energy divergences less severe, and may eventually help in solving such problems as hierarchy and quantum gravity. On the other hand, we do not know what supersymmetry means in simple physical terms. We do not have familiar examples to guide us. In this respect, however, I have a little observation to make.

Recently it has been pointed out by several people that monopoles can catalyze proton decay. This amounts to a violation of baryon number without recourse to the conventional mechanisms in the GUTS. Its enormous implications are of course obvious, but that is not the point here. The reason for such an effect is that fermions can form zero-energy bound states (binding energy equals the mass) with a monopole. So the distinction between particles and holes becomes obscured, and all fermions of different masses become equal in such an environment. Since fermions can be added to a monopole without changing total energy, a kind of supersymmetry is thus created.

Such a phenomenon seems to happen generally in topological excitations, as was first found by Jackiw and others. Occurrence of zero energy bound states are somehow related to spontaneous breaking, because their quantum numbers are similar to those of the Goldstone modes. The Abrikosov flux tube in a superconductor admits such (almost) zero energy states. The empirical supersymmetry in nuclear physics observed by Iachello may also have a similar origin. Although I have not emphasized it before, topology is one of the most interesting aspects of the geometrical principle.

QUANTIZED STRINGS AND QCD

E.S.Fradkin and A.A.Tseytlin
Department of Theoretical Physics, P.N.Lebedev Physical Institute
Leninsky Pr. 53, Moscow 117924, USSR
Unfortunately the authors were unable to attend the Symposium

1. Introduction

It is now generally believed that the $N = \infty$ QCD is equivalent to some string model (see e.g. [1,2]). This equivalence is understood in the sense of the string ansatz for the Wilson loop average:

$$W[C] \equiv \frac{1}{N}\langle tr\, P \exp - \oint_C A_\mu dx_\mu \rangle = \sum_{surfaces,\ \partial S = C} \exp(-I_{string}) \quad (1)$$

However the main question about the type of this string is still open. But suppose we have solved the Makeenko-Migdal equation and thus know the string action I_{string} (which can be an effective one obtained after integrating out some internal string variables). Then the natural problem is how to calculate mesons Green's functions (and thus their spectrum and scattering amplitudes). One may naively think that the old "dual string" approach (see e.g. [3]) may be useful: start with the known string action and define the amplitudes trying to mimic the dual string definitions. However, we will show that the knowledge of I_{string} is not sufficient - one must first calculate $W[C]$ as a functional and then sum over closed paths with appropriate measure. The point is that the QCD string is not a free one but has quarks at the ends (let us stress that only open strings naturally appear from the QCD field formalism if the string ansatz (1) is assumed). In sect. 4 we work out the definition of the mesons scattering amplitudes starting with the QCD path integral and assuming (1). We illustrate our definition on the example of the 2-dimensional QCD.

Lacking the final form for the string ansatz in (1) one can try to study the properties of various possible string models in attempt to establish their common features or to find some distinguished one. This is the topic of sect. 2 where we show that the Brink-Di Vecchia-Howe-Polyakov (BDHP) model [5,6] seems to be the most simple and tractable one among other known bose string ansatze. Sect.3 is devoted to the formal BDHP string theory: we temporarily forget about QCD applications and explicitly calculate the free BDHP string scattering amplitudes, defining them in analogy with the dual string case. The results seem not to support the conjecture that substituting the

Nambu action by the BDHP one we get tachyons free string theory in d=4. A realistic QCD motivated string theory should be the result of the proper account of the quark end point terms as stressed in Sect. 4.

2. String ansatze

Let us list the following three bose string models which may be considered in connection with QCD according to (1):

Nambu model [7] : $\mathcal{D} \subset \mathbb{R}^2$, $h_{ab} = \partial_a x_\mu \partial_b x_\mu$, $h = \det h_{ab}$

$$I_N = M^2 \int_{\mathcal{D}} \sqrt{h}\, d^2 z, \quad M^{-2} = 2\pi\alpha', \quad W_N[c] = \int_{x|_{\partial\mathcal{D}} = c} [\mathcal{D}x]\, e^{-I_N} \quad (2)$$

Eguchi model [8,9] : $I_E = M^{4\nu} \int_{\mathcal{D}} h^\nu d^2 z$, $\quad k(\nu) = \left(\frac{2\nu - 1}{\nu}\right)^{2\nu-1} \cdot \frac{1}{2\nu}$,

$$W_E[c] = \int_0^\infty da\, e^{-a/2} \int_{x|_{\partial\mathcal{D}} = c} [\mathcal{D}x]\, \exp(-k(\nu) I_E), \quad a = \int_{\mathcal{D}} d^2 z .$$

BDHP or "gravitational" model [5,6] :

$$I_G = M^2 \int_{\mathcal{D}} \tfrac{1}{2} g^{ab} \sqrt{g}\, \partial_a x_\mu \partial_b x_\mu\, d^2 z, \quad W_G = \int_{x|_{\partial\mathcal{D}} = c} [\mathcal{D}g\, \mathcal{D}x]\, e^{-I_G} . \quad (3)$$

The study of these models gave the following results [10] : (1) They are equivalent not only at the classical level (minimal surfaces $W \sim \exp(-M^2 A_{min}[c])$) but also in the semiclassical approximation near a minimal surface and thus predict the same long-range $\bar{q}q$ potential $V_{semiclass.} = M^2 R - \gamma/R + const$, $\gamma = \frac{\pi}{12} \cdot \frac{d-2}{2}$ for $C = (R \to \infty) \times (T \to \infty)$; (2) They are equivalent in the leading $1/d$ approximation for the static potential (cf.[9]): $V(d \to \infty) = M^2 R (1 - R_c^2/R^2)^{1/2}$, $R_c^2 = \pi d / 12 M^2$; (3) these models are not equivalent as exact functional integrals (e.g. beyond the semiclassical approximation) being different quantum analogs of the same classical theory; (4) the BDHP model is the most simple and tractable at the quantum level. The main fact is that in order to preserve the $O(d)$ (Lorentz) symmetry of $W[c]$ one is to use $O(d)$-invariant and thus incomplete gauges on x_μ (or g_{ab}). Then additional Weyl symmetry of the BDHP action and its polinomiality in x_μ are essential simplifications. For example, if we consider (instead of W) the formal BDHP string <u>partition function</u> (with $\partial_n x|_{\partial\mathcal{D}} = 0$ boundary conditions) then it is easy to obtain the effective action, integrating the conformal anomaly [6] ; (5) However, one cannot establish the analogous local representation for $W[c]$ due to the nontrivial boundary condition $x/_{\partial\mathcal{D}} = c$: in the gauge $g_{ab} = e^{2\sigma}\delta_{ab}$ we get $W_G = \int d\sigma \exp(-\tilde{I}_G[\sigma,c])$

$\tilde{I}_G = \frac{d}{2} \operatorname{tr} \log \Delta_{0c} - \frac{1}{2} \operatorname{tr} \log \Delta_1$ = non-local functional of σ and c_μ :

$\begin{cases} \Delta_0 = -\nabla_a\nabla^a \\ x_\mu|_{\partial\mathcal{D}} = c_\mu \end{cases}$, $\Delta_{1ab}\xi^b = -(\nabla_c\nabla^c + R/2)\xi_a$. Thus no $(26-d)$ -coefficient arize. As for the formal partition function, it can be evaluated through the anomaly also for locally supersymmetric fermi string models, as was shown for the spinor string in [11] and for the string with spin and charge in [12] . In the latter case one has

$$
\begin{aligned}
Z_{CF} &= \int[\mathcal{D}\varphi_\mu \mathcal{D}\psi_\mu \mathcal{D}A_a \mathcal{D}\chi_a \mathcal{D}g_{ab}]\, e^{-I_{CF}} = \int d\sigma\, d\lambda\, d\varkappa\, e^{-\tilde{I}_{CF}}, \\
I_{CF} &= M^2 \int_{\mathcal{D}} d^2z \sqrt{g} \Big\{ \tfrac{1}{2} g^{ab} \partial_a\varphi \cdot \partial_b\varphi^* + \tfrac{i}{2}\bar{\psi}\gamma^a \overleftrightarrow{\partial_a}\psi + A_a \bar{\psi}\gamma^a\psi + \\
&+ (\partial_a\varphi^* + \bar{\psi}\chi_a)\bar{\chi}_b\gamma^a\gamma^b\psi + \text{c.c.} \Big\}, \qquad (4) \\
\tilde{I}_{CF} &= \frac{2-d}{4\pi}\int d^2z \Big[\tfrac{1}{2}(\partial\sigma)^2 - \tfrac{1}{2}(\partial\varkappa)^2 + i\bar{\lambda}\hat{\partial}\lambda \Big],
\end{aligned}
$$

where $(\varphi_\mu = x_\mu + i y_\mu \ , \psi_\mu)$ form d complex matter multiplets, interacting with N = 2 two-dimensional supergravity fields (g_{ab} , χ_a , A_a) and we used the gauges $g_{ab} = e^{2\sigma}\delta_{ab}$, $\chi_a = \frac{1}{2}\gamma_a\lambda$, $A_a = \frac{1}{2}\varepsilon_{ab}\partial_b\varkappa$. Three terms in (4) are due to the conformal, axial and superconformal anomalies respectively. As was stressed in [12] and [10] the Liouville non-linearities ($\sim e^\sigma$) do not arize in supersymmetrical cases due to the absence of quadratic (L^2) divergences. (Even for the bose BDHP case one can formally neglect L^2 -terms or put the renormalized value of the cosmological constant Λ to zero). However, it is instructive to show how the Λ -term can be introduced in the fermi string theory on the example of quantization of the N=1 2-dimensional supergravity. This is a non-trivial theory with the compact spaces partition functinn given by

$$
Z = \sum_{(\text{topologies})} \int_{(\text{fixed topology})} [\mathcal{D}g_{ab}\,\mathcal{D}\chi_a\,\mathcal{D}A]\, e^{-I(\alpha,\mu)}, \qquad (5)
$$

$$
\begin{aligned}
I &= \alpha\chi + \alpha\int d^2z\sqrt{g}\Big\{ -i\bar{\chi}_a\gamma^a\gamma_5\varepsilon^{mn}\mathcal{D}_m\chi_n + \\
&+ \tfrac{i}{4}\bar{\chi}_a\gamma^b\gamma^a\chi_b - \tfrac{1}{2}A^2\Big\} + \mu\int d^2z\Big(\tfrac{i}{2}\bar{\chi}_a\gamma_5\varepsilon^{ab}\chi_b - A\Big)\sqrt{g},
\end{aligned}
$$

where A is the auxiliary scalar field (cf. [13]) and $\chi = \frac{1}{4\pi}\int R\sqrt{g}\,d^2z$ is the Euler number, while the μ -term is the analog of the Λ -term (cf. [14]). Only taking in account the gauge measure (in the gauge $g_{ab} = e^{2\sigma}\delta_{ab}$, $\chi_a = \frac{1}{2}\partial_a \lambda$) we get the correct counting of degrees of freedom and

$$Z = \sum_{\chi} e^{-\alpha\chi} \int_{(fixed\,\chi)} \mathcal{D}\sigma\,\mathcal{D}\lambda\,\mathcal{D}A\; e^{-\tilde{I}},$$
$$\tilde{I} = \frac{10}{12\pi}\int d^2z\left\{\frac{1}{2}(\partial\sigma)^2 + i\bar{\lambda}\hat{\partial}\lambda - \frac{1}{2}A^2 e^{2\sigma}\right\} + \mu\int d^2z\left\{\bar{\lambda}\lambda e^{\sigma} - A e^{2\sigma}\right\}. \tag{6}$$

Integration over A gives the super-Liouville theory with $\Lambda = \mu^2$ (which probably leads to some space-time foam picture as the correspoinding 2-dim gravity model [15]).

3. Free BDHP string amplitudes

Here we forget about the connection with QCD and study the formal BDHP string theory assuming the Neumann boundary condition $\partial_n x_\mu|_{\partial\mathcal{D}} = 0$ in the path integral. This theory has recently attracted much attention due to a hope that the proper account of the anomalous degree of freedom σ (with the action $\tilde{I} = \frac{26-d}{12\pi}\int d^2z \left[\frac{1}{2}(\partial\sigma)^2 + \mu^2 e^{2\sigma}\right]$ + boundary terms, $\partial_n\sigma|_{\partial\mathcal{D}} \sim \mu\exp(\sigma)$) may help to avoid tachyons and d=26 restriction in loops. As a simplest test of this hope one can put $\mu = 0$ and study the corresponding scattering amplitudes. Still the main problem is the absence of a natural definition of the BDHP strings amplitudes. Thus we are to follow the old dual string definition or try to invent some new one. The realization of this program gave the following results [10] :

I Open strings ($\mathcal{D} = \mathbb{C}^+$, $\partial\mathcal{D} = \mathbb{R}$), on-shell definition a la ref. [4] :

$$V(p_1,\dots,p_n) = \int d\mu_{KN} \prod_{i=1}^{n} |z_{i+1} - z_i|^{-\alpha' m_0^2}\, Z[J_0], \tag{7}$$
$$p_i^2 = -m_0^2,\quad z_i \in \partial\mathcal{D},\quad J_0^\mu = \sum_{i=1}^{n} p_i^\mu \delta^{(2)}(z - z_i),$$
$$Z[J] = \left\langle \exp\left(i\int d^2z\, J_\mu x_\mu\right)\right\rangle_{\partial_n x|_{\partial\mathcal{D}}=0},\quad d\mu_{KN} = \text{"}\prod_{i=1}^{n} dz_i\text{"}$$

, where $d\mu_{KN}$ is the Koba-Nielsen

measure and the averaging is made with the help of the BDHP string path integral. We get: $d=26$; $Z_G=\prod_{i<j}|z_i-z_j|^{2\alpha' p_i p_j}$ i.e. the Veneziano amplitude; $d<26$: $Z_G \sim \prod |z_i-z_j|^{\gamma_{ij}}$, $\gamma_{ij}=2\alpha' p_i p_j - \varkappa$ $\varkappa = 3\alpha'^2 m_0^4/(26-d)$ and the poles are given by $\alpha(0) = -\alpha' m_0^2 + \varkappa/2 = 1 \rightarrow \alpha' m_0^2 \simeq$ 15.6 or -0.94. Thus there is the non-tachyonic ground state solution; the amplitude is dual only for d = 26. Analogous results follow [10] if we start with the heuristic off-shell definition [6, 16]

$$\Gamma(p_1,\dots,p_n) = \langle \prod_{j=1}^{n} \int_{\mathcal{D}} d^2 z_j \sqrt{g(z_j)}\, e^{i p_j \cdot X(z_j)} \rangle , \quad z_j \in \mathcal{D} \qquad (8)$$

Here for d < 26 we get the physical trajectory $\alpha' m_0^2 \simeq 1{,}3n - 0{,}7$ (along with the tachyonic one). However, (8) does not lead to the Veneziano amplitude for d=26 and thus is probably consistent only for closed string case.

II Closed strings ($\mathcal{D} = \mathbb{C}$)

Starting with (8) we get for d=26: Γ has poles at $\frac{\alpha' p_i^2}{2} = 2$, ie describes only the ground state tachyon scattering in the Shapiro-Virasoro model [17] . For d < 26: $\Gamma \sim \int \prod_j d^2 z_j \prod_{i<j} |z_i - z_j|^{\alpha' p_i p_j - \varkappa_{ij}}$ $\varkappa_{ij} = \frac{24}{26-d}\left(1-\frac{\alpha'}{4}p_i^2\right)\left(1-\frac{\alpha'}{4}p_j^2\right)$ and thus the spectrum contains the old tachyon $1-\frac{\alpha'}{4}p_i^2=0$ along with a new physical state $1-\frac{\alpha'}{4}p_i^2 = (26-d)/24$. It remains to be seen if the inclusion of the Liouville term may eliminate tachyons and give unitary and factorizable loop diagrams. However it should be understood that a priori they will have no relation to QCD, where the expression for scattering amplitudes turns to be different from (7) or (8) even if the BDHP ansatz in assumed in (1).

4. $N \rightarrow \infty$ QCD string scattering amplitudes

Trying to derive the string amplitudes from QCD we are to start with the field theoretic definition for meson Green's functions

$$G(x_1,\dots,x_n) = \langle (\bar{q}q)_{x_1} \dots (\bar{q}q)_{x_n} \rangle_{\text{connected}} = \qquad (9)$$

$$= N^{-n/2} \left(\frac{\delta^n \log Z[\mathcal{M}]}{\delta \mathcal{M}(x_1) \dots \delta \mathcal{M}(x_n)} \right)_{\mathcal{M}(x)=m} ,$$

$$Z[\mathcal{M}] \sim \int \mathcal{D}A\, \mathcal{D}\bar{q}\, \mathcal{D}q\, \exp(-I[\mathcal{M}]) ,$$

where $I[m] = \int d^4x \left(\frac{1}{4g^2} F_{\mu\nu}^2 + i\bar{q}(\hat{\mathcal{D}} + m(x))q \right)$.

With the help of the proper time representation for the quark determinant and taking the large N limit (cf.[2]) we finally get

$$G(x_1,\dots,x_n) = N^{1-n/2} \int_0^\infty \frac{dT}{T} \oint_{C(0)=C(T)} \mathcal{D}C_\mu(\tau)\, J_{1/2}[C]\, W[C] \times \tag{10}$$
$$\times \prod_{j=1}^{n} \int_0^T d\tau_j\, \delta^{(4)}[x_j - c(\tau_j)]$$

where W is given by (1) and $J_{1/2}[C] = \int \mathcal{D}\pi\, tz\, P\exp\{i\int_0^T d\tau [\pi_\mu(\dot{x}_\mu + \gamma_\mu) - m]\}$ is the quark end point factor. Using the momentum representation and "covariantizing" the closed path integral we have

$$G(p_1,\dots,p_n) = \int \mathcal{D}e(t) \oint \mathcal{D}c(t) \prod_{j=1}^{n} \int_0^1 dt_j\, e(t_j)\, e^{i\sum_j p_j c(t_j)} \times \tag{11}$$
$$\times J_{1/2}[c,e]\; W[c,e]$$

where e^2 is a one-dimensional metric on C (in the proper time gauge $\dot{e}=0$, $\int_0^1 e\,dt = \int_0^T d\tau$). Suppose now that the string ansatz (1) is valid for W , which can be conviniently written as

$$W[c,e] = \int_{(g_{tt}|_{\partial\mathcal{D}} = e^2(t))} \mathcal{D}g_{ab} \int_{(x|_{\partial\mathcal{D}} = c(t))} \mathcal{D}x(z)\, \exp(-I[g,x]) , \tag{12}$$

where $g_{tt} = g_{ab}\frac{dz^a}{dt}\frac{dz^b}{dt}$ and I is some effective action (e.g. the BDHP one).[It is interesting to note that the final result (11), (12) prompts the open-string analog of the off-shell-closed string amplitude definition (8) for the free BDHP string:

$\Gamma(p_1,\dots,p_n) = \langle \prod_{j=1}^{n} \int_{\partial\mathcal{D}} dz_j\, e(z_j)\, e^{ip_j x(z_j)} \rangle$, where $\langle\dots\rangle =$
$= \int \mathcal{D}e \int \mathcal{D}g\, \delta[g_{tt}|_{\partial\mathcal{D}} - e^2] \int \mathcal{D}x\, \delta[\partial_n x|_{\partial\mathcal{D}} - 0]$. It is this (z_j - reparametrization invariant) definition that should probably be used in the future studies of the BDHP string] . As for $J_{1/2}$, it can be expressed as follows (cf. [18])

$$J_{1/2}[c,e] = \int \mathcal{D}\pi_\mu(t) \int \mathcal{D}\psi(t)\, \mathcal{D}\bar{\psi}(t)\, \exp[i\int_0^1 dt \times \tag{13}$$
$$\times \{\pi_\mu \dot{c}_\mu + i\bar{\psi}\dot{\psi} - e(t)\,\bar{\psi}(\gamma_\mu \pi_\mu - m)\psi\}] ,$$

or $J_0[c,e] = \exp\{-\int_0^1 dt [\frac{1}{2}e^{-1}\dot{x}^2 + m^2 e]$ if we neglect the spins of quarks. The important consequence of (11) and (12) is that we cannot explicitly integrate over $x_\mu(z)$ because of the non-trivial end point factor $J_{1/2}$ (or J_0). Really, if this factor was absent, we could rewrite two integrals $\int \mathcal{D}c \int \mathcal{D}x$ over the "boundary and interiour" as one integral over the whole domain and then assume the Neumann boundary conditions on x_μ , providing the possibility to obtain the explicit expression for the amplitude analogous to those of sect.3 (with t_j playing the role of Koba-Nielsen variables). The conclusion is that contrary to the free string case (sect.3) here we first need to find $W[C]$ as a functional then integrate over C_μ and e . However, the expression for W is difficult to obtain even for the simplest BDHP string case (see sect. 2). In this situation some approximations are needed, for example, the semiclassical one for W ($W \sim \exp(-M^2 A_{min}[C])$) or the semiclassical approximation for the total (string + "ends") action. The second approach was already initiated in a number of papers [19, 20] where it was shown that "ends" are essential to obtain a reasonable spectrum of hadrons. However, these attempts are to be improved (if trying to go beyond the semiclassical approximation) by changing the Nambu action, e.g., by the BDHP (or some fermi string) one and also by using the proper quark end point term (13) instead of the"phenomenological" Bars-Hanson's one (used also in [20]), which actually not follows from QCD

$$I_{BH} = \int_0^T d\tau \left(\frac{i}{2} \bar{\psi} \frac{\dot{c}^\mu}{\sqrt{\dot{c}^2}} \gamma_\mu \overleftrightarrow{\partial_\tau} \psi - m \bar{\psi}\psi \sqrt{\dot{c}^2} \right) \tag{14}$$

(note that (13) and (14) may again be considered as providing different quantum extensions of the same classical theory).

Finally let us illustrate our result for the amplitude (11) on the example of the 2 -dimensional QCD, where the expression for W is explicitly known for the simple curve C [1,21] : $(\gamma = g^2 N)$ $W[C] = \exp\left(-\gamma/2 \int_0^T d\tau\, \varepsilon_{\mu\nu} c_\mu \dot{c}_\nu\right)$. Using the proper time gauge in (11) and assuming that quarks are spinless and have equal masses, we get the following expression for the $\bar{q}q$ - mesons off shell scattering amplitude

$$G(p_1,\dots,p_n) = N^{1-n/2} \int_0^\infty \frac{dT}{T} e^{-\frac{m^2}{2}T} \prod_{j=1}^n d\tau_j \oint dC(\tau) \,\times$$
$$\times \exp\left[-\int_0^T d\tau \left(\frac{1}{2}\dot{c}_\mu^2 + \frac{\gamma}{2}\varepsilon_{\mu\nu} c_\mu \dot{c}_\nu\right) + i\sum_{j=1}^n p_\mu^j c_\mu(\tau_j)\right]. \tag{15}$$

Calculating the path integral we are left with

$$G(p_1,\dots,p_n) = N^{1-n/2} \delta^{(2)}\left(\sum_{j=1}^n p_j\right) \int_0^\infty \frac{ds}{s} e^{-\frac{m^2 s}{\gamma}} \left[\frac{\gamma/2}{\sin s}\right] \times$$

$$\prod_{j=1}^{n}\int_0^1 dt_j\,\exp\Big\{\frac{1}{4\gamma\sin s}\sum_{k,\ell}\Big(p_k^\mu p_\ell^\mu A_{k\ell}(t) - \varepsilon_{\mu\nu}\, p_k^\mu p_\ell^\nu B_{k\ell}(t)\Big)\Big\}\,, \tag{16}$$

$$s = T\gamma/2\,,\quad t_k = \tau_k/T\,,\quad \omega_{k\ell} = s-\gamma\,(t_k-t_\ell)\,,$$

$$A_{k\ell} = \cos\omega_{k\ell}\cdot\theta\,(t_k-t_\ell) + \cos\omega_{\ell k}\cdot\theta\,(t_\ell-t_k)\,,$$

$$B_{k\ell} = \sin\omega_{k\ell}\cdot\theta(t_\ell-t_k) - \sin\omega_{\ell k}\cdot\theta(t_\ell-t_k)\,.$$

This amplitude resembles the dual-like ones with "cos" and "sin" instead of "log's". The spectrum of mesons is given by the poles of the propagator (c.f. with the approach of ref.[22])

$$G(p_1,p_2) = \delta(p_1+p_2)\,G(p_1)\,,$$

$$G(p) = \int_0^\infty \frac{ds}{s}\, e^{-m^2 s/\gamma}\left(\frac{\gamma/2}{\sin s}\right)\int_0^1 d\xi\left(\int_0^{1-\xi} d\eta + \int_0^{\xi} d\eta\right)\times$$

$$\times\exp\Big\{\frac{p^2}{4\gamma}\big[\operatorname{ctg} s\,(1-\cos\gamma\eta) - \sin\gamma\eta\big]\Big\} \equiv \sum_{n=0}^{\infty}\frac{a_n^2}{p^2+m_n^2}\,.$$

It can probably be connected with the longitudinal spectrum of the string with masses at the ends [23] or the t'Hooft spectrum [1].

REFERENCES

1. G. 't Hooft, Nucl. Phys. B72 (1974) 461.
2. Yu.M.Makeenko and A.A.Migdal, Nucl. Phys. B188 (1981) 269; A.A.Migdal, Nucl. Phys. B189 (1981) 253.
3. C.Rebbi, Phys. Rep. 12C (1974)1; S.Mandelstam, Phys. Rep. 13C (1974) 261.
4. C.S.Hsue, B.Sakita and M.A.Virasoro, Phys. Rev. D2 (1970) 2857; J.L.Gervais and B.Sakita, Phys. Rev. D4 (1971) 2291; Phys. Rev. Lett. 30 (1973) 706.
5. L.Brink, P.Di Vecchia and P.S.Howe, Phys. Lett. 65B (1976) 471.
6. A.M.Polyakov, Phys. Lett. 103 B(1981) 211.
7. Y.Nambu, in: Symmetries and Quark Models, ed. by R.Ghand (Gordon and Breach, N.Y., 1960); T.Goto, Progr. Theor. Phys. 46 (1971) 1560.
8. T.Eguchi, Phys. Rev. Lett. 44 (1980) 126.
9. O.Alvarez, Phys. Rev. D24 (1981) 440.
10. E.S.Fradkin and A.A. Tseytlin, Lebedev Inst. preprint N 30(1982), Ann. of Phys. (N.Y.) (to appear).
11. A.M.Polyakov, Phys. Lett. 103B (1981) 207.
12. E.S.Fradkin and A.A.Tseytlin, Phys. Lett. 106B (1981) 63.
13. P.S.Howe, J.Phys. A12 (1979) 393.
14. P.Di Vecchia, B.Durhuus, P.Olesen and J.L.Petersen, N.Bohr Inst. preprint NBI-HE-8279.
15. S.G.Rajeev, Phys. Lett. 113B (1982) 146.
16. B.Durhuus, H.N.Nielsen, P.Olesen and J.L.Petersen, Nucl. Phys. B. 196 (1982) 498.
17. R.I.Nepomechie, Phys. Rev. D25 (1982) 2706.

18. M.B.Halpern, A.Jevicki and P.Senjanovic, Phys. Rev. D16(1977) 2476; R.A.Brandt, F.Neri and D.Zwanziger, Phys. Rev. D19(1979) 1153.
19. I.Bars and A.Hanson, Phys. Rev. D13 (1976) 1744.
20. K.Kikkawa and M.Sato, Phys. Rev. Lett. 38 (1977) 1309; K.Kikkawa, T.Kotani, M.Sato and M.Kenmoku, Phys. Rev. D18(1978) 2606; D19 (1979) 1011.
21. V.A.Kazakov and I.K.Kostov, Nucl. Phys. B176 (1980) 199.
22. A.Strominger, Phys. Lett. B101 (1981) 271.
23. W.A.Bardeen, I.Bars, A.J.Hanson and R.D.Peccei, Phys. Rev. D13 (1976) 2364.

ONE-LOOP DIVERGENCES AND ß-FUNCTIONS IN SUPERGRAVITY THEORIES

E.S.Fradkin and A.A.Tseytlin,
Department of Theoretical Physics,
P.N.Lebedev Physical Institute,
Leninsky Pr. 53, Moscow 117924, USSR
Unfortunately the authors were unable to attend the Symposium

1. Introduction

Supergravity (SG) was invented with a hope to solve the problem of infinities in the Einstein theory. Now came the time of explicite calculations (of counter-terms, ß-functions, off shell and asymptotic behaviour) which are to reveal the structure and the status of quantum SG. Several new results on this way are the topic of this report.

Let us first remind a number of known facts about the infinities in ungauged O(N) SG's (for refs. see [1]): (1) N = 1,...,8 SG's are on-shell finite in L=1,2-order (L is the number of loops): L=1 - diagram calculations of infinities of the S-matrix elements; L=1,2 - general argument of the absence of an on-shell non-vanishing superinvariant; (2) $L \geqslant 3$: there exist superinvariants - candidates for on-shell divergences; (3) N-extended SG's are (off shell) finite (in d-dimensions) for $L < 2(N-1)/(d-2)$, e.g. the N=8, d=4 theory is infinite for $L \geqslant 7$ (some plausable argument based on supergraph power counting rules [2]); (4) N = 8 SG is divergent for $L \geqslant 3$ (implicite argument treating N=8 SG as a $\alpha' \to 0$, $d=10 \to 4$ limit of the superstring theory [3]). Thus different approaches seem to leave the only possibility for finiteness if the actual coefficient of an admissable superinvariant in the (e.g. L=3) infinities is zero. Two examples of such kind of "zeroes" were already found in SG at L=1 order: the absence of topological and gauge field action counter-terms for $N \geqslant 3$ and $N \geqslant 5$ respectively (cf. [1]). A new one - the absence of the off shell L=1 Weyl tensor squared type infinities in the N=8 and N=1, $d=10 \to 4$ theories - will be reported in sect.2, where we discuss the L=1 off shell infinities in gauged O(N) supergravities [4] . Here we make a conjecture that the N=8 SG may be L=1 off-shell finite (cf. [2]) which, if true, may imply an improvement of higher loop behaviour.

Suppose, however, that N=8 SG fails to be finite at 3-loop order. At least two possible modifications of the approach can then be suggested: (i) consider the N=8 SG to be only a low energy manifestation of some fundamental ultraviolet finite superstring theory in d=10 space-time (with six compact dimensions) [3] ; (ii) change the SG lagrangian by adding super-extensions of the curvature squared invariants in order to get a power counting renormalizable theory (just like it can be done already the Einstein theory, see e.g. [5]). It is the second secnnd possibility that we propose here (see sect.4). Sn.3 is devoted to the discussion of the one-loop ß-function [6] (Sect. 3) in conformal supergravities, i.e. the superextensions of the Weyl tensor squared invariant.

2. Off shell one-loop divergences in gauged O(N) supergravities [4]

In order to get a realistic theory one should consider the gauged version of the O(N)-Poincare supergravity (and also try to invent some viable mechanism for a spontaneous supersymmetry breaking). For example, the simplest gauged SG-theory-the O(2)-one-has the following lagrangian [7] $(\Lambda = -3m^2,\ m = 2g/\kappa)$

$$\mathcal{L}_2 = -\frac{1}{\kappa^2}(R-2\Lambda) + \frac{1}{4g^2}F_{\mu\nu}^2 + \frac{1}{2}\epsilon^{\mu\nu\lambda\rho}\bar{\psi}_\mu^i\gamma_5\gamma_\nu\mathcal{D}_\lambda\psi_\rho^i + m\bar{\psi}_\mu^i\sigma_{\mu\nu}\psi_\nu^i + \frac{1}{2}m^{-1}\bar{\psi}_\mu^i F_{\mu\nu}^{+ij}\psi_\nu^j + \ldots, \tag{1}$$

where

$F_{\mu\nu}^{ij} = \varepsilon^{ij}F_{\mu\nu}$, $F_{\mu\nu}^{+} = F_{\mu\nu} + \gamma_5 F_{\mu\nu}^{*}$, i,j = 1,2 and g is dimensionless gauge coupling). This theory is one-loop on-shell renormalizable and one can ask about the value of the ß-function for g. In the first calculation [8] done in the background gravitational sector $\beta(g)$ was implicitly obtained by establishing the Λ-term renormalization and then using the relation $\Lambda\kappa^2 = -12g^2$. The results

$$\begin{array}{cccccccc} N: & 2 & 3 & 4 & 5 & 6 & 7 & 8 \\ \beta: & -\frac{26}{3} & -5 & -2 & 0 & 0 & 0 & 0 \end{array} \tag{2}$$

were then rederived by a "heuristic" calculation in the background gauge field sector[9].The reasoning of [9] contained a number of unjustified assumptions like the validity of the formula $\beta_0(s) = -4\left(\frac{1}{12} - s^2\right)(-1)^{2s}C_2$ (for the cnntribution of spin S field in the gauge field ß-function) for the gravitino ($S = 3/2$) and the possibility to obtain the total result by simple summation of contributions of all spins of the SG multiplet.

To provide understanding of the agreement of these two calculations of $\beta(g)$ one should study the off shell divergences in the combined gravitational-gauge field background sector of the effective action ($g_{\mu\nu} \neq \delta_{\mu\nu}$, $A_\mu \neq 0$, $\psi_\mu = 0$). The one-loop divergences for various fields can be evaluated using the formula

$$(\log \det \Delta)_\infty = \frac{1}{(4\pi)^2} \int d^4x \sqrt{g} \left(\bar{b}_0 L^4 - \frac{1}{2} \bar{b}_2 L^2 - \bar{b}_4 \log \frac{L^2}{\mu^2} \right), \quad L \to \infty , \qquad (3)$$

where $\Delta = -\mathcal{D}^2 + X$ and

$$\bar{b}_0 = \nu , \quad \bar{b}_2 = \rho_1 R + \rho_2 \Lambda + \rho_0 \varkappa F_{\mu\nu}^2 , \quad \varkappa \equiv \frac{\kappa^2}{g^2} ,$$
$$\bar{b}_4 = \beta_1 R^* R^* + \beta_2 W + \frac{1}{3} \beta_3 R^2 + \beta_4 R \Lambda + \beta_5 \Lambda^2 + \qquad (4)$$
$$+ \beta_6 \mathcal{D}^2 R + \gamma_1 \varkappa R_{\mu\nu} T_{\mu\nu} + \gamma_2 \varkappa^2 T_{\mu\nu}^2 + \gamma_3 \varkappa (\mathcal{D}_\mu F_{\mu\nu})^2 +$$
$$+ \beta_0 F_{\mu\nu}^2 , \qquad W = R_{\mu\nu}^2 - \frac{1}{3} R^2 , \; T_{\mu\nu} = F_{\mu\lambda} F_{\nu\lambda} - \frac{1}{4} g_{\mu\nu} F^2 .$$

The central point is establishing the gravitino contribution (in the standart background gauge $\gamma_\mu \psi_\mu^i = \xi^i(x)$)

$$Z_\psi = (\det \Delta_{3/2})^{1/2} (\det \Delta_{gh})^{-3/2} , \quad \Delta_{gh}^{ij} = \hat{\mathcal{D}}^{ij} + 2m \delta^{ij} ,$$
$$\Delta_{3/2\,\mu\nu}^{ij} = \Sigma^\lambda_{\mu\rho} (\mathcal{D}_\lambda)^{ij}_{\rho\nu} - m \delta^{ij} g_{\mu\nu} + m^{-1} F^{+ij}_{\mu\nu} , \quad \Sigma^\lambda_{\mu\nu} = -\frac{1}{2} \gamma_\nu \gamma^\lambda \gamma_\mu .$$

"Squaring" the $\Delta_{3/2}$-operator we get <u>additional</u> F^2-infinities due to mixing of the "mass" and the non-minimal coupling terms. We found that it is this mixing that is essential for the correctness of the $\beta_0(s)$ -formula for $s = 3/2$ used in [9]. The results for different spins contributions in the gravitational and gauge infinities are the following (we also utilize the old off-shell results for the Einstein-Maxwell system [10]):

s	ν	β_0	β_1	β_2	β_3	β_4	β_5	β_6	ρ_1	ρ_2
2	2	$-$	$\frac{53}{45}$	$\frac{7}{10}$	$\frac{3}{4}$	$-\frac{26}{3}$	20	$-\frac{19}{15}$	$-\frac{23}{3}$	20
$\frac{3}{2}$	-2	$-\frac{13}{3}$	$-\frac{233}{720}$	$-\frac{77}{60}$	$-\frac{1}{3}$	$-\frac{4}{9}$	$\frac{44}{9}$	$\frac{1}{60}$	$\frac{1}{6}$	$\frac{16}{3}$
1	2	$\frac{11}{3}$	$-\frac{13}{180}$	$\frac{1}{5}$	0	0	0	$-\frac{1}{10}$	$-\frac{2}{3}$	0
$\frac{1}{2}$	-2	$-\frac{2}{3}$	$\frac{7}{720}$	$\frac{1}{20}$	0	0	0	$\frac{1}{60}$	$\frac{1}{6}$	0
0	1	$-\frac{1}{6}$	$\frac{1}{180}$	$\frac{1}{60}$	$\frac{1}{24}$	$-\frac{1}{9}$	$\frac{2}{9}$	$\frac{1}{30}$	$\frac{1}{6}$	$-\frac{2}{3}$

The important fact is the <u>negative</u> "sign" of the gravitino contribution (β_2) in the "Weyl" infinities, which can be contrasted to the positive ones for S = 0,1/2, 1 fields. It should be stressed that the statement (cf. [10]) about " $\beta_2 > 0$ for any spin" is not actually applicable for S = 2 and 3/2 (in the background gauges). One should take into account that here β_2 is gauge dependent (cf. [11] for S = 2 case and note that the "one-loop" gravitino lagrangian is superinvariant only if the background space is the Einstein space).

The final expression for the one-loop off shell infinities in the gauged O(N) supergravity can be written in the form

$$\bar{b}_0 = 0 , \quad \bar{b}_2 = \rho_1' \bar{R} + \rho_2' \Lambda + 4\kappa^2 \rho_0 \mathcal{L}_2 ,$$

$$\bar{b}_4 = \beta_1 R^* R^* + \beta_2 \left(\bar{R}_{\mu\nu}^2 - \tfrac{1}{3} \bar{R}^2 \right) + \tfrac{1}{3} \beta_3 \bar{R}^2 + \beta_6 \mathcal{D}^2 R + \qquad (5)$$

$$+ \gamma_3 \varkappa (\mathcal{D}_\mu F_{\mu\nu})^2 + \alpha_1 \varkappa \bar{R} + \alpha_2 \varkappa T_{\mu\nu} \bar{R}_{\mu\nu} + 4 g^2 \beta_0 \mathcal{L}_2 ,$$

$$\mathcal{L}_2 = -\frac{1}{\kappa^2} (R - 2\Lambda) + \frac{1}{4g^2} F_{\mu\nu}^2 , \quad \bar{R}_{\mu\nu} = R_{\mu\nu} - \Lambda g_{\mu\nu} - \frac{\kappa^2}{2g^2} T_{\mu\nu} , \bar{R} = R - 4\Lambda ,$$

, where the coefficients are given in the table

N	0	1	2	3	4	5	6	8	$N=1$, $d=10$
β_0	–	–	$-\frac{26}{3}$	-5	-2	0	0	0	–
β_1	$\frac{53}{45}$	$41/48$	$11/24$	0	0	0	0	0	0
β_2	$\frac{7}{10}$	$-7/12$	$-5/3$	$-5/2$	-3	-3	-2	0	0
β_3	$\frac{3}{4}$	$5/12$	$1/12$	$-1/4$	$-1/2$	$-1/2$	0	1	$13/12$
β_6	$-19/15$	$-5/4$	$-4/3$	$-3/2$	$-21/15$	$-1/3$	$11/3$	$26/3$	$22/5$
γ_3	–	–	$1/3$	$1/2$	$2/3$	$11/12$	$4/3$	$13/6$	–
α_1	–	40	64	88	114	146	190	178	–
α_2	–	–	-2	$-5/2$	$-17/6$	$-7/2$	-6	-5	-3
ρ_1'	$-23/3$	$-15/2$	-8	-9	-10	-10	-8	-4	$-31/3$
ρ_2'	$-32/3$	$-14/3$	$-4/3$	0	0	0	0	0	0
ρ_0	–	–	0	0	0	0	0	0	–

Several conclusions follow from these results:

(1) we explicitly demonstrate that <u>gauged</u> SG's are on-shell (L=1) renormalizable up to topological infinity (i.e. after the use of

$$R_{\mu\nu} - \Lambda g_{\mu\nu} - \frac{\kappa^2}{2g^2} T_{\mu\nu} = 0 , \quad R - 4\Lambda = 0, \quad \mathcal{D}_\mu F_{\mu\nu} = 0);$$

(2) the on shell renormalizations of Λ and g^2 are given by the <u>same</u> coefficient β_0, explaining the agreement of the results of refs.[8] and [9] ;
(3) there is no on-shell quadratic divergences in all theories with $N \geqslant 3$;
(4) N = 8 SG and the theory, obtained by a reduction of the N=1,d=10 SG (the last column of the table)are off shell finite in the gravitational 2^+-sector (have $\beta_2 = 0$). Thus the N=8 SG is distinguished by having a maximal degree of the off shell finiteness: $\beta_0 = \beta_1 = \beta_2 = 0$. One may even conjecture that it is completely off shell finite (for L=1) when treated in a suitable background supergauge where $\alpha_1 = \alpha_2 = 0$ (it may turn out that also $\beta_3 = \gamma_3 = 0$ in this gauge if the N=8 superextension of R^2 does not exist).

3. <u>One-loop ß-function in pure conformal supergravities</u> [6]

Conformal supergravities are $U(N)$ ($N = 1, \dots, 4$) superconformal extensions of the Weyl invariant $W = R_{\mu\nu}^2 - \frac{1}{3} R^2$. The lagrangian has the following structure [12]

$$\mathcal{L}_N = \alpha^{-2} (W)_{ss} = \alpha^{-2} \Big[W - \frac{4-N}{4N} F_{\mu\nu}^2(A) - F_{\mu\nu}^{ij\,2}(V) + \tag{6}$$
$$+ 4 \epsilon^{\mu\nu\rho\sigma} \bar{\phi}_\rho^i \gamma_5 \gamma_\sigma \mathcal{D}_\nu^+ \phi_\mu^i + \dots \Big] ,$$

where $\phi_\mu = \frac{1}{3}\gamma^\nu (\mathcal{D}_\nu \psi_\mu - \mathcal{D}_\mu \psi_\nu + \frac{1}{2}\gamma_5 \epsilon_{\mu\nu\rho\sigma} \mathcal{D}_\sigma \psi_\rho)$, $\mathcal{D}^+ = \partial + A + V + \dots$, while A_μ and V_μ are $U(1)$ and $SU(N)$ gauge fields and ψ_μ^i is "conformal gravitino". Why these theories are interesting: (1) they are gauge theories with the maximal known group -superconformal group, including the ordinary and conformal supersymmetries, scale and chiral $U(N)$ transformations; 2) they are power counting renormalizable due to higher derivatives in the kinetic terms, $L_{linear} = h \Box^2 h + \bar{\psi} \partial^3 \psi + A \Box A + \dots$ (Note that for the correct counting of degrees of freedom one should properly account for the "averaging over gauges" operators, e.g. for the N = 1 theory we have $\nu_{tot} = (6)_h + (-8)_\psi + (2)_A = 0$); 3) N = 1,2,3-theories are asymptotically free, while the N=4 theory is finite (in one-loop). More explicitly, one can obtain the following results for the ß-function for the dimensionless coupling α :

$$N = 1 \; : \quad \beta_I = \beta_g + \beta_\psi + \beta_A = \frac{17}{2} \quad , \tag{7}$$

where $\beta_g = \frac{199}{15}$ is the value for the pure Weyl theory [5], $\beta_\psi = -149/30$ is the conformal gravitino W-infinities in the gravitational sector established in [6] with the use of the algorithms for the divergences of the 4-th and 3-d order differential operators and β_A = 1/5 is the axial vector field contribution.

N=2: $\beta_{II} = \beta_g + 2\beta_\psi + \beta_A + \beta_V$ (SU_2 -gauge field V_μ^{ij}) + + β_χ (2 spinors χ^i) + β_T (1 antisymmetric tensor field $T_{\mu\nu}^{ij}$, $\mathcal{L}_T \sim T\Box T$) = $\frac{13}{3}$;

N=3: $\beta_{III} = \beta_g + 3\beta_\psi + \beta_A + \beta_V(SU_3) + \beta_\chi(9\chi^{ij}) +$ $+\beta_E$ (3 complex scalars E_i) + $\beta_T(3T) + \beta_\Lambda$ (1 spinor Λ, $\mathcal{L}_\Lambda \sim \bar{\Lambda}\hat{\partial}^3\Lambda$) = 1 ;

N = 4:

$$\beta_{IV} = \beta_g + 4\beta_\psi + \beta_V(SU_4) + \beta_\chi(20\chi^{ij}_k) + \beta_E(10E_{(ij)})$$
$$+ \beta_T(6T) + \beta_\Lambda(4\Lambda_i) + \beta_C(1 \text{ complex scalar } C, \mathcal{L}_C \sim$$
$$\sim C^*\Box^2 C) = 0 .$$

(where we assumed an appropriate gravitational coupling of the scalar C). Interpreting $-\alpha^2$ as the $U(N)$ gauge coupling, we get the following sequence (cf. with the O(N) Poincare SG sequence (2))

$$\begin{array}{cccccc} N & : & 1 & 2 & 3 & 4 \\ \beta(-\alpha^2) & : & -\frac{17}{2} & -\frac{13}{3} & -1 & 0 \end{array} \tag{8}$$

However, the conformal supergravities lack a low energy correspondence with the Einstein theory. That is why in order to get a viable theory one should add also the ordinary (linear in curvature) supergravity term in the lagrangian.

4. Renormalizable supergravity models

Let us consider the following lagrangian

$$\mathcal{L} = -\frac{1}{K^2}(R)_{ss} + \frac{1}{\alpha^2}(W)_{ss} - \frac{1}{3\sigma^2}(R^2)_{ss} , \tag{9}$$

where the brackets denote the corresponding superextensions. This theory is renormalizable, possesses the correct Einstein limit but lacks perturbative unitarity due to the presence of ghosts. However, ghosts here fill a supermultiplet and thus may decouple in some non-perturbative way. The physical spectrum contains gauge fields of Poincare SG, while $U(N)$ gauge fields of conformal SG are in fact auxiliaries for Poincare SG. Some generalization of (9) probably exists for N=4, ..., 8 with higher spin fields being auxiliary (propagating) in P_0incare (conformal) SG-parts.

One can prove that the inclusion of the $(R+\ldots)$-term in the conformal SG lagrangian increases the value of the Weyl coupling $\beta(\alpha)$-function. Thus we get the asymptotically free behaviour for α also for N=4 theory and may hope for $\beta(\alpha)=0$ for some $N > 4$. In turn, the addition of conformal SG term to the P_0incare one (1) changes the renormalization of the physical gauge field coupling g: all negative gravitino and matter fields conributions in $\beta(g)$ are suppressed due to higher derivative terms in the conformal SG part. Therefore the value of $\beta(g)$ is the same as for the free O(N) gauge field in the flat space-time, i.e. corresponds to the asymptotic freedom in contrast to the non-asymptotically free results in pure Poincare SG case (2). Now let us mention a possibility that a superextension of the R^2-term may not exist for some $N > 2$. Then the superconformal theory

$$\mathcal{L}=\left(-\frac{1}{12}R\phi^2-\frac{1}{2}\partial_\mu\phi\partial_\mu\phi+\lambda\phi^4+\frac{1}{\alpha^2}W\right)_{ss} \qquad (10)$$

is an attractive candidate for a fundamental theory if it has $\beta(\alpha)=0$ (no superconformal anomalies and possible solution of the problem of ghosts). The presence of conformal supergravity term may also help to provide a spontaneous supersymmetry breaking. The important fact is that one can add some matter multiplets to (9) or (10) without destroing renormalizability. Obtaining in this way a sufficient spectrum of particles (or taking in account that additional particles may appear as monopoles after a spontaneous supersymmetry breaking) we get a power counting renormalizable asymptotically free (or finite) unified theory.

In conclusion we want to point out that the renormalizable supergravity (9) can be considered as an "induced supergravity" theory. Suppose we start with the lagrangian

$$\mathcal{L}=\left(\frac{1}{4g^2}F_{\mu\nu}^2+i\bar{\psi}\hat{\mathcal{D}}\psi\right)_{sc}+\frac{1}{\alpha^2}(W)_{ss} \quad , \qquad (11)$$

containing massless gauge and spinor matter fields interacting with the external conformal supergravity fields and also the pure conformal SG part. Assuming that the regularization breaks the conformal symmetry but preserves the general covariance and local supersymmetry, we get (according to the ideas of "induced gravity" approach

[13]) the following effective lagrangian

$$\mathcal{L} = -\frac{1}{k_{ind}^2}\left(R-2\Lambda_{ind}\right)_{ss} - \frac{1}{36_{ind}^2}\left(R^2\right)_{ss} + \frac{1}{\alpha_{ren}^2}\left(W\right)_{ss} + \ldots \quad , \tag{12}$$

where k_{ind} , Λ_{ind} and σ_{ind} are finite calculable constants. One can probably induce the Poincare SG term in (12) even without matter terms in (11) (i.e. starting only with conformal SG term). Thus the conformal supergravity itself may be a true theory on some fundamental level.

References

1. P. van Nieuwenhuizen, Phys. Repts.68 (1981) 189; M.J.Duff, in:"Supergravity 81", S.Ferrara and J.G.Taylor eds, (Cambridge Univ. Press 1982); R.E.Kallosh, ibid.
2. M.T.Grisaru and W.Siegel, Nucl. Phys. B187 (1981) 149; preprint CALT-68-892.
3. M.B.Green, J.H.Schwarz and L.Brink, Nucl. Phys. B198 (1982) 474.
4. E.S.Fradkin and A.A.Tseytlin, Lebedev Inst. preprint N 114 (1982); Phys. Lett. B (to be published).
5. E.S.Fradkin and A.A.Tseytlin, Phys. Lett. 104B (1981) 377; Nucl. Phys. B201 (1982) 469.
6. E.S.Fradkin and A.A.Tseytlin, Phys. Lett. 110B (1982) 117; Nucl. Phys. B203 (1982) 157.
7. D.Z.Freedman and A.Das, Nucl. Phys. B120 (1977) 317.
8. S.M.Christensen, M.J.Duff, G.W.Gibbons and M.Roček, Phys. Rev. Lett. 45 (1980) 161.
9. T.L.Curtright, Phys. Lett. 102B (1981) 17.
10. S.Deser and P. Van Nieuwenhuizen, Phys. Rev. D10 (1974) 401.
11. R.E.Kallosh, O.V.Tarasov and I.V.Tyutin, Nucl. Phys. B137 (1978) 145.
12. M.Kaku, P.K.Townsend and P. van Nieuwenhuizen, Phys. Rev. D17 (1978) 3179; E.Bergshoeff, M. De Roo and B.De Wit, Nucl. Phys. B182 (1981) 173.
13. S.L.Adler, Revs. Mod. Phys. (July 1982).

A Geometrical Foundation of a Unified Field Theory

Nathan Rosen
Technion, Israel Institute of Technology, Haifa
and
Gerald E. Tauber
Tel Aviv University, Tel Aviv

Unfortunately the authors were unable to attend the Symposium

I. Introduction

In a series of two not well known papers Einstein and Mayer [1] proposed a formalism by which they were able to obtain a theory of gravitation and electromagnetism similar to that of Kaluza and Klein. [2] Instead of assuming, as these authors did, the existence of a five-dimensional continuum they assumed that at each point of space-time, regarded as a Riemannian space, there exists a five-dimensional vector space. The purpose of this work is to generalize the approach of Einstein and Mayer to N-dimensions, and to lay the geometrical foundation of a possible unified field theory of gravitation with other fields. [3]

Accordingly, we assume the existence of a four-dimensional Riemannian base space, characterized by coordinates x^i $(i = 1,..4)$ and metric g_{ij}. At each point (of the base space) there is a linear vector space of N dimensions (N 5), vectors in which would have components a_μ , b^ν $(\mu,\nu = 1,2..N)$. Quantities in the two spaces are connected a mixed tensor or projector $h_\mu{}^k$ so that

$$a^k = h_\mu{}^k a^\mu \tag{1.1}$$

For a given a^μ (1.1) determines a^k uniquely, but the reverse is not the case. In particular, for a vector $A^k = 0$ we can write

$$A^k = h_\mu{}^k A^\mu = 0 \tag{1.2}$$

which will have $n = N-4$ independent solutions, if the matrix $(h_\mu{}^k)$ is of rank 4. Labelling these solutions with an index P $(P = 1,2,..n)$ we can define a metric g_{PQ}

$$A_P{}^\mu A_{Q\mu} = g_{PQ} \tag{1.3}$$

In general, g_{PQ} will be functions of coordinates y^m $(m = 1,2,..N-4)$ in this subspace, but for the present discussion[4] we shall assume that g_{PQ} are constants, which we can take as

$$g_{PQ} = \delta_{PQ} \tag{1.4}$$

However, to keep the notation more uniform we shall replace the A's by quantities sudh as $h_\mu^{\ P}$, $h_{\mu P}$, $h^\mu_{\ P}$ etc. In the N-dimensional space we define a metric tensor $f_{\mu\nu}$ related to g_{ij} through

$$g_{ij} = f_{\mu\nu}\, h^\mu_{\ i}\, h^\nu_{\ j} \quad \text{or} \quad g^{ij} = f^{\mu\nu}\, h_\mu^{\ i}\, h_\nu^{\ j} \tag{1.5}$$

and by raising (or lowering) indices

$$h_\lambda^{\ i}\, h^\lambda_{\ j} = \delta^i_j \tag{1.6}$$

A simple calculation using (1.2) and (1.6) gives

$$f_{\mu\nu} = h_\mu^{\ i}\, h_\nu^{\ j}\, g_{ij} + h_\mu^{\ P}\, h_\nu^{\ Q}\, g_{PQ} \tag{1.7}$$

which can be considered to be the inverse of the relation given by (1.5).

II. Curvature tensor

Let us now consider covariant differentiation, which involves the ordinary Christoffel symbols in base space, but a number of connections or three-index symbols in vector-space. For example, the covariant derivative of $S^\mu_{\ k}$ is

$$S^\mu_{\ k\|j} = S^\mu_{\ k,j} + \Gamma^\mu_{\ \lambda j}\, S^\lambda_{\ k} - \left\{{m \atop k\,j}\right\} S^\mu_{\ m} \tag{2.1}$$

Thus, in particular

$$a^k_{\ \|j} = a^k_{\ ;j} \tag{2.2}$$

where a semi-colon denotes a Riemannian covariant derivative. Furthermore,

$$g_{ij\|k} = g_{ij;k} = 0 \tag{2.3}$$

as usual, and we shall also assume

$$f_{\mu\nu\|k} = 0 \tag{2.4}$$

In order to determine the form of the three-index symbol $\Gamma^\lambda_{\ \mu k}$ consider a simpler quantity $\mathring{\Gamma}^\lambda_{\ \mu k}$ involved in covariant derivatives denoted by a single bar | and defined such that

$$h_\mu^{\ j}{}_{|k} = h_\mu^{\ J}{}_{|k} = 0 \tag{2.5}$$

One finds then from (2.5)

$$\mathring{\Gamma}^\lambda_{\ \mu k} = h^\lambda_{\ j}\, h_\mu^{\ j}{}_{,k} + h^\lambda_{\ P}\, h_\mu^{\ P}{}_{,k} + h^\lambda_{\ j}\, h_\mu^{\ m} \left\{{j \atop m\,k}\right\} + h^\lambda_{\ P}\, h_\mu^{\ Q}\, \Gamma^P_{\ Qk} \tag{2.6}$$

where $\Gamma^P_{\ Qk}$ is still undetermined.[5] If we now write

$$\Gamma^\lambda_{\ \mu k} = \mathring{\Gamma}^\lambda_{\ \mu k} + V^\lambda_{\ \mu k} \tag{2.7}$$

it follows from (2.4) that

$$V_{\mu\nu k} + V_{\nu\mu k} = 0 \qquad (2.8)$$

Consequently, we can take $V_{\mu\nu k}$ in the form

$$V_{\mu\nu k} = h_\mu{}^i h_\nu{}^j W_{ijk} + (h_\mu{}^P h_\nu{}^j - h_\nu{}^P h_\mu{}^j) F_{Pjk} + h_\mu{}^P h_\nu{}^Q U_{PQk} \qquad (2.9)$$

where

$$W_{ijk} = - W_{jik} \quad \text{and} \quad U_{PQk} = - U_{QPk}$$

and the functions appearing in (2.9) are to be determined.

The curvature tensor in the base-space is just the Riemann-Christoffel tensor $R^i{}_{jkn}$. In the vector space we can define the curvature tensor

$$P^\lambda{}_{\mu jk} = - \Gamma^\lambda{}_{\mu j,k} + \Gamma^\lambda{}_{\mu k,j} + \Gamma^\lambda{}_{\sigma j} \Gamma^\sigma{}_{\mu k} - \Gamma^\lambda{}_{\sigma k} \Gamma^\sigma{}_{\mu j} \qquad (2.10)$$

From $P^\lambda{}_{\mu jk}$ one can form tensors of lower order

$$P_{\mu j} = P^\lambda{}_{\mu jk} h_\lambda{}^k \quad \text{and} \quad P = P_{\mu j} h^{\mu j} \qquad (2.11)$$

being the analogues of the Ricci tensor and invariant curvature.

From the usual anti-commutation relations we obtain

$$K_{\mu j} = h_\mu{}^k{}_{\| j \| k} - h_\mu{}^k{}_{\| k \| j} = P_{\mu j} - h_\mu{}^m R_{mj}$$

Multiplication by $h^{\mu j}$ then gives

$$P - R = h^{\mu j} K_{\mu j} \qquad (2.12)$$

The right hand side can be evaluated by making use of (2.5), (2.7) and

$$h_\mu{}^i{}_{\| k} = - h_\sigma{}^i V^\sigma{}_{\mu k}$$

where $V^\sigma{}_{\mu k}$ is given by (2.9). Carrying out the indicated calculations we find

$$P = R + W^{ijk} W_{ijk} + F^{Pjk} F_{Pjk} \qquad (2.13)$$

III. Field equations

To obtain the field equations it is convenient to make use of a variational principle. Since for constant $g_{PQ} = \delta_{PQ}$ the only scalar at our disposal in vector space is P given by (2.13) we shall take the variational functional as

$$\delta I = \int P \, (-g)^{\frac{1}{2}} \, d^4x = 0 \qquad (3.1)$$

where $g = \det|g_{ij}|$. Varying (3.1) with respect to g_{ab} we then obtain

$$\begin{aligned} R^{ab} - \tfrac{1}{2} g^{ab} R = & - 2(F_Q{}^{aj} F^{Qb}{}_j - \tfrac{1}{4} g^{ab} F^{QPq} F_{QPq}) \\ & - 3(W^{arp} W^b{}_{rp} - \tfrac{1}{6} g^{ab} W^{rpq} W_{rpq}) \end{aligned} \qquad (3.2)$$

Varying (3.1) with respect to W_{ijk} and F_{Pjk} gives $W^{ijk} = F^{Pjk} = 0$, as can be seen from (2.13). In order to avoid this, let us express these functions as potentials. Let us first assume [6]

$$F_{Pjk} = -F_{Pkj} \quad \text{and} \quad F_{Pjk} = F_{Pj|k} - F_{Pk|j} \tag{3.3}$$

Varying now the potentials F_{Pj} we obtain the field equations

$$F_P{}^{jk}{}_{|k} = F_P{}^{jk}{}_{;k} - \Gamma^Q{}_{Pk} F_Q{}^{jk} = 0 \tag{3.4}$$

Assume also

$$W_{ijk} = -W_{ikj}$$

so that now W_{ijk} is completely antisymmetric upon interchange of any two indices. Two possibilities now suggest themselves:

a) For

$$W_{ijk} = (-g)^{\frac{1}{2}} e_{ijkm} g^{mn} \Psi_{,n} \tag{3.5}$$

where e_{ijkm} is the completely antisymmetric Levi-Civita symbol. Varying now Ψ in (3.1) results in the wave equation

$$g^{jk} \Psi_{;jk} = 0 \tag{3.6}$$

b) Alternately,

$$W_{ijk} = w_{ij,k} + w_{jk,i} + w_{ki,j} \quad \text{with } w_{ij} = -w_{ji} \tag{3.5'}$$

If we now vary with respect to w_{ij} we find from (3.1)

$$W^{ijk}{}_{;k} = 0 \tag{3.6'}$$

It should also be noted that no equations have been obtained for U_{PQk}.

IV. Gauge fields

So far we have considered the vectors $h_\mu{}^P$ satisfying (1.2) and (1.3) as having permanent identities. However, we can get a further generalization by taking into account the possibility of replacing them by linear combinations. If under the transformation

$$h_\mu{}^P \longrightarrow h_\mu{}^{P'} = S^P{}_Q h_\mu{}^Q \tag{4.1}$$

where $S_{PQ} = (S^{-1})_{QP}$ is an orthogonal matrix, the vector Ψ_μ is invariant

$$\Psi_\mu = \Psi_P h_\mu{}^P = \Psi_P{}' h_\mu{}^{P'} \tag{4.2}$$

then

$$\Psi_P{}' = S_P{}^Q \Psi_Q \quad \text{or, in matrix notation,} \quad \underset{\sim}{\Psi}' = \underset{\sim}{S} \underset{\sim}{\Psi} \tag{4.3}$$

If we now define the covariant derivative

$$\underset{\sim}{\Psi}_{|j} = \underset{\sim}{\Psi}_{,j} - \underset{\sim}{B}_j \underset{\sim}{\Psi} \quad \text{so that} \quad \underset{\sim}{\Psi}'_{|j} = \underset{\sim}{S} \underset{\sim}{\Psi}_{|j} \tag{4.4}$$

we obtain the transformation law for B_j

$$\underset{\sim}{B}'_j = \underset{\sim}{S}\,\underset{\sim}{B}_j\,\underset{\sim}{S}^{-1} + \underset{\sim}{S}_{,j}\,\underset{\sim}{S}^{-1} \tag{4.5}$$

Thus,

$$\underset{\sim}{B}_{jk} = \underset{\sim}{B}_{j,k} - \underset{\sim}{B}_{k,j} + [\underset{\sim}{B}_j\,,\,\underset{\sim}{B}_k] \tag{4.6}$$

transforms according to the relation

$$\underset{\sim}{B}'_{jk} = \underset{\sim}{S}\,\underset{\sim}{B}_{jk}\,\underset{\sim}{S}^{-1} \tag{4.7}$$

We see that we have here the gauge-field formalism. Writing out (4.4) in terms of matrix elements gives

$$\Psi_{P|j} = \Psi_{P,j} - B_P{}^Q{}_j\,\Psi_Q = \Psi_{P,j} - \Gamma^Q{}_{Pj}\,\Psi_Q \tag{4.8}$$

which shows that the matrix elements $B_P{}^Q{}_j$ are nothing else than the three-index symbols $\Gamma^Q{}_{Pj}$ we met previously (cf. 2.6), and thus our formalism does contain the seed of the gauge transformation. In particular, (4.6) written out is just the tensor (apart from an overall sign)

$$B^P{}_{Qjk} = \Gamma^P{}_{Qj,k} - \Gamma^P{}_{Qk,j} + \Gamma^P{}_{Rj}\,\Gamma^R{}_{Qk} - \Gamma^P{}_{Rk}\,\Gamma^R{}_{Qj} \tag{4.9}$$

We also note that the three-index symbol $\overset{o}{\Gamma}{}^\lambda{}_{\mu k}$ (2.6) is gauge-invariant as it stands. Moreover, the field equations for F_{Pjk} are gauge-invariant as can be seen from (3.4). Also, from (3.3) we obtain

$$F_{Pjk} = F_{Pj|k} - F_{Pk|j} = F_{Pj,k} - F_{Pk,j} - \Gamma^Q{}_{Pk}\,F_{Qj} + \Gamma^Q{}_{Pj}\,F_{Qk} \tag{4.10}$$

We, then, note that varying (3.1) with respect to $\Gamma^Q{}_{Pj}$ would impose a restriction on F_{Qj}. It is, therefore, suggestive to add to the Lagrangian in (3.1) a term in volving these connections. Thus, we replace (3.1) by

$$\delta \int (P + B^P{}_{Qjk}\,B^Q{}_{Pmn}\,g^{mj}\,g^{nk})(-g)^{\frac{1}{2}}\,d^4x = 0 \tag{4.11}$$

Varying (4.11) with respect to g_{ab} adds to (3.2) a term on the r.h.s. of the form

$$-\,(B^P{}_Q{}^{ka}\,B^Q{}_{Pk}{}^b + B^P{}_Q{}^{kb}\,B^Q{}_{Pk}{}^a - \tfrac{1}{2}g^{ab}\,B^P{}_Q{}^{mn}\,B^Q{}_{Pmn}) \tag{4.12}$$

while varying with respect to $\Gamma^P{}_{Qk}$ gives

$$-\,4B_P{}^{Qjk}{}_{|k} + 2(F_P{}^{jk}\,F^Q{}_k - F^{Qjk}\,F_{Pk}) = 0 \tag{4.13}$$

---

1) A. Einstein and W. Mayer, Sitzber. Preuss. Akad. Wiss. 1931, p. 541; 1932, p.130

2) Th. Kaluza, Sitzber. Preuss. Akad. Wiss. 1921, p. 966

3) Work along these lines is now in progress.

4) The general case will be presented in a separate publication elsewhere.

5) These symbols play the role of gauge fields (see section IV)

6) For P = 1 we get the Maxwell fields considered by Einstein and Mayer.

AUTHOR INDEX

List of Participants

Akama, K.
Saitama Medical School
Japan

Aoki, K.
RIFP, Kyoto University
Japan

Arafune, J.
Inst. CRR, University of Tokyo
Japan

Arisue, H.
Department of Physics
Kyoto University
Japan

Arnowitt, R.L.
Department of Physics
Northeastern University
USA

Azuma, T.
College of General Education
University of Tokyo
Japan

Baba, K.
Nara Women's University
Japan

Banks, T.
Department of Physics
Tel Aviv University
Israel

Bedding, S.P.
Department of Physics
University of Newcastle upon
Tyne, England

Candelas, P.
Department of Physics
University of Texas at Austin
USA

DeWitt, B.S.
Department of Physics
University of Texas at Austin
USA

DeWitt-Morette, C.
Department of Astronomy
University of Texas at Austin
USA

Doi, M.
Department of Physics
Kobe University
Japan

Eguchi, T.
Department of Physics
University of Tokyo
Japan

Freund, P.G.
Enrico Fermi Institute
University of Chicago
USA

Fujii, Y.
College of General Education
University of Tokyo
Japan

Fujikawa, K.
INS, University of Tokyo
Japan

Fujiwara, T.
RIFP, Kyoto University
Japan

Fukuda, R.
RIFP, Kyoto University
Japan

Fukuda, T.
RITP, Hiroshima University
Japan

Fukuhara, A.
Hitachi Central Research
Japan

Fukui, M.
Department of Fundamental
Engineering
Osaka University
Japan

Fukui, T.
College of General Education
Dokkyo University
Japan

Fukuyama, T.
Department of Physics
Osaka University
Japan

Fulling, S.A.
Department of Mathematics
Texas A&M University
USA

Furusawa, T.
Department of Physics
Osaka University
Japan

Halpern, L.E.
Department of Physics
Florida State University
USA

Hara, O.
Atomic Energy Research Institute
of Nihon University
Japan

Harada, K.
Department of Physics
Tôhoku College of Pharmacy
Japan

Hashimoto, T.
Department of Fundamental
Engineering
Osaka University
Japan

Hata, H.
Department of Physics
Kyoto University
Japan

Hayashi, K.
Inst. CRR, University of Tokyo
Japan

Hayashi, M.
Department of Physics
Kyoto University
Japan

Hehl, F.W.
Institut für Theoretische Physik
Universität zu Köln
West Germany

Higuchi, A.
Department of Physics
Kyoto University
Japan

Horibe, M.
Faculty of Education
Fukui University
Japan

Horwitz, G.
Lyman Physics Laboratory
Harvard University
USA

Hosoda, M.
Department of Physics
Tokyo Metropolitan University
Japan

Hosoya, A.
Department of Physics
Osaka University
Japan

Ichinose, S.
Institute of Physics
University of Tsukuba
Japan

Inoue, K.
Department of Physics
Kyushu University
Japan

Ishihara, H.
RITP, Hiroshima University
Japan

Ishikawa, K.
Department of Physics
Osaka University
Japan

Itoh, C.
Department of Physics
Meiji-Gakuin University
Japan

Ito, H.
Department of Physics
Nagoya University
Japan

Ito, K.
Department of Physics
Kyoto University
Japan

Iwao, S.
Department of Physics
College of Liberal Arts
Kanazawa University
Japan

Iwasaki, Y.
Institute of Physics
University of Tsukuba
Japan

Jackiw, R.
MIT
USA

Kanaya, K.
Department of Physics
Nagoya University
Japan

Kasuya, M.
INS, University of Tokyo
Japan

Kato, M.
Department of Physics
Kyoto University
Japan

Kawai, T.
Department of Physics
Osaka City University
Japan

Kawasaki, S
Department of Physics
Chiba University
Japan

Kayama, Y.
Department of Physics
Kobe University
Japan

Kazama, Y.
Department of Physics
Kyoto University
Japan

Kenmoku, M.
Nara Women's University
Japan

Kikkawa, K.
Department of Physics
Osaka University
Japan

Kikuchi, Y.
Department of Physics
Tôhoku University
Japan

Kim, S.W.
Department of Physics
KAIST, Korea

Kim, Y.
Department of Physics
Sogang University
Korea

Kimura, K.
Department of Physics
Kobe University
Japan

Kimura, T.
Department of Physics
Chiba University
Japan

Kitakaze, K.
Department of Physics
Kobe University
Japan

Kirii, K.
College of General Education
Osaka University
Japan

Kobayashi, M.
KEK, Japan

Kodama, H.
Department of Physics
Kyoto University
Japan

Kohado, A
Department of Physics
Kwansei Gakuin University
Japan

Koh, I.G.
Department of Physics
Sogang University
Korea

Koide, Y.
Shizuoka Women's University
Japan

Koikawa, T.
RIFP, Kyoto University
Japan

Kondo, K.
Department of Physics
Nagoya University
Japan

Konisi, G.
Department of Physics
Kwansei Gakuin University
Japan

Konuma, M.
RIFP, Kyoto University
Japan

Kubo, R.
RITP, Hiroshima University
Japan

Kugo, T.
Department of Physics
Kyoto University
JAPAN

Kuramoto, T.
Department of Physics
Kyoto University
Japan

Kurimoto, T.
College of General Education
Osaka University
Japan

Kuroda, Y.
RITP, Hiroshima University
Japan

Kuwabara, Y.
Department of Physics
Kwansei Gakuin University
Japan

Lee, C.H.
Department of Physics
Hanyang University
Korea

Lee, H.Y.
Department of Physics
Tokyo Metropolitan University
Japan

Liu, Y.Y.
Department of Modern Physics
University of Science and
Technology of China
China

Maki, Z.
RIFP, Kyoto University
Japan

Masukawa, J.
Department of Fundamental Engineering
Osaka University
Japan

Matsukura, D.
Department of Physics
Kobe University
Japan

Midorikawa, S.
RIFP, Kyoto University
Japan

Minakata, H.
Department of Physics
Tokyo Metropolitan University
Japan

Minamikawa, T.
Tokyo University of Mercantile Marine
Japan

Mishima, T.
Department of Physics
Nagoya University
Japan

Morita, M.
Department of Physics
Osaka University
Japan

Murai, Y.
Department of Physics
Saitama University
Japan

Nagamachi, S.
Technical College
Tokushima University
Japan

Naito, S.
Department of Physics
Osaka City University
Japan

Naka, S.
College of Science and Technology
Nihon University
Japan

Nakanishi, N.
RIMS, Kyoto University
Japan

Nakano, T.
Department of Physics
Osaka City University
Japan

Nakariki, S.
Department of Applied Physics
Okayama College of Science
Japan

Nakatani, H.
Department of Physics
Nagoya University
Japan

Nakazawa, N.
Kogakuin University
Japan

Nambu, Y.
Enrico Fermi Institute
University of Chicago
USA

Nariai, H.
RITP, Hiroshima University
Japan

Nishijima, K.
Department of Physics
University of Tokyo
Japan

Nishino, H.
College of General Education
University of Tokyo
Japan

Nishioka, M.
Faculty of Liberal Arts
Yamaguchi University
Japan

Nishiura, H.
RIFP, Kyoto University
Japan

Nitta, S.
Department of Physics
Gifu University
Japan

Nojiri, S.
Department of Physics
Kyoto University
Japan

Nonoyama, T.
Department of Physics
Nagoya University
Japan

Odaka, K.
KEK
Japan

Ohkuwa, Y.
Department of Physics
Osaka University
Japan

Ohta, N.
Department of Physics
University of Tokyo
Japan

Ohta, T.
Department of Physics
Miyagi University of Education
Japan

Ohtani, T.
Kansai University of Foreign Studies
Japan

Ohtsubo, N.
Kanazawa Technical College
Japan

Ohta, T.
Miyagi University of Education
Japan

Ojima, I.
RIMS, Kyoto University
Japan

Okada, J.
Department of Physics
Kobe University
Japan

Okada, Y.
Department of Physics
University of Tokyo
Japan

Okamura, H.
Department of Physics
Kogakuin University
Japan

Omote, K.
Department of Physics
Osaka University
Japan

Omote, M.
University of Tsukuba
Japan

Otokozawa, J.
College of Science and Technology
Nihon University
Japan

Ozaki, K.
Osaka Institute of Technology
Japan

Parker, L.
Department of Physics
University of Wisconsin-Milwaukee
USA

Pi, S.Y.
Department of Physics
Harvard University
USA

Ryang, S.
Department of Physics
Osaka University
Japan

Saito, T.
Kyoto Prefectural University
of Medicins
Japan

Sakagami, M.
Department of Physics
Osaka University
Japan

Sakai, N.
KEK
Japan

Sakamoto, J.
Department of Physics
Shimane University
Japan

Sasaki, R.
RITP, Hiroshima University
Japan

Sato, H.
Hyogo University of Education
Japan

Sato, H.
RIFP, Kyoto University
Japan

Sato, M.
New York University
USA

Sato, Y.
Department of Physics
Kyoto University
Japan

Sawada, S.
Department of Physics
Osaka University
Japan

Shigemitsu, J.
Department of Physics
Ohio State University
USA

Shinohara, Y.
Fujita-Gakuen University
Japan

Shintani, M.
RITP, Hiroshima University
Japan

Shirafuji, T.
Department of Physics
Saitama University
Japan

So, H.
Department of Physics
Kyushu University
Japan

Strominger, A.E.
Institute for Advanced Study
Princeton,
USA

Sugano, R.
Department of Physics
Osaka City University
Japan

Sugawara, H.
KEK
Japan

Suura, H.
School of Physics
University of Minnesota
USA

Suzuki, T.
Department of Physics
Kanazawa University
Japan

Taguchi, Y.
College of Engineering
University of Osaka Prefecture
Japan

Takahasi, W.
Kwansei Gakuin University
Japan

Takao, M.
Department of Physics
Osaka University
Japan

Takasuji, E.
College of General Education
Osaka University
Japan

Takeno, H.
Hiroshima Institute of Technology
Japan

Takeshita, S
Department of Physics
Kyushu University
Japan

Tanaka, A.
Osaka Unstitute of Technology
Japan

Tanaka, K.
Department of Electrical
Engineering
Gifu University
Japan

Tanaka, S.
Department of Physics
Kyoto University
Japan

Terazawa, H.
INS, University of Tokyo
Japan

Tomimatsu, A.
RITP, Hiroshima University
Japan

Tomita, K.
RITP, Hiroshima University
Japan

Townsend, P.K.
Lab. de Physique Théorique,
Ecole Normale Supérieure
France

Uehara, S.
Department of Physics
Kyoto University
Japan

Ukawa, A.
INS, University of Tokyo
Japan

Utiyama, R.
Tezukayama University
Japan

Viallet, C.M.
LPTHE, Université de Paris
France

Watanabe, T.
Asia University
Japan

Watanabe, T.
Department of Physics
Kobe University
Japan

Watamura, S.
Department of Physics
University of Tokyo
Japan

Yabuki, H.
Hyogo University of Education
Japan

Yamagishi, K.
Department of Physics
Osaka University
Japan

Yamaguchi, Y.
Department of Physics
University of Tokyo
Japan

Yamamoto, K.
RIFP, Kyoto University
Japan

Yamamoto, K.
College of General Education
Osaka University
Japan

Yamamoto, N.
Department of Physics
Osaka University
Japan

Yamazaki, M.
Department of Physics
Osaka University
Japan

Yokoyama, K.
RITP, Hiroshima University
Japan

Yoneya, T.
College of General Education
University of Tokyo
Japan

Yoshii, I.
Department of Physics
Nagoya University
Japan

Yotsuyanagi, I.
Kanazawa Medical University
Japan

Lecture Notes in Physics

Vol. 136: The Role of Coherent Structures in Modelling Turbulence and Mixing. Proceedings 1980. Edited by J. Jimenez. XIII, 393 pages. 1981.

Vol. 137: From Collective States to Quarks in Nuclei. Edited by H. Arenhövel and A. M. Saruis. VII, 414 pages. 1981.

Vol. 138: The Many-Body Problem. Proceedings 1980. Edited by R. Guardiola and J. Ros. V, 374 pages. 1981.

Vol. 139: H. D. Doebner, Differential Geometric Methods in Mathematical Physics. Proceedings 1981. VII, 329 pages. 1981.

Vol. 140: P. Kramer, M. Saraceno, Geometry of the Time-Dependent Variational Principle in Quantum Mechanics. IV, 98 pages. 1981.

Vol. 141: Seventh International Conference on Numerical Methods in Fluid Dynamics. Proceedings. Edited by W. C. Reynolds and R. W. MacCormack. VIII, 485 pages. 1981.

Vol. 142: Recent Progress in Many-Body Theories. Proceedings. Edited by J. G. Zabolitzky, M. de Llano, M. Fortes and J. W. Clark. VIII, 479 pages. 1981.

Vol. 143: Present Status and Aims of Quantum Electrodynamics. Proceedings, 1980. Edited by G. Gräff, E. Klempt and G. Werth. VI, 302 pages. 1981.

Vol. 144: Topics in Nuclear Physics I. A Comprehensive Review of Recent Developments. Edited by T.T.S. Kuo and S.S.M. Wong. XX, 567 pages. 1981.

Vol. 145: Topics in Nuclear Physics II. A Comprehensive Review of Recent Developments. Proceedings 1980/81. Edited by T. T. S. Kuo and S. S. M. Wong. VIII, 571-1.082 pages. 1981.

Vol. 146: B. J. West, On the Simpler Aspects of Nonlinear Fluctuating. Deep Gravity Waves. VI, 341 pages. 1981.

Vol. 147: J. Messer, Temperature Dependent Thomas-Fermi Theory. IX, 131 pages. 1981.

Vol. 148: Advances in Fluid Mechanics. Proceedings, 1980. Edited by E. Krause. VII, 361 pages. 1981.

Vol. 149: Disordered Systems and Localization. Proceedings, 1981. Edited by C. Castellani, C. Castro, and L. Peliti. XII, 308 pages. 1981.

Vol. 150: N. Straumann, Allgemeine Relativitätstheorie und relativistische Astrophysik. VII, 418 Seiten. 1981.

Vol. 151: Integrable Quantum Field Theory. Proceedings, 1981. Edited by J. Hietarinta and C. Montonen. V, 251 pages. 1982.

Vol. 152: Physics of Narrow Gap Semiconductors. Proceedings, 1981. Edited by E. Gornik, H. Heinrich and L. Palmetshofer. XIII, 485 pages. 1982.

Vol. 153: Mathematical Problems in Theoretical Physics. Proceedings, 1981. Edited by R. Schrader, R. Seiler, and D.A. Uhlenbrock. XII, 429 pages. 1982.

Vol. 154: Macroscopic Properties of Disordered Media. Proceedings, 1981. Edited by R. Burridge, S. Childress, and G. Papanicolaou. VII, 307 pages. 1982.

Vol. 155: Quantum Optics. Proceedings, 1981. Edited by C.A. Engelbrecht. VIII, 329 pages. 1982.

Vol. 156: Resonances in Heavy Ion Reactions. Proceedings, 1981. Edited by K.A. Eberhard. XII, 448 pages. 1982.

Vol. 157: P. Niyogi, Integral Equation Method in Transonic Flow. XI, 189 pages. 1982.

Vol. 158: Dynamics of Nuclear Fission and Related Collective Phenomena. Proceedings, 1981. Edited by P. David, T. Mayer-Kuckuk, and A. van der Woude. X, 462 pages. 1982.

Vol. 159: E. Seiler, Gauge Theories as a Problem of Constructive Quantum Field Theory and Statistical Mechanics. V, 192 pages. 1982.

Vol. 160: Unified Theories of Elementary Particles. Critical Assessment and Prospects. Proceedings, 1981. Edited by P. Breitenlohner and H.P. Dürr. VI, 217 pages. 1982.

Vol. 161: Interacting Bosons in Nuclei. Proceedings, 1981. Edited by J.S. Dehesa, J.M.G. Gomez, and J. Ros. V, 209 pages. 1982.

Vol. 162: Relativistic Action at a Distance: Classical and Quantum Aspects. Proceedings, 1981. Edited by J. Llosa. X, 263 pages. 1982.

Vol. 163: J. S. Darrozes, C. Francois, Mécanique des Fluides Incompressibles. XIX, 459 pages. 1982.

Vol. 164: Stability of Thermodynamic Systems. Proceedings, 1981. Edited by J. Casas-Vázquez and G. Lebon. VII, 321 pages. 1982.

Vol. 165: N. Mukunda, H. van Dam, L.C. Biedenharn, Relativistic Models of Extended Hadrons Obeying a Mass-Spin Trajectory Constraint. Edited by A. Böhm and J.D. Dollard. VI, 163 pages. 1982.

Vol. 166: Computer Simulation of Solids. Edited by C.R.A. Catlow and W.C. Mackrodt. XII, 320 pages. 1982.

Vol. 167: G. Fieck, Symmetry of Polycentric Systems. VI, 137 pages, 1982.

Vol. 168: Heavy-Ion Collisions. Proceedings, 1982. Edited by G. Madurga and M. Lozano. VI, 429 pages. 1982.

Vol. 169: K. Sundermeyer, Constrained Dynamics. IV, 318 pages. 1982.

Vol. 170: Eighth International Conference on Numerical Methods in Fluid Dynamics. Proceedings, 1982. Edited by E. Krause. X, 569 pages. 1982.

Vol. 171: Time-Dependent Hartree-Fock and Beyond. Proceedings, 1982. Edited by K. Goeke and P.-G. Reinhard. VIII, 426 pages. 1982.

Vol. 172: Ionic Liquids, Molten Salts and Polyelectrolytes. Proceedings, 1982. Edited by K.-H. Bennemann, F. Brouers, and D. Quitmann. VII, 253 pages. 1982.

Vol. 173: Stochastic Processes in Quantum Theory and Statistical Physics. Proceedings, 1981. Edited by S. Albeverio, Ph. Combe, and M. Sirugue-Collin. VIII, 337 pages. 1982.

Vol. 174: A. Kadić, D. G. B. Edelen, A Gauge Theory of Dislocations and Disclinations. VII, 290 pages. 1983.

Vol. 175: Defect Complexes in Semiconductor Structures. Proceedings, 1982. Edited by J. Giber, F. Beleznay, J. C. Szép, and J. László. VI, 308 pages. 1983.

Vol. 176: Gauge Theory and Gravitation. Proceedings, 1982. Edited by K. Kikkawa, N. Nakanishi, and H. Nariai. X, 316 pages. 1983.